1,001 Algebra II Practice Problems

FOR DUMMIES®

A Wiley Brand

by Mary Jane Sterling

1,001 Algebra II Practice Problems For Dummies®

Published by
John Wiley & Sons, Inc.
111 River St.
Hoboken, NJ 07030-5774
www.wiley.com

Published by John Wiley & Sons, Inc., Hoboken, New Jersey

Published simultaneously in Canada

For general information on our other products and services, please contact our Customer Care Department within the U.S. at 877-762-2974, outside the U.S. at 317-572-3993, or fax 317-572-4002.

For technical support, please visit www.wiley.com/techsupport.

Wiley publishes in a variety of print and electronic formats and by print-on-demand. Some material included with standard print versions of this book may not be included in e-books or in print-on-demand. If this book refers to media such as a CD or DVD that is not included in the version you purchased, you may download this material at http://booksupport.wiley.com. For more information about Wiley products, visit www.wiley.com.

Library of Congress Control Number: 2013933946

ISBN 978-1-118-44662-1 (pbk); ISBN 978-1-118-44657-7 (ebk); ISBN 978-1-118-44661-4 (ebk);
ISBN 978-1-118-44658-4 (ebk)

Manufactured in the United States of America

10 9 8 7 6 5 4 3 2

About the Author

Mary Jane Sterling is the author of several *For Dummies* titles: *Algebra I For Dummies, Algebra II For Dummies, Trigonometry For Dummies, Math Word Problems For Dummies, Business Math For Dummies,* and *Linear Algebra For Dummies.* She has also written many supplementary workbooks and study aids.

Mary Jane continues doing what she loves best: teaching mathematics. When not teaching or writing *For Dummies* books, she loves spending her time reading murder mysteries and fishing for her dinner.

Dedication

The author dedicates this book to her son, Sgt. James T. Sterling, USMC, and the other members of the 4th Air/Naval Gunfire Liaison Company, Det Juliet (part of Operation Enduring Freedom 2012). Jim and the others in his unit have our utmost respect and appreciation.

Author's Acknowledgments

The author issues a big thank you to project editor Tim Gallan, who has taken on the huge challenge of creating this new product. He has taken a lot of raw material and made it into this wonderful, finished workbook. Thank you so much for your hard work and patience.

Also, a heartfelt thank you to the math editors, Suzanne Langebartels and Amy Nicklin. As careful as I may be to do all the problems correctly, there is always that chance of a silly error. The editors keep me honest!

And, of course, a grateful thank you to acquisitions editor Lindsay Lefevere, who again found me another interesting project.

Publisher's Acknowledgments

We're proud of this book; please send us your comments at http://dummies.custhelp.com. For other comments, please contact our Customer Care Department within the U.S. at 877-762-2974, outside the U.S. at 317-572-3993, or fax 317-572-4002.

Some of the people who helped bring this book to market include the following:

Acquisitions, Editorial, and Vertical Websites

Senior Project Editor: Tim Gallan

Executive Editor: Lindsay Sandman Lefevere

Copy Editors: Suzanne Langebartels, Christine Pingleton

Assistant Editor: David Lutton

Editorial Program Coordinator: Joe Niesen

Technical Editor: Amy Nicklin

Editorial Manager: Michelle Hacker

Editorial Assistants: Rachelle S. Amick, Alexa Koschier

Cover Photos: © Nadia/iStockphoto.com

Composition Services

Senior Project Coordinator: Kristie Rees

Layout and Graphics: Erin Zeltner

Proofreaders: Lindsay Amones, Joni Heredia Language Services

Indexer: Potomac Indexing, LLC

Publishing and Editorial for Consumer Dummies

Kathleen Nebenhaus, Vice President and Executive Publisher

David Palmer, Associate Publisher

Kristin Ferguson-Wagstaffe, Product Development Director

Publishing for Technology Dummies

Andy Cummings, Vice President and Publisher

Composition Services

Debbie Stailey, Director of Composition Services

Contents at a Glance

Table of Contents

Introduction

One-thousand-one algebra problems: You must wonder what inspired such an endeavor.

One possibility for the inspiration is *1,001 Arabian Nights*. (Okay, I'm really stretching here, but why not?) *1,001 Arabian Nights* is a collection of stories and folk tales, compiled over several centuries. Likewise, *1,001 Algebra II Practice Problems For Dummies* is a collection of math problems and explanations, and some of them involve interesting stories. This book wasn't compiled over centuries (though my editors say it seemed to take that long), but the concepts and ideas involved took mathematicians centuries to develop.

Practice makes perfect. Unlike other subjects where you can just read or listen and absorb the information sufficiently, mathematics takes practice. The only way to figure out how the different algebraic rules work and interact with one another is to get into the problems — get your hands dirty, so to speak. Many problems appear to be the same, on the surface, but different aspects and challenges have been inserted to make the different problems unique. The concepts become more set in your mind when you work with the problems and have the properties confirmed with your solutions.

Yes, whether it's 1,001 algebra problems or 1,001 mathematical adventures, you'll find them here. Enjoy!

What You'll Find

This book contains 1,001 algebra problems, their answers, and complete solutions to each. There are 17 problem chapters, and each chapter has many different sets of questions. The sets of questions are sometimes in a logical, sequential order, going from one part of a topic to the next and then to the next. Other times the sets of questions represent the different ways a topic can be presented. In any case, you're given brief instructions on doing the problems. And sometimes you're given a particular formula or format to use. Feel free to refer to other algebra books, such as *Algebra II For Dummies,* to give you ideas on how to solve some of the problems.

Instead of just having answers to the problems, you find a worked-out solution for each and every one. Flip to the last chapters of the book for the step-by-step processes needed to solve the problems. The solutions include verbal explanations inserted in the work where necessary. Sometimes an alternate procedure may be offered. Not everyone does algebra exactly the same way, but this book tries to provide the most understandable and success-promoting process to use when solving the algebra problems presented.

How This Workbook Is Organized

This workbook is divided into two main parts: questions and answers. But you probably figured that out already.

Part I: Questions

The questions chapters cover many different topics:

- **Review of basic operations:** The chapter takes you through some of the main concepts from Algebra I that are essential to working in Algebra II. You'll find problems on powers of binomials and patterns in those powers. Solving linear equations and linear inequalities are fairly straightforward tasks, but it doesn't hurt to review these types of problems to bring your skill level up to speed. Two other topics covered here are radicals and complex numbers. Each topic is special in its own way, but similarities crop up — such as the use of conjugates when simplifying the expressions.
- **Solving nonlinear equations and inequalities:** Quadratic equations are by far one of the most used and referred-to equation types in secondary mathematics. That's why you'll find techniques such as factoring (with the multiplication property of zero), the square root rule, the quadratic formula, and completing the square to be the main methods covered here.

  Throw in some radical equations (square roots, cube roots) and rational equations (fractional expressions), and you have more tricks and techniques to practice to increase your repertoire.
- **Graphing lines:** A line is a very basic structure and is easy to graph when you have two points. You'll also find lines to graph when you're given their equations and then equations to write when you're given information about the line. You get to consider lines that are parallel to one another and others that are perpendicular. The graphing part is simplified when you recognize the basics: a point on the line and the line's slope.
- **Functions:** A function in mathematics has a very specific definition. You can have a function when you have a relationship between sets of numbers and the relationship is described with mathematical operations. The operations take input values and produce output values based on the rules created with the operations. What's particularly special about functions is that there's only one output for every input. The functions you'll find include linear, quadratic, polynomial, rational, exponential, and logarithmic. There are more functions out there, but you get a really good start right here.
- **Systems of equations and inequalities:** When you have two or more statements or equations and want to know whether there are any solutions common to both or all of them at the same time, you're talking about solving systems. The equations can be linear, quadratic, exponential, and so on. You'll use algebraic techniques as well as matrices to solve some of the linear systems.
- **Sequences, sets, and counting techniques:** Getting ready for future studies in probability and statistics, you'll want to practice problems involving lists, counting, sets, and set notation. You'll use Venn diagrams and make lists of elements in sets. You'll also write the terms in sequences and add up the terms in series.

Part II: Answers

This part provides not only the answers to all the questions but explanations of the answers as well. So you get the solution, and you see how to arrive at that solution.

Beyond the Book

This book gives you plenty of Algebra II problems to work on. But maybe you want to track your progress as you tackle the problems, or maybe you're having trouble with certain types of problems and wish they were all presented in one place where you could methodically make your way through them. You're in luck. Your book purchase comes with a free one-year subscription to all 1,001 practice problems online. You get on-the-go access any way you want it — from your computer, smartphone, or tablet. Track your progress and view personalized reports that show where you need to study the most. Study what, where, when, and how you want.

What you'll find online

The online practice that comes free with this book offers you the same 1,001 questions and answers that are available here, presented in a multiple-choice format. The beauty of the online problems is that you can customize your online practice to focus on the topic areas that give you the most trouble. So if you need help with graphing or solving quadratic functions, then select these problem types online and start practicing. Or, if you're short on time but want to get a mixed bag of a limited number of problems, you can specify the quantity of problems you want to practice. Whether you practice a few hundred problems in one sitting or a couple dozen, and whether you focus on a few types of problems or practice every type, the online program keeps track of the questions you get right and wrong so that you can monitor your progress and spend time studying exactly what you need.

You can access this online tool using a PIN code, as described in the next section. Keep in mind that you can create only one login with your PIN. Once the PIN is used, it's no longer valid and is nontransferable. So you can't share your PIN with other users after you've established your login credentials.

How to register

Purchasing this book entitles you to one year of free access to the online, multiple-choice version of all 1,001 of this book's practice problems. All you have to do is register. Just follow these simple steps:

1. **Find your PIN code.**
   - **Print book users:** If you purchased a hard copy of this book, turn to the back of this book to find your PIN.
   - **E-book users:** If you purchased this book as an e-book, you can get your PIN by registering your e-book at `dummies.com/go/getaccess`. Go to this website, find your book and click it, and then answer the security question to verify your purchase. Then you'll receive an e-mail with your PIN.

2. **Go to `http://onlinepractice.dummies.com`.**
3. **Enter your PIN.**
4. **Follow the instructions to create an account and establish your own login information.**

That's all there is to it! You can come back to the online program again and again — simply log in with the username and password you choose during your initial login. No need to use the PIN a second time.

If you have trouble with the PIN or can't find it, please contact Wiley Product Technical Support at 800-762-2974 or `http://support.wiley.com`.

Your registration is good for one year from the day you activate your PIN. After that time frame has passed, you can renew your registration for a fee. The website gives you all the important details about how to do so.

Where to Go for Additional Help

The written directions given with the individual problems are designed to tell you what you need to do to get the correct answer. Sometimes the directions may seem vague if you aren't familiar with the words or the context of the words. Go ahead and look at the solution to see whether that helps you with the meaning. But if the vocabulary is still unrecognizable, you may want to refer to the glossaries in algebra books, such as *Algebra I For Dummies* or *Algebra II For Dummies,* written by yours truly and published by John Wiley & Sons, Inc.

This book is designed to provide you with enough practice to become very efficient in algebra, but it isn't intended to give the step-by-step explanation on how and why each step is necessary. You may need to refer to *Algebra II For Dummies* or *Algebra II Essentials For Dummies* (also written by me) to get more background on a problem or to understand why a particular step is taken in the solution of the problem.

Algebra is sometimes seen as being a bunch of rules without a particular purpose. Why do you have to solve for the solutions of a quadratic equation? Where will you use that again? The answers to all these questions are more apparent when you see them tied together and when more background information is available. Don't be shy about seeking out that kind of information.

You may become intrigued with a particular topic or particular type of problem. Where do you find more problems like those found in a section? Where do you find the historical background of a favorite algebra process? There are many resources out there, including a couple that I wrote:

- Do you like the applications? Try *Math Word Problems For Dummies*.
- Are you more interested in the business-type uses of algebra? Take a look at *Business Math For Dummies.*

If you're ready for a another area of mathematics, look for a couple more of my titles: *Trigonometry For Dummies* and *Linear Algebra For Dummies.*

Part I

The Questions

Visit www.dummies.com for great *For Dummies* content online.

In this part . . .

You get to tackle 1,001 Algebra II problems. Have fun! Here are the general types of questions you'll be dealing with:

- Algebra basics, quadratic equations, and graphing lines (Chapters 1 through 4)
- All kinds of functions (Chapters 5 through 9)
- Conic sections and linear and nonlinear equations (Chapters 10 through 12)
- Complex numbers, matrices, sequences, series, and sets (Chapters 13 through 17)

Chapter 1

Reviewing Algebra Basics

The basics of Algebra II consist of the processes learned in earlier exposures to algebra — in this case, all lumped together in one chapter. It's hard to cover every little thing in this one book that you'll need to continue your algebra study, but this is a really good place to start. Under the guise of solving some equations and inequalities, you get to review many of the most important properties and procedures needed to be successful. Anything missing in this discussion is covered in later chapters, as part of the problems' explanations.

The Problems You'll Work On

In this chapter, you'll work with simplifying expressions and solving equations and inequalities in the following ways:

- Multiplying binomials and trinomials
- Expanding higher powers of binomials
- Solving linear equations and absolute value equations
- Solving linear inequalities
- Simplifying radical expressions
- Rewriting expressions involving imaginary numbers

What to Watch Out For

Don't let common mistakes trip you up; watch for the following when working with simplifying expressions and solving equations and inequalities:

- Distributing the factor over every term in the parentheses
- Multiplying terms by a negative factor
- Reversing the inequality sense when multiplying or dividing by a negative factor
- Correctly multiplying a binomial and its conjugate
- Simplifying expressions involving powers of i

Multiplying Binomials and Trinomials

1–8 Simplify the expressions by performing the operations and combining like terms.

1. $(2x + 3)(4x - 2) =$

2. $3x^2 + (x + 4)(x - 1) =$

3. $(3x + 1)(x - 3) + (x + 2)(5x - 4) =$

4. $5(x - 3)(x + 2) + 3(x - 3)(x - 2) + 1 =$

5. $(x + 4)(x^2 - 3x + 5) =$

6. $(x - 1)(3x^2 + 2x - 1) =$

7. $(2x + 1)(x - 3)(x + 4) =$

8. $(x - 3)(x + 3)(7x + 11) =$

Using Pascal's Triangle to Multiply Binomials

9–12 Use Pascal's Triangle to expand the binomial powers.

9. $(x - 3)^3 =$

10. $(x + 2)^5 =$

11. $(3x - 2y)^4 =$

12. $(a^2 + b)^6 =$

Solving Linear Equations

13–18 Solve the linear and absolute value equations for x.

13. $4x + 2 = 3(x - 3)$

14. $5x + 2(x + 7) = 3(x - 2)$

15. $|3x-2|=14$

16. $|4x+1|-2=3$

17. $4|x-6|=8$

18. $3|2x-5|+5=8$

Solving Linear Equations for Variables

19–24 Solve for the indicated variable.

19. Solve for l in $P = 2l + 2w$.

20. Solve for s_1 in $P = 2s_1 + s_2$.

21. Solve for b_2 in $A = \frac{1}{2}h(b_1 + b_2)$.

22. Solve for F in $C = \frac{5}{9}(F-32)$.

23. Solve for t in $A = P + Prt$.

24. Solve for n in $a_n = a_1 + (n-1)d$.

Solving Linear Inequalities

25–34 Solve the inequalities.

25. $3x - 4 \leq 5x + 6$

26. $4(x-3) > x + 6$

27. $-3 \leq 2x + 7 < 9$

28. $0 < 7 - 3x < 13$

29. $|x+6| < 4$

30. $|2x-3|\geq 5$

31. $|4x-5|+1\leq 4$

32. $2|6x-5|>20$

33. $4x-9<2x+1\leq 3x-1$

34. $2x+6\leq x+3<3x+11$

Making Radical Expressions Simpler

35–44 Simplify the radical expressions.

35. $\sqrt{50}$

36. $\sqrt{300}$

37. $\sqrt{180}$

38. $\sqrt{960}$

39. $\left(1+\sqrt{2}\right)^2$

40. $\left(\sqrt{3}-\sqrt{5}\right)^2$

41. $\dfrac{2}{2+\sqrt{6}}$

42. $\dfrac{10}{5-\sqrt{5}}$

43. $\dfrac{4+\sqrt{10}}{4-\sqrt{10}}$

44. $\dfrac{12-\sqrt{3}}{4-2\sqrt{3}}$

Working with Complex Expressions

45–50 Simplify the complex numbers.

45. i^{138}

46. i^{1001}

47. $(4-i)^2$

48. $(3+2i)^2$

49. $i(2i)^3$

50. $4i^{21}(1+i)^2$

Chapter 2

Solving Quadratic Equations and Nonlinear Inequalities

A quadratic expression is one containing a term raised to the second power. When a quadratic expression is set equal to 0, you have an equation that has the possibility of two real solutions; for example, you may have an equation for which the answers are $x = 1$ or $x = 3$. Nonlinear inequalities can have an infinite number of solutions, so those answers are written with expressions such as $x > 8$ or $x > -2$; these solutions can also be written using interval notation.

The Problems You'll Work On

In this chapter, you'll work with quadratic equations and inequalities in the following ways:

- Solving simple equations using the *square root rule*
- Rewriting quadratics as the product of two binomials in order to solve
- Applying the *quadratic formula*
- Completing the square
- Solving quadratic-like equations
- Finding the solutions of quadratic and other nonlinear inequalities

What to Watch Out For

Don't let common mistakes like the following ones trip you up when working with quadratic equations and inequalities:

- Forgetting to consider $\pm x$ when using the *square root rule*
- Reducing the fraction incorrectly when applying the *quadratic formula*
- Stopping too soon when solving quadratic-like equations
- Eliminating values as solutions when they create a 0 in the denominator of a fraction

Applying the Square Root Rule on Quadratic Equations

51–60 Solve the equations using the square root rule.

51. $x^2 = 81$

52. $x^2 - 144 = 0$

53. $3y^2 - 75 = 0$

54. $5z^2 - 125 = 0$

55. $9x^2 = 4$

56. $98x^2 = 18$

57. $x^2 = 11$

58. $y^2 = 20$

59. $4x^2 = 200$

60. $3z^2 - 726 = 0$

Solving Quadratic Equations Using Factoring

61–76 Solve the quadratic equations by factoring and applying the Multiplication Property of Zero.

61. $x^2 + 2x - 35 = 0$

62. $y^2 - 4y - 96 = 0$

63. $z^2 + 14z + 48 = 0$

64. $8x^2 - 10x - 3 = 0$

65. $30x^2 + 11x - 30 = 0$

66. $12y^2 + 17y + 6 = 0$

67. $125z^2 + 70z - 3 = 0$

68. $12x^2 - x - 11 = 0$

69. $400y^2 - 528y + 35 = 0$

70. $32z^2 - 46z + 15 = 0$

71. Solve for y: $15y^2 + 4yz - 32z^2 = 0$

72. Solve for z: $90z^2 + 129wz + 28w^2 = 0$

73. $ax^2 + (a + c)x + c = 0$

74. $acx^2 + (bc - a)x - b = 0$

75. $2x^2 + 5\sqrt{2}x + 6 = 0$

76. $12z^2 + 5\sqrt{2}z - 6 = 0$

Using the Quadratic Formula to Solve Equations

77–86 Solve the quadratic equations using the quadratic formula.

77. $x^2 - 6x + 4 = 0$

78. $y^2 + 8y + 10 = 0$

79. $x^2 + 9x - 9 = 0$

80. $2x^2 - 3x - 14 = 0$

81. $8x^2 + 43x + 15 = 0$

82. $2y^2 + 15y + 25 = 0$

83. $3x^2 - 5x + 3 = 0$

84. $4y^2 + 3y + 1 = 0$

85. $5z^2 - 8z + 4 = 0$

86. $9z^2 + 2z + 1 = 0$

Recognizing and Solving Quadratic-Like Equations

87–96 Solve the quadratic-like equations.

87. $4x^4 - 13x^2 + 9 = 0$

88. $8y^6 + 19y^3 - 27 = 0$

89. $5z^4 - 45 = 0$

90. $w^6 - 1001w^3 + 1000 = 0$

91. $x^{-4} - 4x^{-2} - 21 = 0$

92. $4z^{-2} - 24x^{-1} + 27 = 0$

93. $8x^{2/3} - 63x^{1/3} - 8 = 0$

94. $4x^{1/2} - 29x^{1/4} + 25 = 0$

95. $x^{-1/2} - 5x^{-1/4} + 4 = 0$

96. $64y^{-6} - 65y^{-3} + 1 = 0$

Completing the Square to Solve Quadratic Equations

97–104 Solve the quadratic equations by completing the square.

97. $x^2 - 8x - 9 = 0$

98. $3x^2 - 14x + 8 = 0$

99. $x^2 + 10x + 7 = 0$

100. $x^2 - 8x - 5 = 0$

101. $3x^2 - x - 6 = 0$

102. $2x^2 + 5x - 1 = 0$

103. $x^2 + bx + c = 0$

104. $ax^2 + bx + c = 0$

Combining Methods When Solving Quadratic Equations

105–114 Solve using the most efficient methods.

105. $5x^2 - 45 = 0$

106. $-18y^2 + 6y + 3 = 0$

107. $x^4 - 1 = 0$

108. $16y^4 - 72y^2 + 81 = 0$

109. $z^4 - 13z^2 + 36 = 0$

110. $4y^4 - 101y^2 + 25 = 0$

111. $27z^4 - 39z^2 + 12 = 0$

112. $25x^4 - 626x^2 + 25 = 0$

113. $x^8 - 17x^4 + 16 = 0$

114. $y^8 - 256y^4 = 0$

Solving Quadratic Inequalities Using Number Lines

115–130 Solve the quadratic inequalities.

115. $(x - 4)(x + 3) \geq 0$

116. $(2x + 1)(x - 5) > 0$

117. $(8 - x)(x + 6) \leq 0$

118. $x(x + 3) > 0$

119. $8 + 7x - x^2 \geq 0$

120. $x^2 + 7x + 12 < 0$

121. $\frac{x+2}{x-4}>0$

122. $\frac{3x-2}{x+5}<0$

123. $\frac{x-9}{x+3}\geq 0$

124. $\frac{x+1}{2x-7}\leq 0$

125. $\frac{x^2-4}{x+3}>0$

126. $\frac{x-1}{x^2-25}\leq 0$

127. $\frac{x^2-9}{x^2-16}\geq 0$

128. $\frac{x}{x^2-1}<0$

129. $\frac{x^2-5x+6}{x^2-5x-6}>0$

130. $\frac{6x^2+x-1}{20x^2-x-1}\leq 0$

Chapter 3

Solving Radical and Rational Equations

A *radical* equation is one that starts out with a square root, cube root, or some other root and gets changed into another form to make the solving process easier. The new form may have solutions that don't work in the original equation, but this method is still the easiest. A *rational* equation is one that involves a fractional expression — usually with a polynomial in the numerator and denominator. These equations are also changed in order to solve them, and they also carry the concern of an *extraneous* or false root.

The Problems You'll Work On

In this chapter, you'll work with radical and rational equations in the following ways:

- Solving radical equations with just one radical term
- Solving radical equations with two or more radical terms
- Checking answers for extraneous roots
- Solving rational equations by forming proportions
- Solving rational equations by finding a common denominator

What to Watch Out For

Don't let common mistakes like the following ones trip you up when working with radical or rational equations:

- Forgetting to check for extraneous solutions
- Squaring a binomial incorrectly when squaring both sides to get rid of the radical
- Distributing incorrectly when writing equivalent fractions using a common denominator
- Eliminating solutions that create a 0 in the denominator

Solving Rational Equations

131–150 Solve the rational equations for x.

131. $\frac{x}{40}=\frac{18}{24}$

132. $\frac{12}{x+1}=\frac{8}{x-1}$

133. $\frac{x+3}{10}=\frac{x+5}{6}$

134. $\frac{3x+1}{8}=\frac{2}{x+1}$

135. $\frac{7}{x+2}+\frac{5}{x}=\frac{14}{x+2}$

136. $\frac{x}{4}-\frac{8}{x}=\frac{x+5}{3}$

137. $\frac{5}{x+1}+\frac{3}{x-1}=\frac{x-16}{x^2-1}$

138. $\frac{6}{x+4}+\frac{2}{x-3}=\frac{6}{x^2+x-12}$

139. $\frac{9}{x}-\frac{2}{x-2}=\frac{3}{x^2-2x}$

140. $\frac{x}{6}+\frac{1}{x}=\frac{5}{3x}$

141. $\frac{5}{2x-1}-\frac{1}{x+3}=\frac{4}{2x^2+5x-3}$

142. $\frac{4}{x}-\frac{1}{x+1}=\frac{5}{x^2+x}$

143. $3-\frac{4}{x+1}=\frac{10}{x+2}$

144. $4+\frac{3}{x-6}=\frac{1}{x-4}$

145. $10-\frac{x}{x+3}=\frac{6}{x+5}$

146. $9+\frac{x+6}{x}=\frac{8}{x+4}$

147. $\frac{x}{x-3}=\frac{3}{x+3}+\frac{9}{x^2-9}$

148. $\frac{2x}{x+1}=\frac{5}{x-3}+\frac{1}{x^2-2x-3}$

149. $\frac{3x}{x-4}=\frac{x}{x+2}-\frac{8}{x^2-2x-8}$

150. $\frac{x}{x-5}=\frac{2x}{x+3}+\frac{4x}{x^2-2x-15}$

Taking on Radical Equations Involving One Radical Term

151–180 Solve the radical equations.

151. $\sqrt{x+1}=4$

152. $\sqrt{3x-2}=5$

153. $\sqrt{2x+1}-4=1$

154. $5+\sqrt{1-3x}=9$

155. $\sqrt{x-1}+3=2$

156. $1-\sqrt{3x-2}=3$

157. $\sqrt{3x+4}+2=x$

158. $\sqrt{4x-4}+4=x$

159. $\sqrt{6-5x}-6=x$

160. $\sqrt{1-2x}-7=x$

161. $x-\sqrt{5x-1}=3$

162. $x+\sqrt{7x+4}=8$

163. $x-\sqrt{3x-6}=2$

164. $x-\sqrt{2x+11}=2$

165. $x+\sqrt{8-4x}=2$

166. $x-\sqrt{13-2x}=5$

167. $\sqrt{x+4}-x=4$

168. $\sqrt{x+7}-x=7$

169. $\sqrt[3]{x-3}=2$

170. $\sqrt[4]{2x+1}=3$

171. $\sqrt[5]{x-8}=2$

172. $\sqrt[4]{5-x}=1$

173. $\sqrt[3]{x+4}-x=4$

174. $\sqrt[3]{3x+11}-x=3$

175. $\sqrt[3]{2x+7}+7=x$

176. $\sqrt[3]{4-4x}-3=x$

177. $\sqrt{5+\sqrt{x+3}}=3$

178. $\sqrt{\sqrt{x-9}-8}=1$

179. $\sqrt{1+\sqrt{x+2}}=2$

180. $\sqrt{x-\sqrt{x+2}}=2$

Solving Radical Equations with Multiple Radical Terms

181–190 Solve the radical equations involving two or more radical terms.

181. $\sqrt{3x+1}-4=\sqrt{x-7}$

182. $\sqrt{x+4}+1=\sqrt{2x+6}$

183. $\sqrt{5-4x}-3=\sqrt{x+9}$

184. $\sqrt{2-2x}+1=\sqrt{4-3x}$

185. $\sqrt{x+5}+\sqrt{5-x}=4$

186. $\sqrt{3x+1}+\sqrt{2x-1}=7$

187. $\sqrt{8-4x}-\sqrt{x+6}=2$

188. $\sqrt{5x+4}-\sqrt{2x+1}=3$

189. $\sqrt{8-x}-\sqrt{x+2}=\sqrt{x+5}$

190. $\sqrt{11x+3}-\sqrt{x+7}=\sqrt{x+2}$

Chapter 4

Graphs and Equations of Lines

Lines are graphs of linear expressions. When graphing lines or writing their equations, you have the slope intercept form, $y = mx + b$, or you can use the point-slope form, $y - y_1 = m(x - x_1)$. Slopes of parallel lines are equal in value, and slopes of perpendicular lines are negative reciprocals of one another.

The Problems You'll Work On

In this chapter, you'll work with the graphs and equations of lines in the following ways:

- Graphing a line given a point and a slope
- Graphing lines given two points
- Graphing a line parallel or perpendicular to a particular line
- Writing the equation of a line given a point and a slope
- Writing the equation of a line given two points
- Writing the equations of lines either parallel or perpendicular to a given line

What to Watch Out For

Don't let common mistakes like those that follow trip you up when working with the graphs and equations of lines:

- Drawing a line with a positive slope rather than a negative slope
- Not using the slope formula correctly when inserting the coordinates of points
- Forgetting to change the sign when determining the slope of a line perpendicular to a given line
- Distributing incorrectly when simplifying the point-slope form

Sketching Lines Using a Point and Slope

191–210 Sketch a graph of the line described.

191. Through (3, 2) with $m = 2$

192. Through (–2, –3) with $m = -3$

193. Through (0, 3) with $m = \frac{3}{5}$

194. Through (3, –2) with $m = -\frac{5}{2}$

195. Through (1, 3) and parallel to the line $y = 3x - 2$

196. Through (5, 1) and parallel to the line $y = -\frac{1}{3}x + 4$

197. Through (–2, –3) and parallel to the line $y = \frac{4}{7}x - 1$

198. Through (–3, –1) and parallel to the line $y = -2x + 1$

199. Through (4, 1) and parallel to the line $y = \frac{1}{2}x + 1$

200. Through (0, 0) and parallel to the line $y = -7x - 2$

201. Through (2, 4) and perpendicular to the line $y = 2x - 3$

202. Through (–3, –3) and perpendicular to the line $y = -\frac{1}{3}x - 2$

203. Through (–2, 7) and perpendicular to the line $y = -3x + 1$

204. Through (4, –1) and perpendicular to the line $y = \frac{1}{4}x$

205. Through (2, 3) and perpendicular to the line $y = 4$

206. Through (–2, 4) and perpendicular to the line $y = -2$

207. Through (2, 5) and perpendicular to the line $x = 1$

208. Through (1, –1) and perpendicular to the line $x = 3$

209. Through (4, 3) and perpendicular to the x-axis

210. Through (3, –5) and perpendicular to the y-axis

Writing Equations of Lines Given Point and Slope

211–220 Write an equation of the line with the given point and slope.

211. Through (3, –1); $m = 2$

212. Through (3, –1); $m = -2$

213. Through (3, –1); $m = \frac{1}{3}$

214. Through (3, –1); $m = -\frac{3}{4}$

215. Through (3, –1); $m = \frac{4}{5}$

216. Through (3, –1); $m = -5$

217. Through (3, –1); $m = \frac{2}{7}$

218. Through (3, –1); $m = -\frac{5}{3}$

219. Through (3, –1); $m = 0$

220. Through (3, –1); m is undefined

Writing Equations of Lines Given Two Points

221–228 Write an equation of the line through the two points.

221. (3, –2) and (4, 2)

222. (–5, 1) and (–3, 7)

223. (4, –3) and (1, 6)

224. (–2, –3) and (3, 4)

225. (5, 6) and (–1, 12)

226. (–3, 5) and (4, –4)

227. (6, 3) and (6, –8)

228. (4, –2) and (5, –2)

Finding Equations of Parallel Lines

229–234 Write an equation of the line parallel to the given line through the point.

229. Parallel to $y = 3x - 7$ through (1, 1)

230. Parallel to $y = -\frac{4}{3}x + 1$ through (2, 3)

231. Parallel to $x + 3y = 4$ through (3, 0)

232. Parallel to $4x - 2y = 7$ through (0, 5)

233. Parallel to $y = 6$ through (4, 8)

234. Parallel to $x = -3$ through (4, –7)

Writing Equations of Perpendicular Lines

235–240 Write an equation of the line perpendicular to the given line through the point.

235. Perpendicular to $y = -2x + 1$ through (2, 3)

236. Perpendicular to $y = \frac{5}{8}x - 2$ through (10, –7)

237. Perpendicular to $x - 5y = 10$ through (–2, 3)

238. Perpendicular to $3x + 2y = 4$ through (–9, –1)

239. Perpendicular to $x = 4$ through (–6, 2)

240. Perpendicular to $y = -8$ through (5, –3)

Chapter 5

Functions

A *function* is a relationship in which there is exactly one output value for each input value. The functions used in algebra incorporate all sorts of operations — from addition and subtraction to absolute value and factorial. Functions might be restricted as to their input values; the input values constitute the *domain*. And a function may have a limited number of output values (its *range*) due to the way the operations are performed on the input values.

The Problems You'll Work On

In this chapter, you'll work with functions in the following ways:

- Determining the domain and range from the function equation
- Recognizing *odd* and *even* functions
- Categorizing some functions as *one-to-one*
- Performing function composition
- Finding inverses of functions
- Working with piecewise functions

What to Watch Out For

Don't let common mistakes trip you up; watch for the following when working with functions:

- Making the correct exclusions in the domains of functions that have radicals or fractions in their function rule
- Distributing factors correctly when performing function compositions
- Squaring and cubing binomials correctly when performing function compositions
- Working through the correct steps when solving for a function's inverse

Determining a Function's Domain and Range

241–250 Find the domain and range of the function.

241. $f(x) = x^2 + 5$

242. $g(x) = -x^2 + 7$

243. $h(x) = \frac{x+2}{x-3}$

244. $k(x) = \frac{2x-1}{x+4}$

245. $f(x) = \sqrt{x-3}$

246. $g(x) = \sqrt{5-x}$

247. $h(x) = \sqrt{16-x^2}$

248. $k(x) = \frac{\sqrt{x+4}}{x-1}$

249. $f(x) = |x-2|$

250. $g(x) = 2|x+3|+8$

Finding Inverses of Functions

251–260 Find the inverse of the function.

251. $f(x) = x - 3$

252. $f(x) = 3x + 4$

253. $f(x) = x^3 + 7$

254. $f(x) = 3x^3 - 5$

255. $f(x) = \frac{x^5 - 4}{2}$

256. $f(x) = \frac{2}{x^3 - 1}$

257. $f(x)=2^x$

258. $f(x)=e^x$

259. $f(x)=\frac{x+3}{2x+1}$

260. $f(x)=\frac{x}{x-1}$

Recognizing Even and Odd Functions

261–270 Determine if the function is even or odd and whether it's one-to-one.

261. $f(x) = x^3$

262. $g(x) = x^2 + 2$

263. $h(x) = x^3 + 5$

264. $k(x)=\frac{x^2}{x^4-3}$

265. $f(x)=\sqrt[3]{x+1}$

266. $g(x)=\frac{x+3}{x-2}$

267. $h(x)=\frac{1}{x}$

268. $k(x)=\sqrt{9-x^2}$

269. $f(x)=\frac{1}{3}x^5$

270. $g(x) = 6$

Performing the Composition Operation on Functions

271–280 Find the composition of $(f \circ g)(x)$ given f and g.

271. $f(x) = x^2 + 3, g(x) = x + 1$

272. $f(x) = 2x^2 + 5x, g(x) = x - 2$

273. $f(x) = -x + 3$, $g(x) = x^2 - 2$

274. $f(x) = 4 - 5x^2$, $g(x) = 3 - x$

275. $f(x) = \frac{1-x}{2+x}$, $g(x) = x - 5$

276. $f(x) = \frac{x^2-9}{x^2-1}$, $g(x) = x + 4$

277. $f(x) = \sqrt{x-4}$, $g(x) = x^2 + 1$

278. $f(x) = \sqrt{9-x^2}$, $g(x) = x - 2$

279. $f(x) = x^3 + 2x + 1$, $g(x) = x + 1$

280. $f(x) = x^3 - 8$, $g(x) = x + 2$

Creating a Difference-Quotient and Simplifying

281–290 Given f(x), find the difference-quotient $\frac{f(x+h)-f(x)}{h}$.

281. $f(x) = x + 3$

282. $f(x) = 4x - 2$

283. $f(x) = x^2 - 2x + 3$

284. $f(x) = 5x^2 + 4x - 5$

285. $f(x) = x^3 - 3x + 4$

286. $f(x) = 2x^3 + 4x^2$

287. $f(x) = \frac{1}{x+3}$

288. $f(x)=\frac{3}{x-2}$

289. $f(x)=\sqrt{x+4}$

290. $f(x)=\sqrt{1-x}$

Evaluating Piecewise Functions for Particular Inputs

291–300 Given the piecewise function, evaluate as requested.

291. Evaluate $f(1)$ in $f(x)=\begin{cases} 2x, & x<0 \\ x+3, & x\geq 0 \end{cases}$

292. Evaluate $f(-1)$ in $f(x)=\begin{cases} x^2, & x<-1 \\ x-4, & x\geq -1 \end{cases}$

293. Evaluate $f(2)$ in $f(x)=\begin{cases} 4x, & x\leq 3 \\ -3, & x>3 \end{cases}$

294. Evaluate $f(0)$ in $f(x)=\begin{cases} 3-x, & x\leq 0 \\ x+5, & x>0 \end{cases}$

295. Evaluate $f(-1)$ in $f(x)=\begin{cases} 3-x^2, & x\leq -3 \\ x-1, & x>-3 \end{cases}$

296. Evaluate $f(-4)$ in $f(x)=\begin{cases} 2x^2, & x<-5 \\ 5-x, & x\geq -5 \end{cases}$

297. Evaluate $f(-5)$ in $f(x)=\begin{cases} |x|, & x \text{ is even} \\ -x^2, & x \text{ is odd} \end{cases}$

298. Evaluate $f(51)$ in $f(x)=\begin{cases} 2x-1, & x \text{ is prime} \\ -x, & x \text{ is composite} \end{cases}$

299. Evaluate $f(0)$ in $f(x)=\begin{cases} \sqrt{x}, & x\geq 9 \\ x!, & 0\leq x<9 \\ |x|, & x<0 \end{cases}$

300. Evaluate $f(4)$ in

$$f(x)=\begin{cases} \frac{x-4}{(x-3)(x-4)}, & x\neq 3,4 \\ 3x+7, & x=3 \\ 1, & x=4 \end{cases}$$

Chapter 6

Quadratic Functions and Relations

A quadratic function is created from a *quadratic expression* — an expression with a variable raised to the second power. The graphs of quadratic functions look like U-shaped curves that open upward or downward. A quadratic *relation* may open left or right. The key to graphing a quadratic function or relation is to find its vertex, determine which way it opens, and find a point or two that can be used to sketch the curve.

The Problems You'll Work On

In this chapter, you'll work with quadratic curves in the following ways:

- Determining the vertex and intercepts from the function rule
- Rewriting quadratic functions in the standard form for a parabola
- Sketching the graphs of parabolas
- Using quadratic functions and their properties to solve applications

What to Watch Out For

Don't let common mistakes trip you up; watch for the following when working with quadratic curves:

- Finding the *opposite* of the coefficient of *b* when solving for the coordinates of the vertex
- Watching for the correct direction of the parabola's opening when sketching
- Performing *completing the square* correctly when rewriting the parabola's equation in standard form
- Using the correct property of a parabola when solving an application

Determining the Vertex and Intercepts of a Parabola

301–310 Find the intercept(s) and vertex of the parabola.

301. $y = x^2 - 1$

302. $y = 9 - x^2$

303. $y = 10x - x^2$

304. $y = x^2 - 6x + 5$

305. $y = x^2 + 4x - 32$

306. $y = x^2 - 6x + 9$

307. $y = x^2 + 7$

308. $y = x^2 + 4x - 6$

309. $y = 2x^2 + 5x - 3$

310. $y = 3x^2 + 2x - 8$

Writing Equations of Parabolas in a Standard Form

311–320 Write the equation of the parabola in the standard form $y - k = a(x - h)^2$. Then identify the vertex.

311. $y = x^2 + 4x - 1$

312. $y = x^2 - 6x - 7$

313. $y = 2x^2 - 12x + 3$

314. $y = x^2 + 8x + 16$

315. $y = 3x^2 - 30x - 4$

316. $y = -x^2 + x - 3$

317. $y = x^2 - 12x$

318. $y = -2x^2 + 3x + 2$

319. $y = -10x^2 + 60x - 80$

320. $y = 6x^2 + 4x + 3$

Sketching Graphs of Parabolas

321–330 Sketch the graph of the parabola.

321. $y = x^2 - 16$

322. $y = 25 - x^2$

323. $y = x^2 + 2x - 15$

324. $y = -x^2 - 2x + 8$

325. $y = x^2 + 4x + 3$

326. $y = 6x^2 - 7x - 3$

327. $y = -6x^2 + 2x + 20$

328. $y = x^2 + 2x + 5$

329. $y = -x^2 + 4x - 8$

330. $y = 3x^2 + 6x + 3$

Using Quadratic Equations in Applications

331–340 Solve the following quadratic applications.

331. The height of a rocket (in feet), t seconds after being shot upward in the air, is given by $h(t) = -16t^2 + 128t + 320$. How high does the rocket rise before returning to the ground?

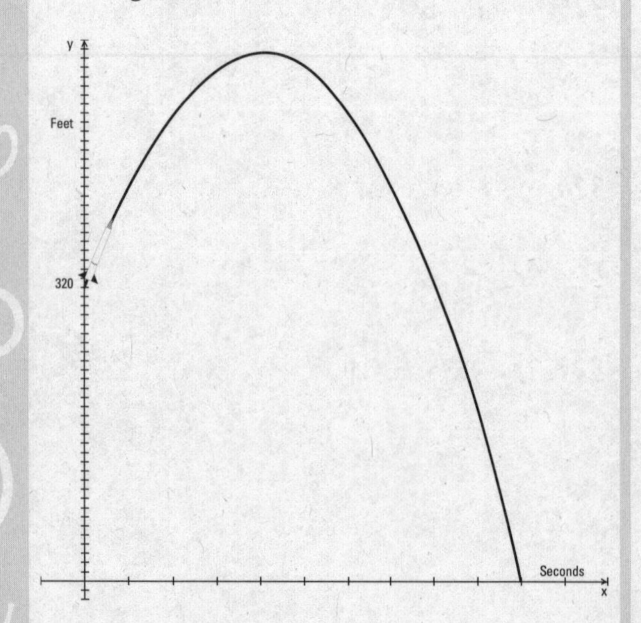

332. The height of a rocket (in feet), t seconds after being shot upward in the air, is given by $h(t) = -16t^2 + 128t + 320$. How long does it take before hitting the ground?

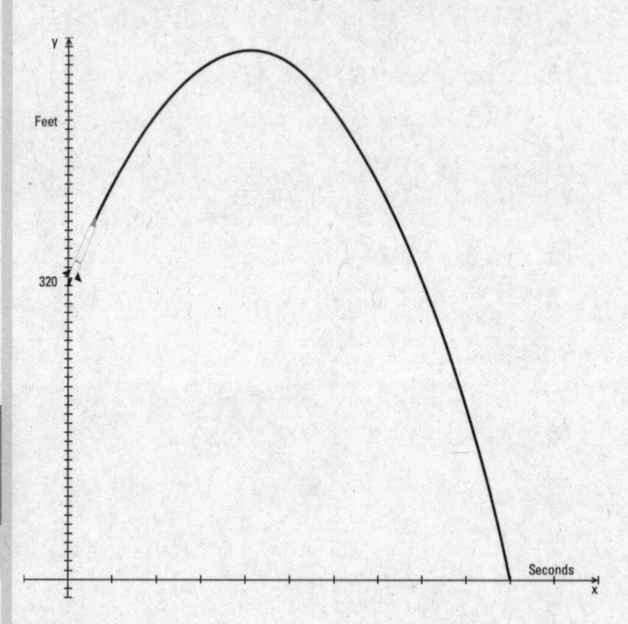

333. The height of a ball, t seconds after being shot upward in the air, is given by $h(t) = -16t^2 + 80t$. How high does the ball get before returning to the ground?

334. The height of a ball, t seconds after being shot upward in the air, is given by $h(t) = -16t^2 + 80t$. How long does it take before hitting the ground?

335. The amount of profit (in dollars) made when x items are sold is determined with the profit function $p(x) = -10x^2 + 1{,}000x - 9{,}000$. How many items must be sold before the "profit" is positive?

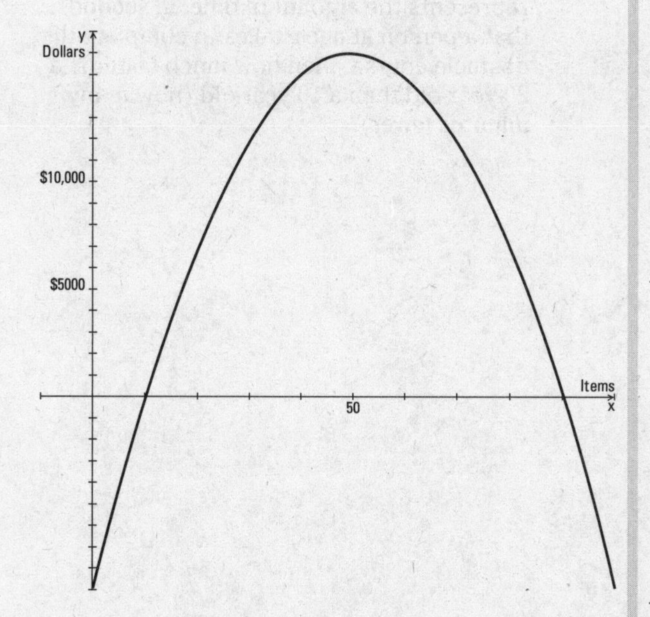

336. The amount of profit (in dollars) made when x items are sold is determined with the profit function $p(x) = -10x^2 + 1{,}000x - 9{,}000$. What is the greatest possible profit?

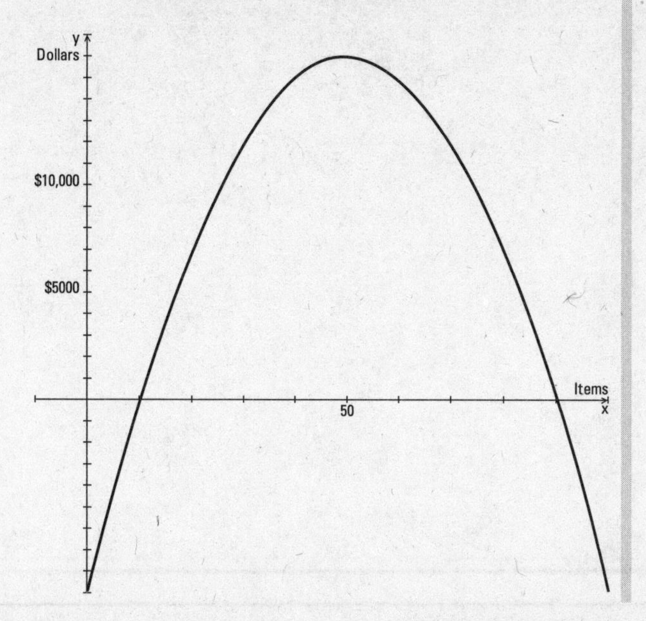

337. The average number of skis per day sold at a sports store during the month of January is projected to be $k(n) = -\frac{2}{15}n^2 + \frac{10}{3}n + 100$, where n corresponds to the day of the month. On what day is the greatest number of skis expected to be sold?

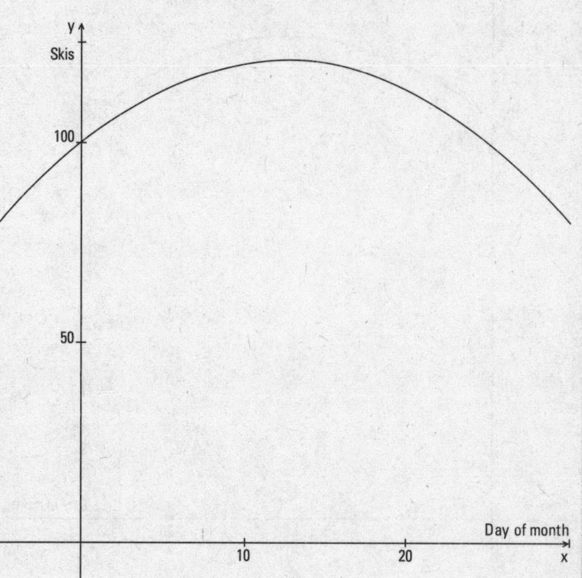

338. The average number of skis per day sold at a sports store during the month of January is projected to be $k(n) = -\frac{2}{15}n^2 + \frac{10}{3}n + 100$, where n corresponds to the day of the month. When will the least number of skis be sold?

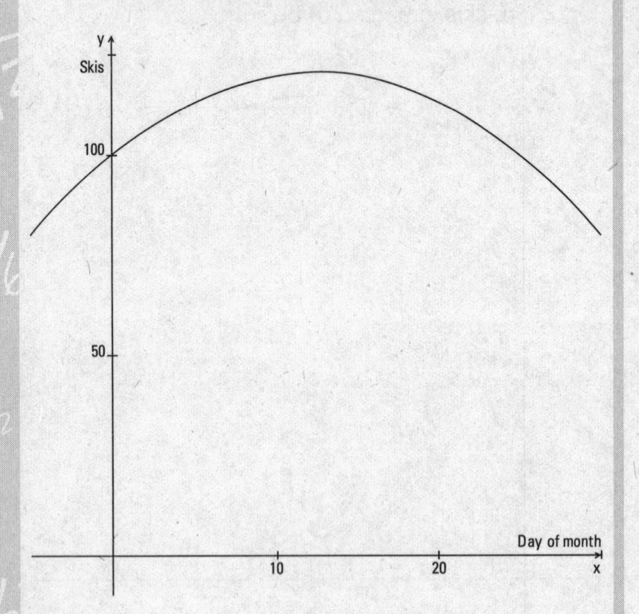

339. The average amount of time (in seconds) it takes a person to complete an obstacle course depends on the person's age. If the function $t(g) = \frac{1}{7}\left(40g^2 - 2{,}400g + 36{,}070\right)$ represents the amount of time, in seconds, that a person at age g takes to complete this obstacle course, then at what age is a person expected to be the fastest (take the least amount of time)?

340. The average amount of time (in seconds) it takes a person to complete an obstacle course depends on the person's age. If the function $t(g) = \frac{1}{7}\left(40g^2 - 2{,}400g + 36{,}070\right)$ represents the amount of time, in seconds, that a person at age g takes to complete the obstacle course, then how much faster is a 20-year-old than a 10-year-old (how many minutes fewer)?

Chapter 7

Polynomial Functions and Equations

A polynomial function is one in which the coefficients are all real numbers, and the exponents on the variables are all whole numbers. A polynomial whose greatest power is 2 is called a *quadratic polynomial;* if the highest power is 3, then it's called a *cubic polynomial.* A highest power of 4 earns the name *quartic* (not to be confused with quadratic), and a highest power of 5 is called *quintic.* There are more names for higher powers, but the usual practice is just to refer to the power rather than to try to come up with the Latin or Greek prefix.

The Problems You'll Work On

In this chapter, you'll work with polynomial functions and equations in the following ways:

- Determining the x and y intercepts from the function rule (equation)
- Solving polynomial equations using grouping
- Applying the *rational root theorem* to find roots
- Using *Descartes' rule of sign* to count possible real roots
- Making use of *synthetic division*
- Graphing polynomial functions

What to Watch Out For

Don't let common mistakes trip you up; watch for the following ones when working with polynomial functions and equations:

- Forgetting to change the signs in the factored form when identifying x-intercepts
- Making errors when simplifying the terms in $f(-x)$ applying Descartes' rule of sign
- Not changing the sign of the divisor when using synthetic division
- Not distinguishing between curves that cross from those that just touch the x-axis at an intercept
- Graphing the incorrect *end-behavior* on the right and left of the graphs

Recognizing the Intercepts of Polynomials

341–350 Find the intercepts of the polynomial.

341. $y = x(x - 1)(x + 1)$

342. $y = x^2(x + 2)(x + 3)^2$

343. $y = (x + 4)(x - 3)(x - 2)$

344. $y = (x + 3)^2(x - 2)^2$

345. $y = x^2 + 6x$

346. $y = x^3 + 2x^2$

347. $y = x^2 + x + 1$

348. $y = -x^2 - 2x - 4$

349. $y = x^3 + 4x^2 + 3x$

350. $y = x^3 - 1$

Factoring by Grouping to Solve for Intercepts

351–360 Find the intercepts of the polynomial. To find the x-intercepts, use factoring by grouping.

351. $y = x^3 + 3x^2 - x - 3$

352. $y = x^3 - 2x^2 - 4x + 8$

353. $y = x^3 + 10x^2 - 100x - 1{,}000$

354. $y = -x^3 - x^2 + 9x + 9$

355. $y = -2x^3 - 3x^2 + 32x + 48$

356. $y = 4x^3 + 28x^2 - x - 7$

357. $y = -3x^3 + x^2 + 108x - 36$

358. $y = x^3 - 7x^2 - 49x + 343$

359. $y = x^4 + 3x^3 - 8x - 24$

360. $y = x^4 - 4x^3 + x - 4$

Applying the Rational Root Theorem to Find Roots

361–370 Use the Rational Root Theorem and Descartes' Rule of Signs to list the possible rational roots of the polynomial.

361. $0 = x^4 + 3x^3 + 2x + 6$

362. $0 = x^5 - 2x^4 + 3x^2 + 2x - 8$

363. $0 = x^5 - 2x^4 + 3x^2 + 2x + 36$

364. $0 = x^5 - 2x^4 - 3x^2 + 2x - 50$

365. $0 = 2x^4 + 3x^2 - 2x + 5$

366. $0 = 3x^4 + 3x^2 - 2x - 6$

367. $0 = 4x^4 + 3x^2 - 2x - 6$

368. $0 = 2x^5 - 2x^4 + 3x^2 + 2x + 10$

369. $0 = 6x^5 - 2x^4 + 3x^2 + 2x + 2$

370. $0 = 8x^4 + 3x^2 - 2x + 12$

Performing Synthetic Division to Factor Polynomials

371–380 Factor the polynomial expressions using synthetic division.

371. $y = x^3 - x^2 - 32x + 60$

372. $y = x^3 + 2x^2 - 55x - 56$

373. $y = x^3 + 11x^2 + 15x - 27$

374. $y = x^3 + 9x^2 + 20x + 12$

375. $y = x^3 + x^2 - 33x + 63$

376. $y = x^3 + 2x^2 - 32x - 96$

377. $y = x^4 + 3x^3 - 15x^2 - 19x + 30$

378. $y = x^4 + 8x^3 + 17x^2 - 2x - 24$

379. $y = 2x^4 + 11x^3 - 3x^2 - 44x - 20$

380. $y = -3x^4 + 5x^3 + 25x^2 - 45x + 18$

Evaluating Polynomials for Input Values

381–390 Evaluate the functions for the given input using the remainder theorem.

381. Given $f(x) = x^3 - x^2 - 32x + 60$, find $f(1)$.

382. Given $f(x) = x^3 + 2x^2 - 55x - 56$, find $f(1)$.

383. Given $f(x) = x^3 + 11x^2 + 15x - 27$, find $f(1)$.

384. Given $f(x) = x^3 + 9x^2 + 20x + 12$, find $f(-1)$.

385. Given $f(x) = x^3 + x^2 - 33x + 63$, find $f(2)$.

386. Given $f(x) = x^3 + 2x^2 - 32x - 96$, find $f(-2)$.

387. Given $f(x) = x^4 + 3x^3 - 15x^2 - 19x + 30$, find $f(-1)$.

388. Given $f(x) = x^4 + 8x^3 + 17x^2 - 2x - 24$, find $f(-2)$.

389. Given $f(x) = 2x^4 + 11x^3 - 3x^2 - 44x - 20$, find $f(1)$.

390. Given $f(x) = -3x^4 + 5x^3 + 25x^2 + 9x + 18$, find $f(2)$.

Investigating End-Behavior of Polynomials

391–400 Determine the end-behavior of the polynomials.

391. $y = x^3 + 10x^2 + 20x + 12$

392. $y = -x^3 + x^2 - 33x - 63$

393. $y = 11x^3 - 3x^2 - 44x - 20$

394. $y = -2x^4 + 11x^3 - 3x^2 - 44x - 20$

395. $y = -3x^4 + 5x^3 + 18$

396. $y = 4x^4 + 3x^2 - 2x - 6$

397. $y = 2x^5 - 2x^4 + 3x^2 + 2x + 10$

398. $y = -x^5 - 2x^4 + 2x + 10$

399. $y = 2x^6 - 10$

400. $y = -x^6 + 3x^2 + 2x + 10$

Sketching the Graphs of Polynomial Functions

401 – 410 Sketch the graph of the polynomial.

401. $y = x^3 - 2x^2 - 15x$

402. $y = x^3 - 6x^2 + 8x$

403. $y = x^4 - 17x^2 + 16$

404. $y = -x^4 + 29x^2 - 100$

405. $y = x^3 - 14x^2 + 63x - 90$

406. $y = x^4 - 2x^3 - 19x^2 + 8x + 60$

407. $y = x^4 + 9x^3 + 5x^2 - 81x - 126$

408. $y = 2x^4 - 3x^3 - 34x^2 + 48x + 32$

409. $y = 4x^4 - 40x^3 + 91x^2 + 90x - 225$

410. $y = 9x^4 - 36x^3 + 35x^2 + 4x - 4$

Chapter 8

Rational Functions

A *rational* function is special in that the function rule involves a fraction with a polynomial in both the numerator and the denominator. Rational functions have restrictions in their domain; any value creating a 0 in the denominator has to be excluded. Many of these exclusions are identified as vertical asymptotes. The x-intercepts of rational functions can be solved for by setting the numerator equal to 0; this is done after you've determined that there are no common factors in the numerator and denominator. A rational function can have a horizontal asymptote — as long as the highest power in the numerator is not greater than that in the denominator.

The Problems You'll Work On

In this chapter, you'll work with rational functions in the following ways:

- Determining the domain and range of the function
- Removing discontinuities, when possible
- Finding limits at infinity and infinite limits
- Writing equations of vertical, horizontal, and slant asymptotes
- Solving for intercepts
- Graphing rational functions

What to Watch Out For

Don't let common mistakes trip you up; watch out for the following ones when working with rational functions:

- Categorizing a discontinuity as a vertical asymptote rather than removable
- Not dividing correctly when solving for the horizontal asymptote
- Sketching the curve on the wrong side of the horizontal asymptote

Investigating the Domain of a Rational Function

411–420 Determine the domain of the rational function.

411. $y = \frac{x-2}{x+1}$

412. $y = \frac{x-3}{x-5}$

413. $y = \frac{2x+3}{x+4}$

414. $y = \frac{3x-5}{x-2}$

415. $y = \frac{x-4}{x^2-3x+2}$

416. $y = \frac{x+6}{x^2-25}$

417. $y = \frac{x^2+3x-4}{x^2-5x}$

418. $y = \frac{2x^2+3x-1}{x^2+2x+1}$

419. $y = \frac{4x^2-25}{x^2-9}$

420. $y = \frac{3x^2-4}{4x^2+1}$

Determining a Function's Removable Discontinuity

421–430 Find the removable discontinuity of the function.

421. $y = \frac{x^2-1}{x^2-3x+2}$

422. $y = \frac{x+4}{x^2-16}$

423. $y = \frac{x^2+5x+6}{x^2-4}$

424. $y = \frac{2x^2+9x+9}{2x^2-5x-12}$

425. $y = \frac{2x^2 - 5x - 3}{x^2 + x - 12}$

426. $y = \frac{x^2 - 5x}{x^2 - 25}$

427. $y = \frac{x^2 - 3x - 4}{5x^2 - 22x + 8}$

428. $y = \frac{x^3 - x^2 + 3x - 3}{x^3 + 2x^2 - x - 2}$

429. $y = \frac{x^4 - 4x^3 + x - 4}{x^4 - 10x^2 + 9}$

430. $y = \frac{x^4 - 9x^2 + 20}{x^4 + 7x^3 + 8x + 56}$

Evaluating a Function When x Is Infinitely Large or Small

431–440 Find the limit of the function as x approaches ∞.

431. $y = \frac{x + 3}{x - 2}$

432. $y = \frac{2x - 4}{x + 3}$

433. $y = \frac{3x^2 - 4x + 1}{x^2 - 2x + 3}$

434. $y = \frac{x^2 - 1}{3x^2 + 7}$

435. $y = \frac{4x^2 + 3x - 2}{2x^2 - 1}$

436. $y = \frac{x^2 + x - 3}{2x^2 + 5x + 3}$

437. $y = \frac{x + 1}{x^2 - 2x + 4}$

438. $y = \frac{4x + 3}{x^3 - 4}$

439. $y = \frac{x^2 + 4}{7x + 3}$

440. $y = \frac{x^3 - 4x^2 + 2}{x^2 - 4x + 1}$

Determining a Function's Infinite Limit

441–450 Find the indicated limit.

441. $\lim_{x\to 4^+} \frac{x+1}{x-4}$

442. $\lim_{x\to -6^+} \frac{5-x}{x+6}$

443. $\lim_{x\to -3^-} \frac{2-x}{x+3}$

444. $\lim_{x\to 2^-} \frac{x+1}{x-2}$

445. $\lim_{x\to 3^-} \frac{x}{(x-3)(x+2)}$

446. $\lim_{x\to 3^+} \frac{x^2}{(x-3)(x+3)}$

447. $\lim_{x\to -4^+} \frac{x+3}{x(x-1)(x+4)}$

448. $\lim_{x\to 5^+} \frac{x^2-3x+2}{x^2-25}$

449. $\lim_{x\to -4^-} \frac{x+3}{4x+x^2}$

450. $\lim_{x\to 0^+} \frac{x+3}{x^2-6x}$

Writing Equations of Asymptotes

451–460 Identify all vertical and horizontal asymptotes of the rational function.

451. $y=\frac{x+3}{x-2}$

452. $y=\frac{2x-3}{x+7}$

453. $y=\frac{x}{x+2}$

454. $y=\frac{x^2-3x}{x^2-4}$

455. $y = \frac{3x^2 + 2x - 1}{x^2 + 4x - 5}$

456. $y = \frac{x^2 + 3x - 4}{x^2 + 7x + 12}$

457. $y = \frac{x^2 - 5x + 4}{x^2 - 1}$

458. $y = \frac{2x^2}{x^2 + 3}$

459. $y = \frac{x^2 + 3}{2x^2}$

460. $y = \frac{x^3 - 16x}{x^3 - 9x}$

Adding Oblique Asymptotes to the Mix

461–465 Identify all vertical and oblique asymptotes.

461. $y = \frac{x^2 + x - 6}{x - 1}$

462. $y = \frac{x^2 - 2x - 3}{x + 4}$

463. $y = \frac{x^2 + 4x}{x - 5}$

464. $y = \frac{x^3 - 3x^2 + 1}{x^2 - 4}$

465. $y = \frac{x^3}{x^2 + 1}$

Sketching the Graphs of Rational Functions

466–480 Sketch the graph of the rational function, indicating all intercepts and asymptotes.

466. $y = \frac{x}{x - 3}$

467. $y = \frac{2x}{x + 1}$

468. $y = \frac{x - 4}{x - 2}$

469. $y = \frac{x + 7}{x - 1}$

470. $y=\frac{x^2-5x}{x^2-1}$

471. $y=\frac{x^2+3x}{x^2-4}$

472. $y=\frac{2x^2-5x-12}{x^2-5x-6}$

473. $y=\frac{3x^2-3}{x^2+2x-8}$

474. $y=\frac{x}{x^2-9}$

475. $y=\frac{4x}{1-x^2}$

476. $y=\frac{x^2-3x-4}{x^2+4x+3}$

477. $y=\frac{x^2+5x}{x^2+3x-10}$

478. $y=\frac{x^2+x}{x-1}$

479. $y=\frac{x^2+3x-5}{x+3}$

480. $y=\frac{x}{x^2+1}$

Chapter 9

Exponential and Logarithmic Functions

Exponential functions are inverses of logarithmic functions — and vice versa. Working with exponential and logarithmic functions requires facility with the correspondence and interrelationship between them. The base of an exponential function must be positive; many mathematicians also exclude the number 1 from being a base. A logarithmic function can have any positive base except the number 1, and the *argument* (input value) of a logarithmic function must also be positive.

The Problems You'll Work On

In this chapter, you'll work with exponential and logarithmic functions in the following ways:

- Evaluating exponential and logarithmic expressions
- Using the laws of logarithms
- Sketching exponential and logarithmic functions
- Solving for the function's inverse
- Solving exponential equations
- Solving logarithmic equations and checking for extraneous roots

What to Watch Out For

Don't let common mistakes trip you up; remember the following ones when working with exponential and logarithmic functions:

- Not using the order of operations properly when evaluating exponential expressions
- Writing a binomial argument (input statement) incorrectly — splitting it up
- Using laws of logarithms incorrectly

Evaluating Exponential Functions for Input

481–490 Evaluate the exponential functions for the input values indicated.

481. If $f(x) = 4^x$, $f(2) =$

482. If $g(x) = 3^x$, $g(-3) =$

483. If $h(x) = 5^{-2x}$, $h\left(-\frac{1}{2}\right) =$

484. If $k(x) = 2^{x^2+1}$, $k(-1) =$

485. If $f(x) = \left(\frac{1}{4}\right)^{x-1}$, $f(0) =$

486. If $g(x) = \left(\frac{1}{2}\right)^{-x}$, $g(4) =$

487. If $k(x) = 5(2)^{x+1}$, $k(3) =$

488. If $h(x) = 10\left(\frac{1}{5}\right)^{2x}$, $h(-3) =$

489. If $m(x) = 4(3)^{2x} + 5$, $m(2) =$

490. If $n(x) = 100(0.01)^{-2x} - 3$, $n(2) =$

Evaluating Exponential Functions in Base e

491–500 Evaluate the exponential functions for the input values indicated. Write your answer as a non-negative power of e.

491. If $f(x) = e^{x-1}$, $f(1) =$

492. If $g(x) = e^{2x}$, $g(-1) =$

493. If $h(x) = e^{x^2-2}$, $h(-2) =$

494. If $k(x) = e^{4x-2}$, $k(1) =$

495. If $m(x) = 2e^x$, $m(-4)$ =

496. If $n(x) = -3e^{-3x}$, $n(-1)$ =

497. If $p(x) = \left(\frac{1}{e}\right)^{x-1}$, $p(-2)$ =

498. If $q(x) = \left(\frac{1}{2e}\right)^x$, $q(2)$ =

499. If $r(x) = \left(\frac{3}{e}\right)^{1-x}$, $r(2)$ =

500. If $t(x) = \left(2e^x\right)^2$, $t(-1)$ =

Sketching the Graphs of Exponential Functions

501–505 Sketch the graph of the exponential function.

501. $y = 2^x$

502. $y = 3^{x-1}$

503. $y = \left(\frac{1}{2}\right)^{x+2}$

504. $y = e^x$

505. $y = e^{2-x}$

Finding Values of Logarithmic Expressions for Given Input

506–515 Evaluate the logarithm.

506. $\log_2(8)$

507. $\log_3\left(\frac{1}{9}\right)$

508. $\log_4\left(\frac{1}{2}\right)$

509. $\log_5\left(\frac{1}{125}\right)$

510. $\log_8\left(\sqrt{2}\right)$

511. $\log(0.001)$

512. $\log_9(27)$

513. $\log_{16}(8)$

514. $\log_4(32)$

515. $\log\left(\sqrt{1000}\right)$

Evaluating Natural Logarithms in Base e

516–525 Evaluate the natural logarithm.

516. $\ln e^2$

517. $\ln e^{-1/2}$

518. $\ln \frac{1}{e}$

519. $\ln \frac{1}{e^5}$

520. $\ln 1$

521. $\ln e$

522. $\ln \sqrt{e}$

523. $\ln \sqrt{e^5}$

524. $\ln \frac{e^2}{\sqrt{e}}$

525. $\ln \frac{\sqrt[3]{e}}{e^5}$

Sketching the Graphs of Log Functions

526–530 Sketch the graph of the log function.

526. $y = \log_2 x$

527. $y = \log_2(x-1)$

528. $y = \log_{10}(x+2)$

529. $y = \ln x$

530. $y = 2\ln x$

Rewriting a Log Function as Its Exponential Inverse

531–535 Find the inverse of the log function.

531. $f(x) = \log_3 x$

532. $f(x) = \log_4 2x$

533. $f(x) = \log(x+1)$

534. $f(x) = \ln(x-2)$

535. $f(x) = 3\ln x$

Changing Exponential Functions to Logarithmic Inverses

536–540 Find the inverse of the exponential function.

536. $f(x) = 2^x$

537. $f(x) = 3^{x+1}$

538. $f(x) = e^{x-1}$

539. $f(x) = 3e^{2x}$

540. $f(x) = \frac{1}{e^x}$

Solving Exponential Equations

541–550 Solve the exponential equation for x.

541. $4^{x+1} = 8^{3-x}$

542. $25^{2x-3} = 125^{x+1}$

543. $9^{x-5} = 27^{2x-3}$

544. $16^{x-2} = 32^{2-x}$

545. $\left(\frac{1}{9}\right)^{x} = 27^{x-3}$

546. $\left(\frac{1}{8}\right)^{2-x} = \left(\frac{1}{4}\right)^{x+3}$

547. $\sqrt{8^{x}} = 4^{3x+1}$

548. $e^{5x+2} = \left(\frac{1}{e}\right)^{x-1}$

549. $\sqrt{e^{x-4}} = e^{5-x}$

550. $e^{x^2} = \frac{1}{e^{3x+2}}$

Solving Logarithmic Equations

551–570 Solve the logarithmic equation for x.

551. $\log_2 x = 3$

552. $\log_3 x = 2$

553. $\log_9 x = \frac{1}{2}$

554. $\log_{16} x = -\frac{1}{4}$

555. $\log_7 x = -1$

556. $\log_6 x = -3$

557. $\ln x = -5$

558. $\ln(x+1) = 2$

559. $\ln(2x-2) = -4$

560. $\ln e^{x+1} = 5$

561. $\log_4 x + \log_4(x+2) = \frac{3}{2}$

562. $\log_3(x-2) + \log_3(x+6) = 2$

563. $\log_2(x+14) - \log_2 x = 3$

564. $\log_7(5-x) + \log_7(2-x) = \log_7 28$

565. $\log_6(x^2-9) - \log_6(x-3) = \log_6 7$

566. $\ln(x+1) + \ln(x-2) = \ln 4$

567. $\ln(x-5) + \ln(x-7) = \ln 3$

568. $\ln x + \ln(x-4) = \ln(x-6)$

569. $\ln x + \ln(x+3) = \ln(2x+12)$

570. $\ln(x+3) - \ln(x-2) = \ln 2$

Chapter 10

Conic Sections

Conic sections get this designation because they can all be described using slices of cones. The most familiar conic section is the circle. Our planets follow elliptical orbits, and comets hyperbolic orbits. That leaves the parabola to reflect upon — it's the shape of the reflector plates in your car's headlights.

The Problems You'll Work On

In this chapter, you'll work with the different conic sections in the following ways:

- Writing the equations of parabolas in their standard form
- Picking out the vertex, focus, directrix, and axis of symmetry from the standard equation of a parabola
- Sketching parabolas
- Writing the equations of circles in their standard form
- Picking out the center and radius from the standard equation of a circle
- Sketching circles
- Writing the equations of ellipses in their standard form
- Picking out the center, foci, and axes from the standard equation of an ellipse
- Sketching ellipses
- Writing the equations of hyperbolas in their standard form
- Picking out the center, foci, and axes from the standard equation of a hyperbola
- Sketching hyperbolas

What to Watch Out For

Don't let common mistakes trip you up; watch for the following when working with conic sections and their standard equations:

- Recognizing which standard form to apply to a general equation
- Completing the square correctly to write the standard equation
- Finding square roots when using the standard form to sketch conic sections

Rewriting Equations of Parabolas in Standard Form

571–580 Write the equation of the parabola in standard form, either $y - k = a(x - h)^2$ or $x - h = a(y - k)^2$.

571. $y = x^2 - 6x + 2$

572. $y = -2x^2 + 8x - 5$

573. $2y = x^2 - 16x + 10$

574. $x^2 + 4x + y = 7$

575. $y + 4x^2 - 24x = 8$

576. $x = y^2 - 10y + 2$

577. $x = 3y^2 + 12y$

578. $2x - 4y^2 + 20y = 5$

579. $2y^2 = 3x$

580. $9x^2 = 5y$

Determining a Vertex, Focus, Directrix, and Axis of Symmetry

581–585 Identify the vertex, focus, directrix, and axis of symmetry of each parabola.

581. $y - 3 = 4(x + 2)^2$

582. $y = -3(x - 1)^2 + 2$

583. $x = -2(y - 4)^2 + 1$

584. $x = \frac{1}{2}y^2 - 8$

585. $3x^2 = 2y$

Sketching the Graph of a Parabola

586–590 Sketch the graph of the parabola.

586. $y-5=-\frac{1}{3}(x+2)^2$

587. $x+3=4(y-1)^2$

588. $y=x^2-8x+1$

589. $x=y^2-4y-3$

590. $3x^2+5y=0$

Rewriting a Circle's Equation in Standard Form

591–600 Write the equation of each circle in the standard form $(x-h)^2+(y-k)^2=r^2$.

591. $x^2+y^2-2y=8$

592. $x^2+4x+y^2=12$

593. $x^2+y^2-8x+10y=23$

594. $x^2+y^2+12x+2y+1=0$

595. $x^2+y^2+2x-y=1$

596. $12x^2+12y^2-72x+60y=9$

597. $5x^2+5y^2-10y=11$

598. $x^2+4x=5-y^2$

599. $y^2-11=10y-x^2$

600. $9x^2=25-9y^2$

Identifying the Center and Radius of a Circle

601–605 Find the center and radius of each circle.

601. $(x-4)^2+(y+3)^2=4$

602. $(x+1)^2+y^2=25$

603. $(x+3)^2+\left(y-\frac{1}{4}\right)^2=\frac{1}{16}$

604. $x^2+y^2-4x-8y=5$

605. $x^2=\frac{1}{9}-y^2$

Sketching the Graph of a Circle

606–610 Sketch the graph of each circle, indicating the center and a point on the circle.

606. $x^2+y^2=16$

607. $(x-3)^2+(y-1)^2=9$

608. $(x+1)^2+y^2=1$

609. $(x-2)^2+(y-2)^2=4$

610. $(x-1)^2+y^2=\frac{1}{4}$

Writing an Ellipse's Equation in Standard Form

611–620 Write the equation of each ellipse in the standard form $\frac{(x-h)^2}{a^2}+\frac{(y-k)^2}{b^2}=1$.

611. $5x^2+2y^2=10$

612. $3x^2+4y^2=6$

613. $9x^2+4y^2-18x+16y=11$

614. $4x^2+25y^2+40x-50y+25=0$

615. $16x^2 + 9y^2 + 96x = 0$

616. $x^2 + 8y^2 - 32y = 0$

617. $2x^2 + y^2 - 4x + 6y - 9 = 0$

618. $x^2 + 3y^2 + 10x + 30y + 46 = 0$

619. $b^2x^2 + a^2y^2 - 2b^2x - 2a^2y + a^2 + b^2 = a^2b^2$

620. $b^2x^2 + a^2y^2 - 2b^2hx - 2a^2ky + a^2k^2 + b^2h^2 = a^2b^2$

Identifying a Center, the Foci, and Axis Endpoints

621–625 Determine the center, the foci, and the endpoints of the major axis and minor axis of the ellipses.

621. $\frac{x^2}{16} + \frac{y^2}{25} = 1$

622. $\frac{(x-2)^2}{9} + \frac{(y+3)^2}{4} = 1$

623. $25x^2 + 4y^2 = 100$

624. $(x+3)^2 + \frac{(y-5)^2}{9} = 1$

625. $\frac{(x-1)^2}{100} + \frac{(y+3)^2}{4} = 1$

Sketching the Graphs of Ellipses

626–630 Sketch the graph of the ellipse.

626. $\frac{x^2}{4} + \frac{y^2}{36} = 1$

627. $\frac{x^2}{64} + \frac{y^2}{9} = 1$

628. $\frac{(x-3)^2}{25} + \frac{(y+2)^2}{49} = 1$

629. $\frac{x^2}{36} + \frac{(y-1)^2}{25} = 1$

630. $\frac{(x-1)^2}{2} + \frac{(y+3)^2}{8} = 1$

Writing the Equation of a Hyperbola in Standard Form

631–640 Write the equation of each hyperbola in standard form, either $\frac{(x-h)^2}{a^2}-\frac{(y-k)^2}{b^2}=1$ *or* $\frac{(y-k)^2}{b^2}-\frac{(x-h)^2}{a^2}=1$.

631. $9x^2 - 4y^2 = 36$

632. $16y^2 - 25x^2 = 400$

633. $36y^2 - 4x^2 + 216y + 8x + 176 = 0$

634. $9x^2 - 4y^2 - 18x + 40y = 127$

635. $4x^2 - 16y^2 - 16x - 32y = 64$

636. $4x^2 - 25y^2 + 24x - 100y = 164$

637. $3x^2 - y^2 - 30x - 6y + 18 = 0$

638. $y^2 - 9x^2 - 2y - 72x = 152$

639. $4x^2 - 9y^2 + 16x + 15 = 0$

640. $4x^2 - y^2 - 16x - 2y + 11 = 0$

Finding the Center, Foci, and Asymptotes of Hyperbolas

641–645 Determine the center, foci, and equations of the asymptotes of the hyperbola.

641. $\frac{(y+3)^2}{64}-\frac{(x+1)^2}{36}=1$

642. $\frac{x^2}{16}-\frac{(y-3)^2}{9}=1$

643. $\frac{(x-3)^2}{81}-\frac{(y+1)^2}{144}=1$

644. $16y^2 - x^2 = 16$

645. $9x^2 - 36y^2 = 324$

Sketching the Graphs of Hyperbolas

646–650 Sketch the graph of the hyperbola.

646. $\frac{x^2}{4}-\frac{y^2}{49}=1$

647. $\frac{y^2}{16}-\frac{x^2}{1}=1$

648. $\frac{(x-1)^2}{25}-\frac{(y+1)^2}{4}=1$

649. $\frac{(y-2)^2}{9}-\frac{(x-2)^2}{36}=1$

650. $25x^2-y^2=25$

Chapter 11

Systems of Linear Equations

A *linear equation* consists of variable terms whose exponents are always the number 1. When you have two variables, the equation can be represented by a line. With three terms, you can draw a plane to describe the equation. More than three variables is indescribable, because there are only three dimensions. When you have a system of linear equations, you can look for the values of the variables that work for all the equations in the system — the common solutions. Sometimes there's just one solution, sometimes many, and sometimes there's no solution at all.

The Problems You'll Work On

In this chapter on systems of linear equations, you'll see the following:

- Determining the point of intersection of two lines
- Finding a single point of intersection of three planes
- Writing expressions for multiple solutions of systems
- Writing systems in the *echelon* form or *reduced row echelon* form
- *Decomposing* fractions using systems of linear equations

What to Watch Out For

Don't let common mistakes trip you up; watch for the following ones when working with systems of linear equations:

- Trying to eliminate more than one variable when solving a system
- Distributing negative numbers incorrectly over several terms
- Writing the coordinates of ordered pairs and ordered triples in the incorrect positions
- Placing the incorrect numerator over its denominator when decomposing fractions

Solving Systems of Two Linear Equations

651–660 Solve the system of two linear equations for x and y.

651. $\begin{cases} y = 2x + 3 \\ y = 3x - 4 \end{cases}$

652. $\begin{cases} y = 5x + 1 \\ y = 7x + 13 \end{cases}$

653. $\begin{cases} 4x + 3y = 5 \\ 2x - y = -5 \end{cases}$

654. $\begin{cases} 3x - 4y = 10 \\ 2x - 5y = 16 \end{cases}$

655. $\begin{cases} 4x + y = 5 \\ 2x - y = -2 \end{cases}$

656. $\begin{cases} 2x + 9y = -1 \\ 3x + 3y = -5 \end{cases}$

657. $\begin{cases} y = x + 5 \\ 2x - 2y + 10 = 0 \end{cases}$

658. $\begin{cases} 6y = 4 - 2x \\ x = 2 - 3y \end{cases}$

659. $\begin{cases} 2x + y = 6 \\ y = 4 - 2x \end{cases}$

660. $\begin{cases} 3x + 2y = 4 \\ 6x + 4y = 5 \end{cases}$

Performing Row Operations on Linear Equations

661–665 Interpret the operation notation; then perform that operation on the equations.

661. $R_1 \leftrightarrow R_2$ when $\begin{cases} 2x - 3y - z = 12 \\ x + y - 2z = 11 \\ 3x - 2y + z = 3 \end{cases}$

662. $-3R_2 \rightarrow R_2$ when $\begin{cases} 3x + 3y - 2z = 4 \\ x - 2y + z = 7 \\ 2x + 4y + z = 1 \end{cases}$

663. $-3R_1+R_2 \to R_2$ when $\begin{cases} x-3y+2z=7 \\ 3x+y-z=4 \\ 2x-y+z=1 \end{cases}$

664. $-3R_2+R_1 \to R_1$ when $\begin{cases} x+3y+2z=7 \\ y-z=-2 \\ 2y+3z=4 \end{cases}$

665. $-7R_2+R_1 \to R_1$ when $\begin{cases} x+7y+5z=6 \\ y-z=4 \\ z=1 \end{cases}$

Solving Systems of Equations with Row Echelon Form

666-675 Write each system of equations in the row echelon form and substitute back to solve the system.

666. $\begin{cases} x+y+2z=-2 \\ x-2y+3z=-6 \\ 2x-5y+z=-8 \end{cases}$

667. $\begin{cases} x+y-4z=33 \\ 2x+4y-z=10 \\ 3x-y+z=-5 \end{cases}$

668. $\begin{cases} 2x-y-4z=5 \\ x+3y+z=10 \\ x+2y-3z=11 \end{cases}$

669. $\begin{cases} 3x-2y+z=-1 \\ 2x+3y-z=2 \\ x+y+z=2 \end{cases}$

670. $\begin{cases} 2x-3y+z=3 \\ x-4y-2z=9 \\ 3x+y+z=6 \end{cases}$

671. $\begin{cases} x-2y+2z=1 \\ 2x+y-4z=3 \\ 3x-y+6z=8 \end{cases}$

672. $\begin{cases} 3x+3y+z=3 \\ 2x-3y-z=2 \\ x+6y-z=4 \end{cases}$

673. $\begin{cases} 3x+y-z=-2 \\ x-2y+3z=6 \\ 4x+4y+z=2 \end{cases}$

674. $\begin{cases} 3x+y-z=2 \\ x+3y-2z=1 \\ 5x+7y-5z=4 \end{cases}$

675. $\begin{cases} x+2y+4z=9 \\ 2x+3z=4 \\ x-2y-z=-5 \end{cases}$

Using Reduced Row Echelon Form to Solve Systems

676–680 Write each system of equations in the reduced row echelon form and identify the solution.

676. $\begin{cases} x-2y+z=0 \\ 2x-3y+2z=2 \\ x+y-2z=-3 \end{cases}$

677. $\begin{cases} 2x-3y+7z=3 \\ x+y-z=-3 \\ 3x+2y+z=-5 \end{cases}$

678. $\begin{cases} x-y-2z=7 \\ x+3z=-2 \\ 2x+y=9 \end{cases}$

679. $\begin{cases} 2x+y-3z=-9 \\ 4y+z=11 \\ x-2z=-7 \end{cases}$

680. $\begin{cases} 5x+3y-2z=17 \\ x+2y-z=8 \\ 3x-y=1 \end{cases}$

Solving Systems of Linear Equations

681–690 Solve the system of equations.

681. $\begin{cases} x+3y+z=6 \\ 2x+y-z=5 \\ x-y-3z=2 \end{cases}$

682. $\begin{cases} 2x+y+z=3 \\ x-y+4z=11 \\ x+2y+z=8 \end{cases}$

683. $\begin{cases} x-y+2z=4 \\ 3x+2y+z=7 \\ 4x-3y-z=7 \end{cases}$

684. $\begin{cases} 3x+y-z=6 \\ 4x-y-3z=10 \\ x+3y+z=2 \end{cases}$

685. $\begin{cases} x-3y-z=4 \\ 2x+y-4z=5 \\ 3x-y-z=0 \end{cases}$

686. $\begin{cases} x+y+z=2 \\ x-y+z=8 \\ x-y-z=4 \end{cases}$

687. $\begin{cases} x+2y-3z=8 \\ 2x+y-z=5 \\ 3x+3y-4z=13 \end{cases}$

688. $\begin{cases} 4x+2y-z=11 \\ x-2y-2z=8 \\ 6x-2y-5z=27 \end{cases}$

689. $\begin{cases} x+z=5 \\ y-2z=-4 \\ 2x-y=10 \end{cases}$

690. $\begin{cases} x+y=3 \\ 2x-z=-2 \\ 3y-z=7 \end{cases}$

Decomposing Fractions Using Systems of Equations

691–695 Perform decomposition of the fraction using a system of equations.

691. $\frac{x+4}{x(x-2)}$

692. $\frac{x-7}{(x+1)(x-3)}$

693. $\frac{6}{x(x-3)}$

694. $\frac{-x^2+12x-3}{x(x+1)(x-3)}$

695. $\frac{x^2-12x-49}{(x-1)(x+2)(x-5)}$

Chapter 12

Systems of Nonlinear Equations and Inequalities

Systems of *nonlinear* equations usually involve multiple answers. This happens when you have curves that can come back around and intersect one another multiple times. The equations of a circle and parabola can share as many as four different solutions.

A system of inequalities has solutions that cover various areas on the coordinate plane. The best way to describe these solutions is to graph the inequalities and determine where they share points.

The Problems You'll Work On

In this chapter, you'll find the solutions of systems of equations and inequalities such as:

- Intersecting lines and parabolas
- Intersecting parabolas and circles
- Exponential functions intersecting with lines or curves
- Areas defined by lines intersecting with other areas defined by lines
- Curves defining areas intersecting with other lines and curves

What to Watch Out For

Don't let common mistakes trip you up; watch for the following when working with the intersections of lines and curves:

- Applying the elimination process correctly when solving a system of equations
- Distributing correctly when using substitution to solve a system of equations involving curves
- Substituting back into the simpler equation when computing the coordinates of the points of intersection
- Shading in the correct side of a line when solving systems of inequalities

Determining Intersections of Parabolas and Lines

696–705 Find the points of intersection of the parabola and the line.

696. $y = 2x^2 - 3x - 1$ and $y = x + 5$

697. $y = x^2 - 5x - 3$ and $y = -x - 6$

698. $y = x^2 + 2x - 7$ and $y = 2x - 3$

699. $y = -x^2 - x + 5$ and $y = -4x + 1$

700. $y = 3x^2 + 7$ and $y = -6x + 16$

701. $y = -2x^2 + 2x - 4$ and $y = -4x - 12$

702. $y = -x^2 + 6x - 2$ and $y = x + 2$

703. $y = -x^2 - 2x + 12$ and $y = -5x + 12$

704. $y = 4x^2 - 8x$ and $y = -16x + 12$

705. $y = -3x^2 + 4x - 5$ and $y = -8x - 5$

Finding Where Two Parabolas Meet

706–715 Find the points of intersection of the two polynomials.

706. $y = x^2 - 4$ and $y = 4 - x^2$

707. $y = x^2 - 8x + 2$ and $y = -2x^2 + 7x + 2$

708. $y = x^2 - 10x + 5$ and $y = -1.5x^2 + 17.5$

709. $y = x^2 + 3x - 4$ and $y = 0.5x^2 + 2x + 8$

710. $y = x^2 + 5x - 8$ and $y = -0.25x^2 + 3.75x + 7$

711. $y = x^3 - x^2 - 6x - 1$ and $y = 4x^2 - 12x - 1$

712. $y = 2x^3 - 3x^2 - 6x - 4$ and $y = -x^2 - 2x - 4$

713. $y = -x^3 + 5x^2 - 10x + 9$ and $y = -5x^2 + 21x - 21$

714. $y = -x^3 + 2x^2 + 3x + 7$ and $y = -x^2 - 3x + 15$

715. $y = 3x^3 - x^2 + 10x + 4$ and $y = 5x^2 + 13x - 2$

Solving for Intersections of Conics

716–725 Find the points of intersection involving the conics.

716. $x^2 + y^2 = 25$ and $y = 0.2x^2 - 5$

717. $(x - 3)^2 + (y + 4)^2 = 25$ and $y = x - 8$

718. $x^2 + y^2 = 100$ and $y = 0.5x^2 - 26$

719. $\frac{x^2}{25} + \frac{y^2}{9} = 1$ and $y = 0.6x^2 - 7.8$

720. $x^2 + y^2 = 4$ and $y = -x^2 - 2$

721. $x^2 + y^2 = 1$ and $y = 0.25x^2 - 4$

722. $x^2 + y^2 = 25$ and $(x - 8)^2 + y^2 = 25$

723. $x^2 + y^2 = 16$ and $(x - 4)^2 + y^2 = 25$

724. $x^2 + y^2 = 1$ and $x^2 - y^2 = 1$

725. $x^2 + (y + 3)^2 = 25$ and $25x^2 - 16y^2 = 400$

Finding the Intersections of Exponential Functions

726–735 Find the points of intersection of the exponential functions.

726. $y = 2^x, y = 2^{-x}$

727. $y = 3^x, y = 2 - 3^x$

728. $y = 5^x, y = 10 - 5^x$

729. $y = 10^{-x}, y = 10^{x+2}$

730. $y = 6^{x+1}, y = 6^{-x+1}$

731. $y = 4^{x-3}, y = 4^{5-x}$

732. $y = 2^x, y = 8^x$

733. $y = 3^{x+1}, y = 9^x$

734. $y = 4^{x-3}, y = 2^{x-3}$

735. $y = 25^{1-x}, y = 125^{x-1}$

Using Graphing to Solve Systems of Inequalities

736–745 Solve the systems of inequalities by graphing the individual statements and determining their intersection.

736. $\begin{cases} y \geq x \\ y \leq 3 - x \end{cases}$

737. $\begin{cases} y \leq 5 \\ y \geq x + 1 \end{cases}$

738. $\begin{cases} x + y \leq 10 \\ x \geq 3 \end{cases}$

739. $\begin{cases} x+y\leq 1 \\ y\geq -2 \end{cases}$

740. $\begin{cases} y\leq 4 \\ y\geq x^2 \end{cases}$

741. $\begin{cases} y\leq 4x-x^2 \\ y\geq 0 \end{cases}$

742. $\begin{cases} y\leq 4 \\ y\geq 2^x \end{cases}$

743. $\begin{cases} x+y\leq 6 \\ y\leq 5 \\ x\geq 0,\ y\geq 0 \end{cases}$

744. $\begin{cases} 2x+y\leq 12 \\ y\geq x \\ x\geq 0,\ y\geq 0 \end{cases}$

745. $\begin{cases} x+y\leq 7 \\ y\leq 2x+1 \\ x\geq 0,\ y\geq 0 \end{cases}$

Chapter 13

Working with Complex Numbers

The mathematical rules and processes used when working with complex numbers are the same as those used with real numbers — for the most part. Complex numbers were "invented" so that mathematicians could finish their homework. Before complex numbers were created, mathematicians had to stop when they got to a problem involving a negative number under a radical. Consequently, the imaginary number i was born. A complex number is of the form $a + bi$, where a is the real part and bi is the imaginary part. In this chapter, you'll have many opportunities to work with imaginary numbers and see how to deal with their eccentricities.

The Problems You'll Work On

In this chapter, you'll work with complex numbers in the following ways:

- Simplifying powers of i into either i, $-i$, 1, or -1
- Adding and subtracting complex numbers
- Multiplying complex numbers and writing your answers in the standard form
- Dividing complex numbers by multiplying by a conjugate
- Working with equations whose answers require complex numbers

What to Watch Out For

Don't let common mistakes trip you up; watch for the following when working with the real and imaginary parts of complex numbers:

- Finding the correct equivalence for a power of i
- Adding the real parts and imaginary parts of complex numbers separately
- Distributing correctly when subtracting complex numbers
- Using FOIL on the real and imaginary parts when multiplying
- Selecting the correct conjugate when dividing complex numbers
- Simplifying complex answers when they involve radicals

Rewriting Powers of i as i, –i, 1, or –1

746–750 Simplify each power of i.

746. i^{15}

747. i^{402}

748. i^{144}

749. i^{171}

750. i^{57}

Operating on and Simplifying Complex Expressions

751–755 Perform the operations and simplify to a + bi form.

751. $(3 + 2i) + (5 - 3i)$

752. $(-4 - 3i) + (-6 + 5i)$

753. $(9 - 5i) - (-5 - 5i)$

754. $(4 - 6i) - (8 + 6i)$

755. $(7 + 3i) - (2 + 4i) + (5 - 3i)$

Multiplying and Simplifying Complex Numbers

756–770 Perform the multiplication and simplify to a + bi form.

756. $(3 - 6i)(3 + 6i)$

757. $(-4 + 2i)(-4 - 2i)$

758. $(5 - 3i)(4 + 2i)$

759. $(6 - 5i)(-2 + 5i)$

760. $3i(6 - 3i)$

761. $(2 + i)(3 - 2i)$

762. $(7 - 6i)(8 - 6i)$

763. $(4 - 3i)^2$

764. $(-2 + i)^2$

765. $(-4 - 4i)^2$

766. $(2 + 2i)^2$

767. $i^2(3 - 2i)(i + 1)$

768. $i^3(4 - i)(6 + i)$

769. $(3 - 2i^5)(5 + 3i^3)$

770. $2i(3 + 2i)(4 - 2i)$

Using Conjugates to Divide Complex Numbers

771–785 Divide and simplify to a + bi form.

771. $\frac{2+3i}{1-4i}$

772. $\frac{1-10i}{2+9i}$

773. $\frac{9+i}{8-i}$

774. $\frac{2+i}{6-4i}$

775. $\frac{7+i}{4-2i}$

776. $\frac{1-3i}{1+3i}$

777. $\frac{-5-2i}{-5+2i}$

778. $\frac{1+7i}{1-7i}$

779. $\frac{6-8i}{-6+8i}$

780. $\frac{-3-9i}{3+9i}$

781. $\frac{4i}{5-2i}$

782. $\frac{6}{3-i}$

783. $\frac{6-2i}{2i}$

784. $\frac{-6+i}{3i}$

785. $\frac{i}{1+i}$

Solving for Complex Solutions in Quadratic Equations

786–795 Solve the quadratic equation and write your answer in the a + bi form.

786. $x^2 + 2x + 5 = 0$

787. $x^2 - 4x + 5 = 0$

788. $2x^2 - 2x + 1 = 0$

789. $x^2 + 6x + 17 = 0$

790. $2x^2 - 6x + 9 = 0$

791. $x^2 + 8x + 17 = 0$

792. $x^2 + 3x + 5 = 0$

793. $x^2 - x + 7 = 0$

794. $2x^2 - 3x + 8 = 0$

795. $3x^2 + 2x + 2 = 0$

Chapter 14

Matrices

A *matrix* is a rectangular array of numbers or other values, called *elements.* Matrices provide an excellent format for organizing information and performing computations involving that information.

Matrix arithmetic incorporates the operations of addition, subtraction, and multiplication, but in most unique ways. As with complex numbers, there is no division — the process is accomplished by multiplying by an inverse.

One of the first applications of matrices found in the classroom is that of solving systems of equations. It's not usually easier to solve those systems with matrices than to do it with algebraic substitution and elimination, but the process is easily programmed in graphing calculators, making technology the choice for doing those systems problems.

The Problems You'll Work On

In this chapter, you'll work with matrices in the following ways:

- Determining matrix dimensions
- Adding and subtracting matrices with the same dimension
- Performing scalar multiplication
- Multiplying matrices with compatible dimensions
- Finding the inverse of a square matrix
- Dividing matrices using an inverse matrix
- Using matrices to solve systems of linear equations

What to Watch Out For

Don't let common mistakes trip you up; watch for the following when working with matrices:

- Reversing the row and column parts of matrix dimensions
- Placing the results in the correct element positions when multiplying matrices
- Indicating your steps when performing row operations
- Recognizing when a matrix has no inverse

Determining the Dimension and Type of Matrix

796–800 Give the dimension and type of matrix.

796. $\begin{bmatrix} 1 & 2 & 3 \\ 4 & 5 & 6 \end{bmatrix}$

797. $\begin{bmatrix} 1 & 2 \\ 3 & 4 \\ 5 & 6 \\ 7 & 8 \\ 9 & 0 \end{bmatrix}$

798. $\begin{bmatrix} 1 \\ 2 \\ 3 \\ 4 \end{bmatrix}$

799. $\begin{bmatrix} 1 & 2 & 3 \\ 4 & 5 & 6 \\ 7 & 8 & 9 \end{bmatrix}$

800. $\begin{bmatrix} 0 & 0 & 0 \\ 0 & 0 & 0 \\ 0 & 0 & 0 \\ 0 & 0 & 0 \end{bmatrix}$

Adding and Subtracting Matrices

801–805 Perform the matrix addition or subtraction.

801. $\begin{bmatrix} 1 & 0 & 7 \\ -4 & 3 & 3 \end{bmatrix} + \begin{bmatrix} -2 & -1 & 6 \\ 3 & 0 & -5 \end{bmatrix}$

802. $\begin{bmatrix} 4 & 3 \\ -5 & 2 \\ 0 & -7 \end{bmatrix} - \begin{bmatrix} -1 & 0 \\ -2 & 6 \\ -3 & 1 \end{bmatrix}$

803. $\begin{bmatrix} 6 & -4 \\ -1 & 0 \end{bmatrix} + \begin{bmatrix} -6 & 4 \\ 1 & 0 \end{bmatrix}$

804. $\begin{bmatrix} 1 \\ -1 \\ 7 \\ 0 \end{bmatrix} + \begin{bmatrix} -2 \\ 3 \\ -3 \\ 4 \end{bmatrix} - \begin{bmatrix} 5 \\ 2 \\ -1 \\ 6 \end{bmatrix}$

805. $\begin{bmatrix} 3 & 6 \end{bmatrix} - \begin{bmatrix} -2 & 4 \end{bmatrix} - \begin{bmatrix} -1 & -1 \end{bmatrix}$

Performing Scalar Multiplication on Matrices

806–810 Perform the scalar multiplication.

806. $2\begin{bmatrix} -1 & 0 & 4 \\ 2 & -2 & 6 \end{bmatrix}$

807. $-\frac{1}{2}\begin{bmatrix} -2 & 8 \\ 10 & -6 \end{bmatrix}$

808. $-1\begin{bmatrix} 4 \\ -5 \\ 0 \end{bmatrix}$

809. $-0.2\begin{bmatrix} 3 & -2 & 0 \\ 4 & 1 & -1 \end{bmatrix}$

810. $\frac{1}{3}\begin{bmatrix} 3 & 0 & 0 \\ 0 & -9 & 0 \\ 0 & 0 & 6 \end{bmatrix}$

Multiplying Matrices

811–820 Perform the matrix multiplication.

811. $\begin{bmatrix} 1 & -2 \\ 3 & 5 \end{bmatrix} \cdot \begin{bmatrix} -4 & 0 \\ -1 & 2 \end{bmatrix}$

812. $\begin{bmatrix} -1 & 0 & 2 \\ 3 & -2 & 5 \end{bmatrix} \cdot \begin{bmatrix} 6 & 3 \\ -1 & 1 \\ 0 & 4 \end{bmatrix}$

813. $\begin{bmatrix} 1 & -1 \\ 0 & -3 \\ 4 & -2 \end{bmatrix} \cdot \begin{bmatrix} 5 & -5 & 0 \\ -2 & 4 & -1 \end{bmatrix}$

814. $\begin{bmatrix} 1 & -1 & 0 \\ 9 & -1 & 9 \\ 2 & 1 & 6 \\ 4 & -1 & 7 \end{bmatrix} \cdot \begin{bmatrix} 2 \\ 3 \\ -4 \end{bmatrix}$

815. $\begin{bmatrix} -1 & 4 \\ 9 & 0 \end{bmatrix} \cdot \begin{bmatrix} 1 \\ 8 \end{bmatrix}$

816. $\begin{bmatrix} -2 & 3 & 9 & 0 \end{bmatrix} \cdot \begin{bmatrix} 1 \\ -4 \\ 0 \\ -3 \end{bmatrix}$

817. $\begin{bmatrix} 2 & 3 \\ 1 & 2 \end{bmatrix} \cdot \begin{bmatrix} 2 & -1 \\ -3 & 2 \end{bmatrix}$

818. $\begin{bmatrix} 1 & 1 & 1 \\ 1 & 13 & -5 \\ -1 & 5 & -1 \end{bmatrix} \cdot \begin{bmatrix} 2 & 1 & -3 \\ 1 & 0 & 1 \\ 3 & -1 & 2 \end{bmatrix}$

819. $\begin{bmatrix} 5/4 & 7/4 & -1/4 \\ -1/4 & -3/4 & 1/4 \\ -7/4 & -9/4 & 3/4 \end{bmatrix} \cdot \begin{bmatrix} 0 & 3 & -1 \\ 1 & -2 & 1 \\ 3 & 1 & 2 \end{bmatrix}$

820. $\begin{bmatrix} 2 & 3 & -1 & 4 \\ 2 & -3 & 0 & 0 \\ 1 & 0 & -1 & 2 \\ 3 & -3 & 1 & 0 \end{bmatrix} \cdot \begin{bmatrix} 6 & 12 & -12 & -6 \\ 4 & 6 & -8 & -4 \\ -6 & -18 & 12 & 12 \\ -6 & -15 & 15 & 9 \end{bmatrix}$

Determining Inverses of Square Matrices

821–830 Find the inverse of the matrix.

821. $\begin{bmatrix} 3 & -2 \\ -4 & 3 \end{bmatrix}$

822. $\begin{bmatrix} 4 & 7 \\ 3 & 5 \end{bmatrix}$

823. $\begin{bmatrix} -2 & 5 \\ 3 & -7 \end{bmatrix}$

824. $\begin{bmatrix} 3 & -5 \\ -2 & 4 \end{bmatrix}$

825. $\begin{bmatrix} 2 & -1 & 0 \\ 1 & 0 & 0 \\ -1 & 4 & 1 \end{bmatrix}$

826. $\begin{bmatrix} 1 & 0 & -2 \\ 3 & 1 & -6 \\ 2 & -3 & -5 \end{bmatrix}$

827. $\begin{bmatrix} 1 & 2 & 0 \\ 0 & 1 & 3 \\ -2 & 1 & 14 \end{bmatrix}$

828. $\begin{bmatrix} 3 & -1 & 1 \\ -4 & 0 & 2 \\ 2 & -1 & 3 \end{bmatrix}$

829. $\begin{bmatrix} 1 & 2 & -2 \\ 3 & 0 & 6 \\ -1 & 3 & -4 \end{bmatrix}$

830. $\begin{bmatrix} 3/7 & -13/14 & 1/7 \\ 2/7 & 3/14 & -1/14 \\ 2/7 & -2/7 & -1/14 \end{bmatrix}$

Dividing Matrices Using Inverses

831–835 Perform the division.

831. $\dfrac{\begin{bmatrix} 1 & 2 \\ 3 & 4 \end{bmatrix}}{\begin{bmatrix} 3 & 5 \\ -2 & -3 \end{bmatrix}}$

832. $\dfrac{\begin{bmatrix} 1 & 0 \\ -4 & 2 \end{bmatrix}}{\begin{bmatrix} 5 & 8 \\ -3 & -5 \end{bmatrix}}$

833. $\dfrac{\begin{bmatrix} 8 & 10 \\ 12 & 16 \end{bmatrix}}{\begin{bmatrix} 4 & 5 \\ 6 & 8 \end{bmatrix}}$

834. $\dfrac{\begin{bmatrix} 4 & 0 & 2 \\ 3 & 1 & 6 \\ 0 & -2 & 0 \end{bmatrix}}{\begin{bmatrix} 1 & 1 & -1 \\ 1 & 2 & 3 \\ 2 & -1 & -13 \end{bmatrix}}$

835. $\dfrac{\begin{bmatrix} 3 & 1 & 2 \\ 0 & 4 & 0 \\ 1 & -2 & 1 \end{bmatrix}}{\begin{bmatrix} 1 & 0 & 3 \\ 0 & 1 & 4 \\ 2 & -3 & -4 \end{bmatrix}}$

Using Matrices to Solve Systems of Equations

836–840 Solve the system of equations using matrices.

836. $\begin{cases} x-2y=5 \\ 3x-5y=13 \end{cases}$

837. $\begin{cases} 2x+3y=5 \\ 5x+7y=11 \end{cases}$

838. $\begin{cases} 4x+7y=5 \\ 3x+5y=4 \end{cases}$

839. $\begin{cases} x+y+3z=3 \\ x+2y=8 \\ 2x+3y+2z=12 \end{cases}$

840. $\begin{cases} x+2y+z=3 \\ x-3y+2z=17 \\ x+y+2z=9 \end{cases}$

Chapter 15

Sequences and Series

A *sequence* is a list of numbers or terms. When a rule is used to create a numerical sequence, the terms are designated as first, second, third, and so on, and the number of the term is used in the rule. A *series* is the sum of the terms in a sequence.

Many sequences and series can be designated as arithmetic, geometric, Fibonacci, and so on. Sequences and series that fall into certain categories can have special rules used to find sums of terms or particular terms in the listing.

The Problems You'll Work On

In this chapter, you'll work with sequences and series in the following ways:

- Writing specific terms of the sequence when given a rule
- Continuing the listing of terms in a sequence when given four or more existing terms
- Using the format for arithmetic and geometric sequences to write their rules
- Finding the sum of the terms in a series
- Working with other special sequences and series
- Applying sequences and series to practical problems

What to Watch Out For

Don't let common mistakes trip you up; watch for the following ones when working with sequences and series:

- Trying to fit a pattern when not given enough terms in a sequence
- Using incorrect terms when writing recursively defined sequences
- Failing to recognize the notation in the summation symbol
- Subtracting the wrong sum when working with applications involving subtractions of series

Writing Terms of a Sequence Given a Rule

841–845 Write the first five terms of the sequence, given the general term.

841. $a_n = 2n+1$

842. $b_n = n^2 - n + 1$

843. $c_n = \frac{2n-1}{n+1}$

844. $d_n = \frac{n!}{2^n}$

845. $f_n = 3\left(2^{n-1}\right)$

Finding a General Term for an Arithmetic Sequence

846–850 Write the general term for the arithmetic sequence.

846. 2, 4, 6, 8, ...

847. 1, 6, 11, 16, ...

848. 5, 8, 11, 14, ...

849. –6, 0, 6, 12, ...

850. $\frac{1}{4}, \frac{3}{8}, \frac{1}{2}, \frac{5}{8}, \ldots$

Writing a General Term for a Geometric Sequence

851–855 Write the general term for the geometric sequence.

851. 1, 2, 4, 8, ...

852. 2, 6, 18, 54, ...

853. 64, 32, 16, 8, ...

854. 3, –6, 12, –24, ...

855. $\frac{3}{2}, \frac{9}{4}, \frac{27}{8}, \frac{81}{16}, \ldots$

Recognizing Patterns in Sequences

856–860 Find the pattern/rule and write the next four terms of the sequence.

856. 0, 3, 8, 15, 24, ...

857. 1, 3, 7, 15, 31, ...

858. 1, 2, 6, 24, 120, ...

859. 3, 1, 9, 1, 27, 1, 81, ...

860. 3, 10, 29, 66, 127, ...

Creating Recursively Defined Sequences

861–865 Write the first five terms of the recursively defined sequence.

861. $a_1 = 5,\ a_2 = -1$
$a_n = a_{n-1} + a_{n-2}$

862. $a_1 = 1,\ a_2 = 1$
$a_n = a_{n-1} + 2a_{n-2}$

863. $a_1 = 2,\ a_2 = 3$
$a_n = a_{n-2} - a_{n-1}$

864. $a_1 = 0,\ a_2 = 2$
$a_n = 4a_{n-2} - 2a_{n-1}$

865. $a_1 = 1,\ a_2 = 4,\ a_3 = 3$
$a_n = a_{n-1} + a_{n-2} + a_{n-3}$

Writing More Terms of Special Sequences

866–870 Write the next four terms of the special sequence.

866. 1, 1, 2, 3, 5, 8, 13, ...

867. $1, 1, \frac{1}{2}, \frac{1}{6}, \frac{1}{24}, \ldots, \frac{1}{(n-1)!}$

868. $(x-1), \frac{1}{2}(x-1)^2, \frac{1}{3}(x-1)^3, \ldots, \frac{1}{n}(x-1)^n$

869. 1, 3, 6, 10, 15, 21, ...

870. 1, 11, 21, 1211, 111221, ...

Finding the Sum of the Terms in a Series

871–875 Find the sum.

871. $\sum_{i=1}^{3} i^2$

872. $\sum_{i=1}^{4} (2i+1)$

873. $\sum_{i=0}^{3} 3(2^i)$

874. $\sum_{i=0}^{4} 8\left(\frac{1}{2}\right)^i$

875. $\sum_{i=1}^{6} 2^{i-1}$

Summing Arithmetic Series

876–885 Find the sum of the arithmetic series using $S_n = \frac{n(a_1 + a_n)}{2}$ where n is the number of terms in the sequence.

876. $1 + 3 + 5 + 7 + 9 + 11 + 13$

877. $\sum_{i=1}^{6} (2i-1)$

878. $\sum_{i=1}^{100} i$

879. $\sum_{i=0}^{15} (3i+2)$

880. First 20 positive odd integers

881. First 100 positive even integers

882. Number of blocks in a stack; bottom row 16 blocks, each row above has one less block until the top row of one block.

883. Number of people in a picture; first row 19 people, two fewer in each of the next four rows.

884. Number of trees in a triangular plot; three in first row, three more in each of next 39 rows.

885. 13, 16, 19, …, 301

Determining the Sum of a Geometric Series

886–895 Find the sum of the geometric series using $S_n = \frac{g_1(1-r^n)}{1-r}$.

886. $\sum_{i=1}^{10} 2^i$

887. $\sum_{i=0}^{5} 5^i$

888. $\sum_{i=0}^{6} 3^{i+1}$

889. $\sum_{i=0}^{5} 5(2^{i+1})$

890. $\frac{1}{10}+1+10+100+1{,}000+10{,}000$

891. $-2+4-8+16-32+64-128+256$

892. $27+9+3+\cdots+\frac{1}{81}$

893. $\frac{3}{8}+\frac{3}{4}+\frac{3}{2}+\cdots+24$

894. Find the total number of pennies if you get one penny the first day, two pennies the second day, four pennies the third day, eight pennies the fourth day, and so on through the 31st day.

895. You had 364 pieces of candy. You ate 1 piece the first hour, 3 pieces the second hour, 9 pieces the third hour, 27 pieces the fourth hour, and so on. How long did it take to eat all the candy?

Computing the Sum of a Special Series

896–898 Find the sum of the first six terms of the special series.

896. $e = 1 + 1 + \frac{1}{2} + \ldots + \frac{1}{(n-1)!}$

897. $e^2 = 1 + 2 + 2 + \frac{4}{3} + \cdots + \frac{2^{n-1}}{(n-1)!}$

898. $\pi \approx 4\sum_{i=1}^{6} \frac{(-1)^{i+1}}{2i-1}$

Using a Special Formula for the Sum of an Infinite Series

899–900 Find the sum of the infinite series using $S_n = \frac{a_1}{1-r}$.

899. 64 + 32 + 16 + 8 + …

900. A ball is dropped from 256 feet and bounces back $\frac{3}{4}$ of that height. It falls again and bounces back $\frac{3}{4}$ of that height. The ball continues falling and bouncing back and falling and bouncing back. What is the total distance the ball travels?

Chapter 16

Sets

A *set* is a collection of objects called *elements*. A set can be defined by listing all the elements, such as A = {2, 4, 6, 8, 10} or by giving a rule, A = {the first five even natural numbers}. Sets have operations that are performed on them, such as *union* and *intersection*. Some special designations are *empty set* (also known as *null set*} and the *universal set*. Set theory is important in studying number systems as well as answering some practical application problems in many arenas. The biggest challenge in working with sets is knowing the vocabulary so you can perform the task.

The Problems You'll Work On

In this chapter, you'll work with sets and set operations in the following ways:

- Becoming acquainted with set notation and what the symbols represent
- Recognizing equal sets and subsets
- Performing operations on sets
- Using Venn diagrams to analyze problems dealing with sets of numbers or other types of elements
- Solving practical problems using sets

What to Watch Out For

Don't let common mistakes trip you up; watch for the following when working with sets and subsets:

- Confusing union and intersection symbols
- Recognizing which sets are equal
- Using too many elements in intersections of Venn diagrams
- Reading the solutions correctly from Venn diagram work

Creating the Elements of a Set from a Description

901–910 Rewrite the statement using set notation.

901. The set A contains the first five positive even integers.

902. The number two is an element of set B.

903. Set C contains all the positive multiples of seven.

904. Set D has no elements.

905. Sets E and F contain the exact same elements.

906. The number three is not in set G.

907. Set H is a subset of set J and has fewer elements than set J.

908. Do not include set K.

909. Set M consists of all the elements shared by sets P and Q.

910. Set M contains all the elements in set P and all the elements in set Q.

Performing Set Operations

911–920 Use sets A, B, C, D, and E to perform the set operations.

A = {0, 1, 2, 3, 4, 5, 6, 7, 8, 9}

B = {0, 3, 6, 9}

C = {2, 4, 6, 8}

D = {2, 3, 5, 7}

E = {1, 3, 5, 7, 9}

911. $B \cup C$

912. $B \cap E$

913. $A \cap D$

914. $C \cap E$

915. $B \cup D \cup E$

916. $B \cap (C \cup D)$

917. $(B \cap C) \cup D$

918. $n(B)$

919. $n(B \cup C)$

920. $n(B) + n(C) - n(B \cap C)$

Operating on Sets of Special Numbers

921–930 Use sets U, A, B, C, and D to perform the set operations.

U = {all integers}

A = {even integers}

B = {odd integers}

C = {multiples of 6}

D = {multiples of 8}

921. $A \cup B$

922. $A \cap B$

923. $A \cap C$

924. $A \cup C$

925. $C \cap D$

926. B'

927. A'

928. $B \cap C$

929. $A \cup \varnothing$

930. $A \cap \varnothing$

Using Venn Diagrams to Solve Problems Involving Sets

931–940 Use a Venn diagram to solve.

931. If U = {1, 2, …, 14}, A = {odd numbers}, and B = {divisors of 30 that are less than 15}, find $A \cap B'$.

932. If U = {letters in the word *facetious*}, A = {vowels}, and B = {letters in the word *cast*}, find $(A \cap B)'$.

933. If U = {first 19 natural numbers}, A = {even numbers between 1 and 19}, and B = {multiples of 5 between 1 and 19}, find $(A \cup B)'$.

934. If U = {the first 30 natural numbers}, A = {numbers of the form $4k - 1$ that are less than 30}, and B = {all prime numbers less than 30}, find $A' \cap B$.

935. If U = {the letters of the alphabet}, A = {b, c, d, e, g, p, t, v, z}, B = {e, g, y, p, t}, and C = {a, e, i, o, u, y}, find $A \cap B \cap C$.

936. If U = {the letters of the alphabet}, A = {c, a, l, i, f, o, r, n}, B = {w, y, o, m, i, n, g}, and C = {o, r, e, g, n}, find $(A \cup B) \cap C$.

937. If U = {the letters of the alphabet}, A = {s, i, l, v, e, r}, B = {g, r, a, n, i, t, e}, and C = {p, l, a, t, i, n, u, m}, find $(A \cap B) \cup C$.

938. If U = {1, 2, 3, …, 20}, A = {multiples of 4}, B = {multiples of 5}, and C = {odd numbers}, find $(A \cap B) \cap C'$.

939. If U = {a, b, c, ..., k, l}, A = {vowels}, B = {letters in the word *cadillac*}, and C = {letters in the word *hijack*}, find $(A \cup B \cup C)'$.

940. If U = {positive multiples of 3 less than 40}, A = {multiples of 4}, B = {multiples of 5}, and C = {multiples of 7}, find $(A \cup B)' \cap C$.

Enlisting Venn Diagrams to Solve Applications

941–950 Solve the problem using a Venn diagram.

941. Of 100 houses in a subdivision, 45 have wood siding, 50 have brick, and 10 have both wood siding and brick. How many houses have neither wood siding nor brick?

942. This evening 130 orders were placed at a restaurant. Of the orders, 52 included pie and 84 included ice cream. If 15 orders included neither pie nor ice cream, then how many asked for ice cream only?

943. There are 35 fraternity members registering for classes. Seventeen are registering for English, 19 for math, 12 for physics, 10 for both English and math, 7 for math and physics, and 3 for English and physics. Two are registering for all three, and five aren't registering for any of these classes. How many are registering for math, but not physics or English?

944. Fifty pizzas were ordered last night. If 31 contained sausage, 20 had pepperoni, 6 had anchovies, 10 had sausage and pepperoni, 2 had pepperoni and anchovies, 3 had sausage and anchovies, and 2 had all three toppings, then how many had none of the toppings?

945. One hundred people were interviewed as to how they got the daily weather report. Thirty-three said they got it on television, 9 on the radio, 56 on the Internet, 3 on both television and the radio, 5 on the radio and the Internet, 12 on television and the Internet, and 2 got it on all three. How many received the weather by none of these (just went outside)?

946. A florist had 150 flower orders to be delivered on February 14. Of those, 120 included roses, 94 included mums, and 79 included carnations. There were 71 orders including roses and mums, 67 including roses and carnations, and 66 including mums and carnations. Sixty-one orders had all three flowers in them. How many orders had mums and carnations, but no roses?

947. A car dealership placed an order for 100 cars to be made available for sale on the lot. Of those, 55 had GPS, 60 had a roof rack, 60 had 20-inch wheels, 30 had GPS and a roof rack, 45 had a roof rack and 20-inch wheels, and 30 had GPS and 20-inch wheels. Twenty-five cars had all three features. How many had just GPS or none of the three features?

948. A vacation rental agency has 150 units available for a spring week. Of these, 63 have an ocean view, 82 are a block from the beach, and 33 are on the top floor of the building. Twenty-two of the units have the ocean view and are a block from the beach, 3 have the ocean view and are on the top floor, 12 are a block from the beach and are on the top floor, and 2 have all three of these features. How many units have either none of the features or just the ocean view?

949. There are eight different human blood types: A+, A–, B+, B–, AB+, AB–, O+, and O–. The types are determined by which factors occur: A, B, both, or neither, and then positive or negative. Create a Venn diagram, using three intersecting circles, illustrating these eight different factors. Which blood type ends up in the intersection of the three circles?

950. At a recent blood drive, 70 people gave blood. Six of the donors had AB+ type blood, 7 had AB type blood, 20 had A+ type blood, and 8 had B+ type blood. Thirty-two had the A antigen in their blood (either A or AB), and 17 had the B antigen (B or AB). A total of 58 had + blood. How many were the *universal donor,* O–?

Chapter 17

Counting Techniques and Probability

Counting techniques and probability go hand-in-hand. You need the counting techniques to be able to determine how many ways an event can occur, and probability is one of the most common uses of permutations and combinations (counting techniques) in the sciences.

In this chapter, you find many examples of the ways that counting techniques make life much easier — and your answers more accurate. And you see probability in action using these computed numbers.

The Problems You'll Work On

In this chapter, you'll see how to count and how to use the results in the following ways:

- Multiplying the choices together with the *multiplication property*
- Counting the number of arrangements of a select few, chosen from the many, using permutations
- Figuring out how many groupings are possible if you just want some — not all — of the choices, using combinations
- Applying the formula for combinations in the *Binomial Theorem*
- Solving basic probability problems
- Using counting techniques to solve more complex probability problems

What to Watch Out For

Don't let common mistakes trip you up; watch for the following ones when working with counting techniques and probability:

- Selecting the wrong technique for a problem
- Reducing the fractions involving factorial incorrectly
- Failing to recognize the set of all possibilities when doing probability problems
- Writing probabilities as decimals and fractions incorrectly

Solving Problems Using the Multiplication Property

951–955 Use the multiplication property of counting to solve the problem.

951. Steven has ten shirts, seven pairs of slacks, six sweaters, and four ties. How many different outfits can he create (assuming that any combination "works" colorwise)?

952. A local creamery offers either a sugar cone or a waffle cone, six different flavors of ice cream, and five different toppings. How many different desserts can you create if you can have a cone with one scoop of ice cream and one topping?

953. You are going to "name" your car and put that name on the garage door. You're limited, though, to the letters you have in your art box. You can use one vowel (*a, e, i, o,* or *u*), one letter from the last five letters of the alphabet, and one from the letters *j* through *n*. How many different names can be created?

954. A license plate in a nearby state must have the following pattern: a letter of the alphabet (except for *I* or *O*), any digit except 0, and then two more digits — with no restrictions. How many different license plates can be printed?

955. You want to create a code for your bicycle chain lock. The first entry can be any letter from the word *Washington,* the second entry can be any letter from *Alabama,* the third entry can be any letter from *Minnesota,* and the last entry can be any letter from *Illinois*. How many different codes are possible?

Counting How Many Using Permutations

956–965 Use permutations to solve the counting problem. The number of permutations of n things taken r at a time is $P(n,r)=\frac{n!}{(n-r)!}$.

956. How many "words" (arrangements of the letters) can be made from the letters in the word *math?*

957. How many different ways can you arrange seven people for a picture if they have to stand next to each other in a single row?

958. How many ways can you arrange the Seven Dwarfs in a single file if Doc has to be first and Sleepy has to be last?

959. There are 30 people in a club, and it's time to choose officers: president, secretary, and treasurer. If the club members decide to just draw names with the first name drawn becoming president, and so on, how many different arrangements of officers are there?

960. Elliott has to create a four-digit code for his locker. All the digits must be different. How many different codes are there to choose from if he doesn't want to use the digit "0"?

961. Ryan is playing Scrabble and has seven letters to play. If he can find a five-letter word to fit in the corner (and join the letters already played), he'll get a triple-word-score. How many different arrangements of letters does he have to choose from (hoping at least some are actual words)?

962. Jane is creating a display of hair products on a shelf. She has twelve different products to choose from but can only fit eight of them at a time, lining them up in a row. How many different arrangements are possible?

963. A state prints license plates that start with two letters followed by four digits. How many license plates are possible if there are no repeats of the letters or digits?

964. A state prints license plates that start with two letters followed by four digits. How many license plates are possible if neither letter can be O or I, but letters can be repeated, and none of the digits can be repeated?

965. You won a trip on a television game show. You get to choose visits to four different countries — six days in each. You're given 20 countries to choose from and can travel to them in any order you want. How many different trips can you plan?

Incorporating Combinations in Counting Problems

966–975 Use a combination of n things taken r at a time to solve the problem.

$${}_nC_r = \frac{n!}{(n-r)!r!}$$

966. Six people are to be chosen from a group of ten to serve on a committee to organize a trip. How many different committees can be formed?

967. A scout troup has 20 members. Three scouts will be selected to represent the troop in the Thanksgiving parade. In how many ways can these three scouts be chosen?

968. In a state's LOTTO game, you win if you choose the correct five numbers out of the numbers 1 through 40. If you want to buy a ticket for every possible way to choose five numbers, how many tickets do you need to purchase?

969. A multi-state lottery has you choose six numbers from a possible sixty — and then choose a single "bonus" number. How many different ways can the seven numbers be chosen?

970. Tommy has time for eight more rides at the amusement park. He will choose four of those rides from the ten that fling him in the air, and he'll choose the other four from the six rides that send him through water. How many different ways can he choose the rides?

971. Shannon has to read eight books for her sociology class. She can choose any six from the ten in the first group, and then she can choose any two from the three books in the second group. How many different ways can she choose the eight books?

972. Ariel wants to purchase a flower arrangement for her mother. She can choose one of eight vases, pick three types of flowers from ten different flowers available, and pick two types of greenery from the four in the shop. How many different ways can she choose this arrangement?

973. A local pizzeria lets you "build your own pizza." You can choose any two toppings from sausage, pepperoni, Canadian bacon, or ham. You also get any two toppings from onions, mushrooms, green peppers, jalapeño peppers, and olives. The cheese choices are any two from mozzarella, parmesan, cheddar, provolone, and pepper jack. Finally, you can have a thin crust, a thick crust, or a hand-tossed crust. How many different pizzas can be created?

974. The game of bridge is played with a standard deck of 52 cards. How many different bridge hands of 13 cards are possible?

975. A game of 5-card poker is played with a standard deck of 52 cards. How many different hands are possible that contain nothing but face cards and aces?

Performing a Binomial Expansion

976–980 Use combinations to complete the binomial expansion of the binomial. Recall:

$\binom{n}{r}x^n y^{n-r}$

976. $(x+y)^3$

977. $(x+y)^4$

978. $(x+y)^5$

979. $(x+y)^6$

980. $(x+y)^7$

Solving Probability Problems Using Counting Techniques

981–1001 Use counting techniques and the rule for probability, $\frac{n(\text{desired events})}{n(\text{all events available})}$*, to solve the problems.*

981. What is the probability of drawing a vowel (*a*, *e*, *i*, *o*, or *u*) from a set of tiles labeled with the letters of the alphabet?

982. What is the probability of drawing a prime number from a set of disks labeled with the first 20 natural numbers?

983. What is the probability of choosing a multiple of the number 5 if you draw a tile from a set labeled with the first 50 natural numbers?

984. What is the probability of drawing a multiple of 4 or 6 when choosing from a set of disks labeled with the first 40 natural numbers?

985. Your next vacation will be taken in a state chosen at random (you'll throw a dart at a wall map while blindfolded). What is the probability that the state you end up vacationing in has a name starting with the letter *T?*

986. Your next vacation will be taken in a state chosen at random (you'll throw a dart at a wall map while blindfolded). What is the probability that the state you end up vacationing in has a name starting with the letter *B?*

987. You have four Scrabble tiles: *A, H, M,* and *T.* If you turn them face down, mix them up, and then draw them one at a time, what is the probability that you'll draw them in the order *MATH?*

988. You have four Scrabble tiles: *E, O, R,* and *T.* If you turn them face down, mix them up, and then draw just three of them one at a time, what is the probability that you'll draw them in the order *ROT, TOR,* or *ORT?*

989. Adam and Betty are two of three people who belong to a club that has a total of 30 members. If the offices of president and vice president are to be decided by a random drawing of all the names of the club members, then what is the probability that Adam will be chosen as president and Betty will be chosen as vice president?

990. Adam, Betty, Carole, and Dave all belong to a club that has 30 members. If the offices of president, vice president, secretary, and treasurer are to be decided by a random drawing of all the names of the club members, then what is the probability that Adam, Betty, Carole, and Dave will all be chosen as one of the officers?

991. A coin is tossed twice. What is the probability that it will be "heads" both times?

992. A coin is tossed three times. What is the probability that exactly two of the results will be "heads"?

993. What is the probability that, in a family with three children, at least one will be a girl?

994. What is the probability that, in a family with four children, there will be more boys than girls?

995. A jar contains four red marbles, two white marbles, and four blue marbles. What is the probability of drawing two red marbles if you don't replace the first one after drawing it?

996. A jar contains four red marbles, two white marbles, and four blue marbles. What is the probability of drawing two red marbles if you replace the first one after drawing it?

997. The 26 letters of the alphabet are written on separate tiles. You draw one tile at a time and place them in a straight line. What is the probability that the tiles will be drawn in alphabetical order?

998. You plan on a year-long adventure of visiting all 50 states. The names of the states are written on disks. You draw a name, go to that state, and plant the disk in an "interesting" spot. You draw the next name and go to that state. What is the probability that the last state you visit will be Wyoming?

999. You are playing a card game in which you're dealt 5 cards from a standard deck of 52. You glance at your cards and see that they're all red. What is the probability that they're all hearts?

1000. You're playing a card game where you're dealt 5 cards from a standard deck of 52. What is the probability that four of the cards are aces?

1001. You have all the divisors of 1001 written on separate tiles. What is the probability that one of these tiles, drawn at random, will be a prime number?

Part II
The Answers

To access the Cheat Sheet created specifically for this book, go to www.dummies.com/cheatsheet/1001algebra2.

In this part . . .

You get answers and explanations for all 1,001 problems. As you're going over your work, you may realize that you need a little more instruction. Fortunately, the *For Dummies* series offers several excellent resources. The following titles, both written by your humble author, are available at your favorite bookstore or in e-book format:

- *Algebra II For Dummies*
- *Algebra II Workbook For Dummies*

After you've mastered Algebra II and you're ready to move on to more advanced math, you'll find all the help you'll need in these titles:

- *Pre-Calculus For Dummies,* 2nd Edition (Yang Kuang and Elleyne Kase)
- *Calculus For Dummies* (Mark Ryan)

Visit `www.dummies.com` for more information.

Chapter 18

The Answers

1. $\mathbf{8x^2 + 8x - 6}$

Apply "FOIL" to multiply the terms in the binomials.

First: $2x(4x) = 8x^2$

Outer: $2x(-2) = -4x$

Inner: $3(4x) = 12x$

Last: $3(-2) = -6$

Combine these products and simplify:

$$8x^2 - 4x + 12x - 6 = 8x^2 + 8x - 6$$

2. $\mathbf{4x^2 + 3x - 4}$

First, multiply the two binomials together using "FOIL."

$$(x + 4)(x - 1) = x^2 - x + 4x - 4 = x^2 + 3x - 4$$

Now add $3x^2$ to that product.

$$3x^2 + x^2 + 3x - 4 = 4x^2 + 3x - 4$$

3. $\mathbf{8x^2 - 2x - 11}$

First, perform the two multiplications.

$$(3x + 1)(x - 3) = 3x^2 - 9x + x - 3 = 3x^2 - 8x - 3$$

$$(x + 2)(5x - 4) = 5x^2 - 4x + 10x - 8 = 5x^2 + 6x - 8$$

Then add the two products.

$$3x^2 - 8x - 3 + 5x^2 + 6x - 8 = 8x^2 - 2x - 11$$

4. $\mathbf{8x^2 - 20x - 11}$

First, perform the two multiplications.

$$5(x - 3)(x + 2) = 5(x^2 - x - 6) = 5x^2 - 5x - 30$$

$$3(x - 3)(x - 2) = 3(x^2 - 5x + 6) = 3x^2 - 15x + 18$$

Then add the two products and the 1.

$$5x^2 - 5x - 30 + 3x^2 - 15x + 18 + 1 = 8x^2 - 20x - 11$$

5. $x^3 + x^2 - 7x + 20$

Distribute the two terms in the binomial over the terms in the trinomial; then combine like terms.

$$x(x^2 - 3x + 5) + 4(x^2 - 3x + 5)$$
$$= x^3 - 3x^2 + 5x + 4x^2 - 12x + 20$$
$$= x^3 + x^2 - 7x + 20$$

6. $3x^3 - x^2 - 3x + 1$

Distribute the two terms in the binomial over the terms in the trinomial; then combine like terms.

$$x(3x^2 + 2x - 1) - 1(3x^2 + 2x - 1)$$
$$= 3x^3 + 2x^2 - x - 3x^2 - 2x + 1$$
$$= 3x^3 - x^2 - 3x + 1$$

7. $2x^3 + 3x^2 - 23x - 12$

First, multiply the second and third binomials together.

$$(2x + 1)(x - 3)(x + 4) =$$
$$(2x + 1)(x^2 + x - 12)$$

Now distribute the two terms in the binomial over the terms in the trinomial; then combine like terms.

$$2x(x^2 + x - 12) + 1(x^2 + x - 12)$$
$$= 2x^3 + 2x^2 - 24x + x^2 + x - 12$$
$$= 2x^3 + 3x^2 - 23x - 12$$

8. $7x^3 + 11x^2 - 63x - 99$

First, multiply the first and second binomials together.

$$(x - 3)(x + 3)(7x + 11)$$
$$= (x^2 - 9)(7x + 11)$$

Now multiply the remaining two binomials together.

$$(x^2 - 9)(7x + 11) = 7x^3 + 11x^2 - 63x - 99$$

9. $x^3 - 9x^2 + 27x - 27$

When raising a binomial to the third power, use the coefficients in the third row of Pascal's Triangle:

1 3 3 1

Place decreasing powers of x after the coefficients.

$$1x^3 \quad 3x^2 \quad 3x^1 \quad 1x^0$$

Now place increasing powers of –3 after the coefficients.

$$1x^3(-3)^0 \quad 3x^2(-3)^1 \quad 3x^1(-3)^2 \quad 1x^0(-3)^3$$

Finally, simplify each term and combine.

$$x^3 - 9x^2 + 27x - 27$$

10. $\mathbf{x^5 + 10x^4 + 40x^3 + 80x^2 + 80x + 32}$

When raising a binomial to the fifth power, use the coefficients in the fifth row of Pascal's Triangle:

1 5 10 10 5 1

Place decreasing powers of x after the coefficients.

$$1x^5 \quad 5x^4 \quad 10x^3 \quad 10x^2 \quad 5x^1 \quad 1x^0$$

Now place increasing powers of 2 after the coefficients.

$$1x^5 2^0 \quad 5x^4 2^1 \quad 10x^3 2^2 \quad 10x^2 2^3 \quad 5x^1 2^4 \quad 1x^0 2^5$$

Finally, simplify each term and combine.

$$x^5 + 10x^4 + 40x^3 + 80x^2 + 80x + 32$$

11. $\mathbf{81x^4 - 216x^3y + 216x^2y^2 - 96xy^3 + 16y^4}$

When raising a binomial to the fourth power, use the coefficients in the fourth row of Pascal's Triangle:

1 4 6 4 1

Place decreasing powers of $3x$ after the coefficients.

$$1(3x)^4 \quad 4(3x)^3 \quad 6(3x)^2 \quad 4(3x)^1 \quad 1(3x)^0$$

Now place increasing powers of $-2y$ after the coefficients.

$$1(3x)^4(-2y)^0 \quad 4(3x)^3(-2y)^1 \quad 6(3x)^2(-2y)^2 \quad 4(3x)^1(-2y)^3 \quad 1(3x)^0(-2y)^4$$

Finally, simplify each term and combine.

$$81x^4 - 216x^3y + 216x^2y^2 - 96xy^3 + 16y^4$$

12. $\mathbf{a^{12} + 6a^{10}b + 15a^8b^2 + 20a^6b^3 + 15a^4b^4 + 6a^2b^5 + b^6}$

When raising a binomial to the sixth power, use the coefficients in the sixth row of Pascal's Triangle:

1 6 15 20 15 6 1

Place decreasing powers of a^2 after the coefficients.

$$1(a^2)^6 \quad 6(a^2)^5 \quad 15(a^2)^4 \quad 20(a^2)^3 \quad 15(a^2)^2 \quad 6(a^2)^1 \quad 1(a^2)^0$$

Simplify the terms to make the work less "busy."

$$1a^{12} \quad 6a^{10} \quad 15a^{8} \quad 20a^{6} \quad 15a^{4} \quad 6a^{2} \quad 1a^{0}$$

Now place increasing powers of the b after the coefficients.

$$1a^{12}b^{0} \quad 6a^{10}b^{1} \quad 15a^{8}b^{2} \quad 20a^{6}b^{3} \quad 15a^{4}b^{4} \quad 6a^{2}b^{5} \quad 1a^{0}b^{6}$$

Finally, simplify each term and combine.

$$a^{12} + 6a^{10}b + 15a^{8}b^{2} + 20a^{6}b^{3} + 15a^{4}b^{4} + 6a^{2}b^{5} + b^{6}$$

13. $x = -11$

First, distribute the 3 over the terms in the binomial.

$$4x + 2 = 3x - 9$$

Now add –2 to each side of the equation and add $-3x$ to each side.

$$x = -11$$

14. $x = -5$

First, distribute the 2 and 3 over their respective binomials.

$$5x + 2x + 14 = 3x - 6$$

Combine the like terms on the left.

$$7x + 14 = 3x - 6$$

Now add $-3x$ to each side and add –14 to each side.

$$4x = -20$$

Divide each side by 4.

$$x = -5$$

15. $x = \frac{16}{3}$ or $x = -4$

First, rewrite the absolute value equation as its two corresponding linear equations.

$$3x - 2 = 14 \text{ or } 3x - 2 = -14$$

Solve the equations by adding 2 to each side.

$$3x = 16 \text{ or } 3x = -12$$

Now divide each side by 3.

$$x = \frac{16}{3} \text{ or } x = -4$$

16. $x = 1$ or $x = -\frac{3}{2}$

First, add 2 to each side of the equation to isolate the absolute value.

$$|4x + 1| = 5$$

Now rewrite the absolute value equation as its two corresponding linear equations.

$$4x + 1 = 5 \text{ or } 4x + 1 = -5$$

Add –1 to each side of the equations.

$$4x = 4 \text{ or } 4x = -6$$

Divide each side by 4.

$$x = 1 \text{ or } x = \frac{-6}{4} = -\frac{3}{2}$$

17. $\mathbf{x = 8 \text{ or } x = 4}$

First, divide each side of the equation by 4 to isolate the absolute value.

$$|x-6| = 2$$

Now rewrite the absolute value equation as its two corresponding linear equations.

$$x - 6 = 2 \text{ or } x - 6 = -2$$

Add 6 to each side of the equations.

$$x = 8 \text{ or } x = 4$$

18. $\mathbf{x = 3 \text{ or } x = 2}$

First, add –5 to each side of the absolute value equation.

$$3|2x-5| = 3$$

Now divide each side by 3.

$$|2x-5| = 1$$

Next rewrite the absolute value equation as its two corresponding linear equations.

$$2x - 5 = 1 \text{ or } 2x - 5 = -1$$

Add 5 to each side of the equations.

$$2x = 6 \text{ or } 2x = 4$$

Finally, divide each side by 2.

$$x = 3 \text{ or } x = 2$$

19. $\mathbf{l = \frac{P-2w}{2}}$

$P = 2l + 2w$ finds the perimeter of a rectangle.

First, add $-2w$ to each side of the equation.

$$P - 2w = 2l$$

Now divide each side by 2.

$$\frac{P-2w}{2} = l \text{ or } \frac{P}{2} - w = l$$

20. $\mathbf{s_1 = \frac{P - s_2}{2}}$

$P = 2s_1 + s_2$ gives the perimeter of an isosceles triangle.

First, subtract s_2 from each side of the equation.

$$P - s_2 = 2s_1$$

Now divide each side by 2.

$$\frac{P - s_2}{2} = s_1$$

21. $\mathbf{b_2 = \frac{2A}{h} - b_1}$

$A = \frac{1}{2}h(b_1 + b_2)$ is used to find the area of a trapezoid.

Multiply each side by 2.

$$2A = h(b_1 + b_2)$$

Divide each side by h.

$$\frac{2A}{h} = b_1 + b_2$$

Now add $-b_1$ to each side of the equation.

$$\frac{2A}{h} - b_1 = b_2 \text{ or } \frac{2A - b_1h}{h} = b_2$$

22. $\mathbf{F = \frac{9}{5}C + 32}$

$C = \frac{5}{9}(F - 32)$ is used to convert degrees Fahrenheit to degrees Centigrade/Celsius.

Multiply each side of the equation by $\frac{9}{5}$.

$$\frac{9}{5}C = F - 32$$

Now add 32 to each side.

$$\frac{9}{5}C + 32 = F$$

23. $\mathbf{t = \frac{A - P}{Pr}}$

$A = P + Prt$ gives you the total amount in an account earning simple interest.

Add $-P$ to each side.

$$\text{A} - \text{P} = \text{Prt}$$

Now divide each side by Pr.

$$\frac{A - P}{Pr} = t$$

24. $n = \frac{a_n - a_1}{d} + 1$

$a_n = a_1 + (n-1)d$ gives you the *n*th term of an arithmetic sequence.

First, add $-a_1$ to each side.

$$a_n - a_1 = (n-1)d$$

Divide each side by *d*.

$$\frac{a_n - a_1}{d} = n - 1$$

Finally, add 1 to each side.

$$\frac{a_n - a_1}{d} + 1 = n \text{ or } \frac{a_n - a_1 + d}{d} = n$$

25. $x \geq -5$

Add 4 to each side of the inequality.

$$3x \leq 5x + 10$$

Add $-5x$ to each side.

$$-2x \leq 10$$

Divide each side by –2; remember to reverse the sense.

$$x \geq -5$$

In interval notation, this is written $[-5, \infty)$.

26. $x > 6$

Distribute the 4 on the left.

$$4x - 12 > x + 6$$

Add 12 to each side.

$$4x > x + 18$$

Add $-x$ to each side.

$$3x > 18$$

Divide by 3.

$$x > 6$$

In interval notation, this is written $(6, \infty)$.

27. $-5 \leq x < 1$

First, add –7 to each of the three intervals/sections.

$$-10 \leq 2x < 2$$

Now divide each interval/section by 2.

$$-5 \leq x < 1$$

In interval notation, this is written $[-5,1)$.

28. $-2 < x < \frac{7}{3}$

First, add –7 to each interval/section.

$$-7 < -3x < 6$$

Now divide each interval/section by –3. Be sure to reverse both of the senses.

$$\frac{7}{3} > x > -2$$

Rewriting the inequality so that the numbers increase as you read from left to right, you reverse both the numbers and the direction of the senses, which gives you

$$-2 < x < \frac{7}{3}$$

In interval notation, this is written $\left(-2, \frac{7}{3}\right)$.

29. $-10 < x < -2$

First, rewrite the absolute value inequality as its corresponding inequality.

$$-4 < x + 6 < 4$$

Add –6 to each interval/section.

$$-10 < x < -2$$

In interval notation, this is written (–10, –2).

30. $x \le -1$ **or** $x \ge 4$

Rewrite the absolute value inequality as its corresponding inequalities.

$$2x - 3 \ge 5 \text{ or } 2x - 3 \le -5$$

Add 3 to each side of the two inequalities.

$$2x \ge 8 \text{ or } 2x \le -2$$

Divide both sides of the inequalities by 2.

$$x \ge 4 \text{ or } x \le -1$$

In interval notation, this is written $(-\infty, -1] \cup [4, \infty)$.

31. $\frac{1}{2} \le x \le 2$

First, add –1 to each side of the absolute value inequality.

$$|4x - 5| \le 3$$

Now rewrite the absolute value inequality as its corresponding inequality.

$$-3 \le 4x - 5 \le 3$$

Add 5 to each interval/section.

$$2 \le 4x \le 8$$

Now divide each interval/section by 4.

$$\frac{1}{2} \le x \le 2$$

In interval notation, this is written $\left[\frac{1}{2},2\right]$.

32. $x < -\frac{5}{6}$ **or** $x > \frac{5}{2}$

First, divide each side of the absolute value inequality by 2.

$$|6x-5| > 10$$

Next, rewrite the absolute value inequality as its corresponding inequalities.

$$6x-5 > 10 \text{ or } 6x-5 < -10$$

Add 5 to each side of the inequalities.

$$6x > 15 \text{ or } 6x < -5$$

Divide each side of the inequalities by 6.

$$x > \frac{5}{2} \text{ or } x < -\frac{5}{6}$$

In interval notation, this is written $\left(-\infty,-\frac{5}{6}\right)\cup\left(\frac{5}{2},\infty\right)$.

33. $2 \le x < 10$

Create two statements: $4x-9 < 2x+1$ and $2x+1 \le 3x-1$

Solve each and find the intersection of their solutions.

First, with $4x-9 < 2x+1$, add 9 to each side and add $-2x$ to each side.

$$2x < 10$$

Divide each side by 2.

$$x < 5$$

Next, with $2x+1 \le 3x-1$, add -1 to each side and add $-3x$ to each side.

$$-x \le -2$$

Divide each side by -1; remember to reverse the sense.

$$x \ge 2$$

The two solutions determined are $x < 5$ and $x \ge 2$. The intersection of those two statements is $2 \le x < 5$.

In interval notation, this is written $[2,5)$.

34. $-4 < x \le -3$

Create two statements: $2x+6 \le x+3$ and $x+3 < 3x+11$

Solve each and find the intersection of their solutions.

First, with $2x+6 \le x+3$, add $-x$ to each side and add -6 to each side.

$$x \le -3$$

Next, with $x + 3 < 3x + 11$, add –3 to each side and add $-3x$ to each side.

$$-2x < 8$$

Divide each side by –2, reversing the sense.

$$x > -4$$

The two solutions determined are $x \leq -3$ and $x > -4$. The intersection of those two statements is $-4 < x \leq -3$.

In interval notation, this is written (–4, –3].

35. $5\sqrt{2}$

Write the number under the radical as a product with a perfect square factor.

$$\sqrt{50} = \sqrt{25 \cdot 2}$$

Now write the product under the radical as the product of radicals.

$$= \sqrt{25}\sqrt{2}$$

Simplify.

$$= 5\sqrt{2}$$

36. $10\sqrt{3}$

Write the number under the radical as a product with a perfect square factor.

$$\sqrt{300} = \sqrt{100 \cdot 3}$$

Now write the product under the radical as the product of radicals.

$$= \sqrt{100}\sqrt{3}$$

Simplify.

$$= 10\sqrt{3}$$

37. $6\sqrt{5}$

Write the number under the radical as a product with a perfect square factor.

$$\sqrt{180} = \sqrt{36 \cdot 5}$$

Now write the product under the radical as the product of radicals.

$$= \sqrt{36}\sqrt{5}$$

Simplify.

$$= 6\sqrt{5}$$

38. $8\sqrt{15}$

Write the number under the radical as a product with a perfect square factor.

$$\sqrt{960} = \sqrt{64 \cdot 15}$$

Now write the product under the radical as the product of radicals.

$$= \sqrt{64}\sqrt{15}$$

Simplify.

$$= 8\sqrt{15}$$

Sometimes, the larger perfect squares aren't easy to spot in these big numbers. It doesn't hurt to do the problem with more than one go-around. For example, you may have spotted 16 as a factor of 960 and written

$$\sqrt{960} = \sqrt{16 \cdot 60} = \sqrt{16}\sqrt{60} = 4\sqrt{60}$$

This isn't completely simplified, because 60 has a perfect square factor of 4.

$$\text{So } 4\sqrt{60} = 4\sqrt{4 \cdot 15} = 4\sqrt{4}\sqrt{15} = 4 \cdot 2\sqrt{15} = 8\sqrt{15}$$

39. $\mathbf{3 + 2\sqrt{2}}$

Square the binomial.

$$\left(1+\sqrt{2}\right)\left(1+\sqrt{2}\right)$$
$$= 1 + 2\sqrt{2} + \sqrt{4}$$
$$= 1 + 2\sqrt{2} + 2$$
$$= 3 + 2\sqrt{2}$$

40. $\mathbf{8 - 2\sqrt{15}}$

Square the binomial.

$$\left(\sqrt{3}-\sqrt{5}\right)\left(\sqrt{3}-\sqrt{5}\right)$$
$$= \sqrt{9} - 2\sqrt{15} + \sqrt{25}$$
$$= 3 - 2\sqrt{15} + 5$$
$$= 8 - 2\sqrt{15}$$

41. $\mathbf{\sqrt{6} - 2}$

Multiply the numerator and denominator by the conjugate of the denominator.

$$\frac{2}{2+\sqrt{6}} \cdot \frac{2-\sqrt{6}}{2-\sqrt{6}}$$
$$= \frac{2\left(2-\sqrt{6}\right)}{\left(2+\sqrt{6}\right)\left(2-\sqrt{6}\right)}$$

The denominator is the product of the sum and difference of the same two values. The product is the difference of their squares.

$$= \frac{2\left(2-\sqrt{6}\right)}{4-\sqrt{36}}$$
$$= \frac{2\left(2-\sqrt{6}\right)}{4-6}$$
$$= \frac{2\left(2-\sqrt{6}\right)}{-2}$$

Reduce the fraction.

$$=\frac{\not{2}(2-\sqrt{6})}{-\not{2}}=-(2-\sqrt{6})=\sqrt{6}-2$$

42. $\frac{5+\sqrt{5}}{2}$

Multiply the numerator and denominator by the conjugate of the denominator.

$$\frac{10}{5-\sqrt{5}}\cdot\frac{5+\sqrt{5}}{5+\sqrt{5}}$$

$$=\frac{10(5+\sqrt{5})}{(5-\sqrt{5})(5+\sqrt{5})}$$

The denominator is the product of the sum and difference of the same two values. The product is the difference of their squares.

$$=\frac{10(5+\sqrt{5})}{25-\sqrt{25}}$$

$$=\frac{10(5+\sqrt{5})}{25-5}$$

$$=\frac{10(5+\sqrt{5})}{20}$$

Reduce the fraction.

$$=\frac{\not{10}(5+\sqrt{5})}{\not{20}_2}=\frac{5+\sqrt{5}}{2}$$

43. $\frac{13+4\sqrt{10}}{3}$

Multiply the numerator and denominator by the conjugate of the denominator.

$$\frac{4+\sqrt{10}}{4-\sqrt{10}}\cdot\frac{4+\sqrt{10}}{4+\sqrt{10}}$$

$$=\frac{(4+\sqrt{10})(4+\sqrt{10})}{(4-\sqrt{10})(4+\sqrt{10})}$$

The denominator is the product of the sum and difference of the same two values. The product is the difference of their squares.

$$=\frac{16+8\sqrt{10}+\sqrt{100}}{16-\sqrt{100}}$$

$$=\frac{16+8\sqrt{10}+10}{16-10}$$

$$=\frac{26+8\sqrt{10}}{6}$$

Reduce the fraction. Be sure to divide all three terms by 2.

$$=\frac{\not{26}^{13}+\not{8}^{4}\sqrt{10}}{\not{6}^{3}}=\frac{13+4\sqrt{10}}{3}$$

44. $\frac{21+10\sqrt{3}}{2}$

Multiply the numerator and denominator by the conjugate of the denominator.

$$\frac{12-\sqrt{3}}{4-2\sqrt{3}}\cdot\frac{4+2\sqrt{3}}{4+2\sqrt{3}}$$
$$=\frac{\left(12-\sqrt{3}\right)\left(4+2\sqrt{3}\right)}{\left(4-2\sqrt{3}\right)\left(4+2\sqrt{3}\right)}$$

The denominator is the product of the sum and difference of the same two values. The product is the difference of their squares.

$$=\frac{48+24\sqrt{3}-4\sqrt{3}-2\sqrt{9}}{16-4\sqrt{9}}$$
$$=\frac{48+20\sqrt{3}-2(3)}{16-4(3)}$$
$$=\frac{48+20\sqrt{3}-6}{16-12}$$
$$=\frac{42+20\sqrt{3}}{4}$$

Reduce the fraction — dividing both terms in the numerator by 2.

$$=\frac{21+10\sqrt{3}}{2}$$

45. –1

The power on i can be written as the sum of a multiple of 4 and a number between 0 and 3.

$$i^{138}=i^{4(34)+2}$$

Rewrite the power of i as a product.

$$=i^{4(34)}i^{2}$$

Every power of i that is a multiple of 4 is equal to 1. And $i^2=-1$.

$$i^{4(34)}i^{2}=1(-1)=-1$$

46. i

The power on i can be written as the sum of a multiple of 4 and a number between 0 and 3.

$$i^{1001}=i^{4(250)+1}$$

Rewrite the power of i as a product.

$$=i^{4(250)}i^{1}$$

Every power of i that is a multiple of 4 is equal to 1. And $i^1=i$.

$$i^{4(250)}i^{1}=1(i)=i$$

47. **$15 - 8i$**

Square the binomial.

$$(4 - i)(4 - i)$$
$$= 16 - 8i + i^2$$

But $i^2 = -1$, so substitute this value into the expression and simplify.

$$= 16 - 8i + (-1)$$
$$= 15 - 8i$$

48. **$5 + 12i$**

Square the binomial.

$$(3 + 2i)(3 + 2i)$$
$$= 9 + 12i + 4i^2$$

But $i^2 = -1$, so substitute this value into the expression and simplify.

$$= 9 + 12i + 4(-1)$$
$$= 5 + 12i$$

49. **8**

First, cube the term in the parentheses.

$$i(2i)^3 = i(2^3 i^3) = i(8i^3)$$

Multiply by i.

$$= 8i^4$$

And, because $i^4 = 1$,

$$= 8(1) = 8$$

50. **–8**

First, square the binomial.

$$4i^{21}(1+i)^2$$
$$= 4i^{21}(1+i)(1+i)$$
$$= 4i^{21}(1+2i+i^2)$$

Now replace i^2 with –1.

$$= 4i^{21}(1+2i-1) = 4i^{21}(2i)$$

Now multiply.

$$4i^{21}(2i) = 8i^{22}$$

Now simplify the complex numbers by writing the exponents as the sum of a power of 4 and another number between 0 and 3.

$$= 8i^{5(4)+2} = 8i^{5(4)}i^2$$

Every power of i that is a multiple of 4 is equal to 1, and $i^2 = -1$. Substitute these values into the expression and simplify.

$$= 8(1)(-1) = -8$$

51. $x = \pm 9$

Take the square root of both sides of the equation. Be sure to assign the ± sign in front of the numerical part to account for both solutions.

$$\sqrt{x^2} = \pm\sqrt{81}$$

$$x = \pm 9$$

52. $x = \pm 12$

First, add 144 to each side of the equation.

$$x^2 = 144$$

Then take the square root of both sides. Be sure to assign the ± sign in front of the numerical part to account for both solutions.

$$\sqrt{x^2} = \pm\sqrt{144}$$

$$x = \pm 12$$

53. $y = \pm 5$

First, add 75 to each side of the equation.

$$3y^2 = 75$$

Then divide each side by 3.

$$y^2 = 25$$

Then take the square root of both sides. Be sure to assign the ± sign in front of the numerical part to account for both solutions.

$$\sqrt{y^2} = \pm\sqrt{25}$$

$$y = \pm 5$$

54. $z = \pm 5$

First, add 125 to each side of the equation.

$$5z^2 = 125$$

Then divide each side by 5.

$$z^2 = 25$$

Then take the square root of both sides. Be sure to assign the ± sign in front of the numerical part to account for both solutions.

$$\sqrt{z^2} = \pm\sqrt{25}$$

$$z = \pm 5$$

55. $\boldsymbol{x = \pm\frac{2}{3}}$

First, divide each side of the equation by 9.

$$x^2 = \frac{4}{9}$$

Then take the square root of both sides. Be sure to assign the ± sign in front of the numerical part to account for both solutions.

$$\sqrt{x^2} = \pm\sqrt{\frac{4}{9}}$$

$$x = \pm\frac{\sqrt{4}}{\sqrt{9}} = \pm\frac{2}{3}$$

$$x = \pm\frac{2}{3}$$

56. $\boldsymbol{x = \pm\frac{3}{7}}$

First, divide each side of the equation by 98 and reduce.

$$x^2 = \frac{18}{98} = \frac{9}{49}$$

Then take the square root of both sides. Be sure to assign the ± sign in front of the numerical part to account for both solutions.

$$\sqrt{x^2} = \pm\sqrt{\frac{9}{49}}$$

$$x = \pm\frac{\sqrt{9}}{\sqrt{49}} = \pm\frac{3}{7}$$

$$x = \pm\frac{3}{7}$$

57. $\boldsymbol{x = \pm\sqrt{11}}$

Take the square root of both sides of the equation. Be sure to assign the ± sign in front of the numerical part to account for both solutions.

$$\sqrt{x^2} = \pm\sqrt{11}$$

$$x = \pm\sqrt{11}$$

The number 11 isn't a perfect square and has no perfect square factors, so this is the final answer.

58. $y = \pm 2\sqrt{5}$

Take the square root of both sides of the equation. Be sure to assign the ± sign in front of the numerical part to account for both solutions.

$$\sqrt{y^2} = \pm\sqrt{20}$$

$$y = \pm\sqrt{20}$$

The number 20 isn't a perfect square, but it does have a perfect square factor. The answer can be simplified.

$$y = \pm\sqrt{4 \cdot 5} = \pm\sqrt{4}\sqrt{5} = \pm 2\sqrt{5}$$

59. $x = \pm 5\sqrt{2}$

First, divide both sides of the equation by 4.

$$x^2 = 50$$

Take the square root of both sides. Be sure to assign the ± sign in front of the numerical part to account for both solutions.

$$\sqrt{x^2} = \pm\sqrt{50}$$

$$x = \pm\sqrt{50}$$

The number 50 isn't a perfect square, but it does have a perfect square factor. The answer can be simplified.

$$x = \pm\sqrt{25 \cdot 2} = \pm\sqrt{25}\sqrt{2} = \pm 5\sqrt{2}$$

60. $z = \pm 11\sqrt{2}$

First, add 726 to both sides of the equation.

$$3z^2 = 726$$

Then divide both sides by 3.

$$z^2 = 242$$

Take the square root of both sides. Be sure to assign the ± sign in front of the numerical part to account for both solutions.

$$\sqrt{z^2} = \pm\sqrt{242}$$

$$z = \pm\sqrt{242}$$

The number 242 isn't a perfect square, but it does have a perfect square factor. The answer can be simplified.

$$z = \pm\sqrt{121 \cdot 2} = \pm\sqrt{121}\sqrt{2} = \pm 11\sqrt{2}$$

61. $x = 5$ or $x = -7$

First, determine the factor pairs of the last coefficient. These are the possible values you can choose from for the second term in each binomial. The products yielding 35 are 1×35 and 5×7.

You want the difference of the two values to match the coefficient of the middle term, 2, so use 5 and 7 for the second term in each binomial.

$$(x \quad 5)(x \quad 7) = 0$$

The difference is +2, so you want +7 and –5.

$$(x - 5)(x + 7) = 0$$

Setting the binomials equal to 0, you have $x = 5$ or $x = -7$.

62. $y = 12$ **or** $y = -8$

First, determine the factor pairs of the last coefficient. These are the possible values you can choose from for the second term in each binomial. The products yielding 96 are 1×96, 2×48, 3×32, 4×24, 6×16, and 8×12.

You want the difference of the two values to match the coefficient of the middle term, 4, so use 8 and 12 for the second term in each binomial.

$$(y \quad 8)(y \quad 12) = 0$$

The difference is –4, so you want –12 and +8.

$$(y - 12)(y + 8) = 0$$

Setting the binomials equal to 0, you have $y = 12$ or $y = -8$.

63. $z = -6$ **or** $z = -8$

First, determine the factor pairs of the last coefficient. These are the possible values you can choose from for the second term in each binomial. The products yielding 48 are 1×48, 2×24, 3×16, 4×12, and 6×8.

You want the sum of the two values to match the middle term, 14, so use 6 and 8 for the second term in each binomial.

$$(z \quad 6)(z \quad 8) = 0$$

The sum is +14, so you want +6 and +8.

$$(z + 6)(z + 8) = 0$$

Setting the binomials equal to 0, you have $z = -6$ or $z = -8$.

64. $x = -\frac{1}{4}$ **or** $x = \frac{3}{2}$

First, determine the factor pairs of both the first and the last coefficients. These are the possible values you can choose from for the first terms and second terms, respectively, in each binomial.

The products yielding 8 are 1×8 or 2×4.

And the product yielding 3 is just 1×3.

You want the difference of the outer and inner products to be $10x$, so use the 2 and 4 for the two first terms and the 1 and 3 for the two second terms. Align them so you have outer and inner products of $12x$ and $2x$.

$$(4x \quad 1)(2x \quad 3) = 0$$

The difference is $-10x$, so you want the $12x$ product to be negative and the $2x$ product to be positive.

$$(4x + 1)(2x - 3) = 0$$

Setting the binomials equal to 0, you have $x = -\frac{1}{4}$ or $x = \frac{3}{2}$.

65. $x = -\frac{6}{5}$ **or** $x = \frac{5}{6}$

First, determine the factor pairs of both the first and the last coefficients. These are the possible values you can choose from for the first terms and second terms, respectively, in each binomial.

The products yielding 30 are 1×30, 2×15, 3×10, or 5×6.

You want the difference of the outer and inner products to be $11x$, so use the 5 and 6 for the two first terms and the 5 and 6 for the two second terms. Align them so you have outer and inner products of $25x$ and $36x$.

$$(5x \quad 6)(6x \quad 5) = 0$$

The difference is $+11x$, so you want the $25x$ product to be negative and the $36x$ product to be positive.

$$(5x + 6)(6x - 5) = 0$$

Setting the binomials equal to 0, you have $x = -\frac{6}{5}$ or $x = \frac{5}{6}$.

66. $y = -\frac{2}{3}$ **or** $y = -\frac{3}{4}$

First, determine the factor pairs of both the first and the last coefficients. These are the possible values you can choose from for the first terms and second terms, respectively, in each binomial.

The products yielding 12 are 1×12, 2×6, or 3×4.

The products yielding 6 are 1×6 or 2×3.

You want the sum of the outer and inner products to be $17y$, so use the 3 and 4 for the two first terms and the 2 and 3 for the two second terms. Align them so you have outer and inner products of $9y$ and $8y$.

$$(3y \quad 2)(4y \quad 3) = 0$$

The sum in the middle term is $+17y$, so you want both the signs to be positive.

$$(3y + 2)(4y + 3) = 0$$

Setting the binomials equal to 0, you have $y = -\frac{2}{3}$ or $y = -\frac{3}{4}$.

67. $z = -\frac{3}{5}$ **or** $z = \frac{1}{25}$

First, determine the factor pairs of both the first and the last coefficients. These are the possible values you can choose from for the first terms and second terms, respectively, in each binomial.

The products yielding 125 are 1×125 or 5×25.

The product yielding 3 is just 1×3.

You want the difference of the outer and inner products to be $70z$, so use the 5 and 25 for the two first terms and the 1 and 3 for the two second terms. Align them so you have outer and inner products of $5z$ and $75z$.

$$(5z \quad 3)(25z \quad 1) = 0$$

The difference in the middle term is $+70z$, so you want the $75z$ to be positive and the $5z$ to be negative.

$$(5z + 3)(25z - 1) = 0$$

Setting the binomials equal to 0, you have $z = -\frac{3}{5}$ or $z = \frac{1}{25}$.

68. $x = 1$ **or** $x = -\frac{11}{12}$

First, determine the factor pairs of both the first and the last coefficients. These are the possible values you can choose from for the first terms and second terms, respectively, in each binomial.

The products yielding 12 are 1×12, 2×6, or 3×4.

The product yielding 11 is just 1×11.

You want the difference of the outer and inner products to be $1x$, so use the 1 and 12 for the two first terms and the 1 and 11 for the two second terms. Align them so you have outer and inner products of $11x$ and $12x$.

$$(x \quad 1)(12x \quad 11) = 0$$

The difference in the middle term is $-1x$, so you want the $11x$ to be positive and the $12x$ to be negative.

$$(x - 1)(12x + 11) = 0$$

Setting the binomials equal to 0, you have $x = 1$ or $x = -\frac{11}{12}$.

69. $y = \frac{5}{4}$ **or** $y = \frac{7}{100}$

The products yielding 400 are 1×400, 2×200, 4×100, 5×80, 8×50, 10×40, 16×25, or 20×20.

The products yielding 35 are 1×35 or 5×7.

You want the sum of the outer and inner products to be $528y$, so use the 4 and 100 for the two first terms and the 5 and 7 for the two second terms. Align them so you have outer and inner products of $28y$ and $500y$.

$$(4y \quad 5)(100y \quad 7) = 0$$

The sum in the middle term is $-528y$, so you want both signs to be negative.

$$(4y - 5)(100y - 7) = 0$$

Setting the binomials equal to 0, you have $y = \frac{5}{4}$ or $y = \frac{7}{100}$.

70. $z=\frac{1}{2}$ or $z=\frac{15}{16}$

The products yielding 32 are 1×32, 2×16, or 4×8.

The products yielding 15 are 1×15 or 3×5.

You want the sum of the outer and inner products to be $46z$, so use the 2 and 16 for the two first terms and the 1 and 15 for the two second terms. Align them so you have outer and inner products of $30z$ and $16z$.

$$(2z \quad 1)(16z \quad 15)=0$$

The sum in the middle term is $-46z$, so you want both signs to be negative.

$$(2z-1)(16z-15)=0$$

Setting the binomials equal to 0, you have $z=\frac{1}{2}$ or $z=\frac{15}{16}$.

71. $y=\frac{4z}{3}$ or $y=-\frac{8z}{5}$

The products yielding 15 are 1×15 or 3×5.

And the products yielding 32 are 1×32, 2×16, or 4×8.

You want the difference of the outer and inner products to be $4yz$, so use the 3 and 5 for the two first terms and the 4 and 8 for the two second terms. Align them so you have outer and inner products of $24yz$ and $20yz$.

$$(3y \quad 4z)(5y \quad 8z)=0$$

The difference in the middle term is $+4yz$, so you want the $24yz$ to be positive and the $20yz$ to be negative.

$$(3y-4z)(5y+8z)=0$$

Setting the binomials equal to 0 and solving for y, you have $y=\frac{4z}{3}$ or $y=-\frac{8z}{5}$.

72. $z=-\frac{7w}{6}$ or $z=-\frac{4w}{15}$

The products yielding 90 are 1×90, 2×45, 3×30, 5×18, 6×15, or 9×10.

The products yielding 28 are 1×28, 2×14, or 4×7.

You want the sum of the outer and inner products to be $129wz$, so use the 6 and 15 for the two first terms and the 4 and 7 for the two second terms. Align them so you have outer and inner products of $24wz$ and $105wz$.

$$(6z \quad 7w)(15z \quad 4w)=0$$

The sum in the middle term is $+129wz$, so you want both signs to be positive.

$$(6z+7w)(15z+4w)=0$$

Setting the binomials equal to 0 and solving for z, you have $z=-\frac{7w}{6}$ or $z=-\frac{4w}{15}$.

73. $x = -\frac{c}{a}$ **or** $x = -1$

The product yielding a is just $1 \times a$.

The product yielding c is just $1 \times c$.

You want the sum of the outer and inner products to be $(a + c)x$, so align the factors so you have outer and inner products of ax and cx.

$$(ax \quad c)(1x \quad 1) = 0$$

The sum in the middle term is +, so you want both of the signs to be positive.

$$(ax + c)(1x + 1) = 0$$

Setting the binomials equal to 0 and solving for x, you have $x = -\frac{c}{a}$ or $x = -1$.

74. $x = -\frac{b}{a}$ **or** $x = \frac{1}{c}$

The product yielding ac is just $a \times c$.

The product yielding b is just $1 \times b$.

You want the difference of the outer and inner products to be $(bc - a)x$, so align the factors so you have outer and inner products of ax and bcx.

$$(ax \quad b)(cx \quad 1) = 0$$

For the ax to be negative and the bcx positive, so that the combined middle term is $(bc - a)x$, put the negative sign in the second binomial.

$$(ax + b)(cx - 1) = 0$$

Setting the binomials equal to 0 and solving for x, you have $x = -\frac{b}{a}$ or $x = \frac{1}{c}$.

75. $x = -\sqrt{2}$ **or** $x = -\frac{3\sqrt{2}}{2}$

The product yielding 2 is just 1×2, but the radical factor in the middle term suggests you may use $\sqrt{2} \cdot \sqrt{2}$.

The products yielding 6 are 1×6 or 2×3.

You want the sum of the outer and inner products to be $5\sqrt{2}$, so use the 2 and 3 factors of the 6 and the $\sqrt{2}$ factors of the 2. Align them so you have outer and inner products of $3\sqrt{2}$ and $2\sqrt{2}$.

$$\left(\sqrt{2}x \quad 2\right)\left(\sqrt{2}x \quad 3\right) = 0$$

The sum in the middle term is positive, so use the plus sign for both binomials.

$$\left(\sqrt{2}x + 2\right)\left(\sqrt{2}x + 3\right) = 0$$

Setting the binomials equal to 0, you have $x = -\frac{2}{\sqrt{2}}$ or $x = -\frac{3}{\sqrt{2}}$

Rationalizing the fractions,

$$x = -\frac{2}{\sqrt{2}} \cdot \frac{\sqrt{2}}{\sqrt{2}} = -\frac{2\sqrt{2}}{2} = -\sqrt{2}$$

$$x = -\frac{3}{\sqrt{2}} \cdot \frac{\sqrt{2}}{\sqrt{2}} = -\frac{3\sqrt{2}}{2}$$

76. $z = \frac{\sqrt{2}}{3}$ **or** $z = -\frac{3\sqrt{2}}{4}$

The products yielding 12 are just 1×12, 2×6, or 3×4, but the radical factor in the middle term suggests you may use some radicals such as $6\sqrt{2} \cdot \sqrt{2}$ or $3\sqrt{2} \cdot 2\sqrt{2}$.

The products yielding 6 are 1×6 or 2×3, but, again, you may use some radicals such as $3 \cdot \sqrt{2}\sqrt{2}$.

You want the difference between the outer and inner terms to be $5\sqrt{2}$, so use the 3 and 4 as factors of 12 and the $\sqrt{2}$ and $3\sqrt{2}$ as factors of 6. Align the factors so the outer product is $9\sqrt{2}z$ and the inner product is $4\sqrt{2}z$.

$$(3z \quad \sqrt{2})(4z \quad 3\sqrt{2}) = 0$$

The difference in the middle term shows the $5\sqrt{2}z$ as positive, so give the first binomial the negative sign.

$$(3z - \sqrt{2})(4z + 3\sqrt{2}) = 0$$

Setting the binomials equal to 0, you have $z = \frac{\sqrt{2}}{3}$ or $z = -\frac{3\sqrt{2}}{4}$.

77. $x = 3 \pm \sqrt{5}$

The equation is in standard form, so $a = 1$, $b = -6$, and $c = 4$. Substitute the values into the quadratic formula.

$$\begin{aligned} x &= \frac{-b \pm \sqrt{b^2 - 4ac}}{2a} \\ &= \frac{-(-6) \pm \sqrt{(-6)^2 - 4(1)(4)}}{2(1)} \\ &= \frac{6 \pm \sqrt{36 - 16}}{2} = \frac{6 \pm \sqrt{20}}{2} \end{aligned}$$

Simplify the radical and reduce the fraction.

$$\begin{aligned} &= \frac{6 \pm \sqrt{4 \cdot 5}}{2} = \frac{6 \pm 2\sqrt{5}}{2} \\ &= \frac{\cancel{2}(3 \pm \sqrt{5})}{\cancel{2}} = 3 \pm \sqrt{5} \end{aligned}$$

Answers 1–100

78. $y = -4 \pm \sqrt{6}$

The equation is in standard form, so $a = 1$, $b = 8$, and $c = 10$. Substitute the values into the quadratic formula.

$$\begin{aligned} y &= \frac{-b \pm \sqrt{b^2 - 4ac}}{2a} \\ &= \frac{-8 \pm \sqrt{8^2 - 4(1)(10)}}{2(1)} \\ &= \frac{-8 \pm \sqrt{64 - 40}}{2} = \frac{-8 \pm \sqrt{24}}{2} \end{aligned}$$

Simplify the radical and reduce the fraction.

$$\begin{aligned} &= \frac{-8 \pm \sqrt{4 \cdot 6}}{2} = \frac{-8 \pm 2\sqrt{6}}{2} \\ &= \frac{\cancel{2}\left(-4 \pm \sqrt{6}\right)}{\cancel{2}} = -4 \pm \sqrt{6} \end{aligned}$$

79. $x = \frac{-9 \pm 3\sqrt{13}}{2}$

The equation is in standard form, so $a = 1$, $b = 9$, and $c = -9$. Substitute the values into the quadratic formula.

$$\begin{aligned} x &= \frac{-b \pm \sqrt{b^2 - 4ac}}{2a} \\ &= \frac{-9 \pm \sqrt{9^2 - 4(1)(-9)}}{2(1)} \\ &= \frac{-9 \pm \sqrt{81 + 36}}{2} = \frac{-9 \pm \sqrt{117}}{2} \end{aligned}$$

Simplify the radical.

$$= \frac{-9 \pm \sqrt{9 \cdot 13}}{2} = \frac{-9 \pm 3\sqrt{13}}{2}$$

80. $x = \frac{7}{2}$ or $x = -2$

The equation is in standard form, so $a = 2$, $b = -3$, and $c = -14$. Substitute the values into the quadratic formula.

$$\begin{aligned} x &= \frac{-b \pm \sqrt{b^2 - 4ac}}{2a} \\ &= \frac{-(-3) \pm \sqrt{(-3)^2 - 4(2)(-14)}}{2(2)} \\ &= \frac{3 \pm \sqrt{9 + 112}}{4} = \frac{3 \pm \sqrt{121}}{4} \\ &= \frac{3 \pm 11}{4} \end{aligned}$$

The fact that the value under the radical was a perfect square means that the quadratic can be factored. Continuing with naming the solutions:

$$\frac{3\pm 11}{4}=\frac{3+11}{4} \text{ or } \frac{3-11}{4}$$

$$x=\frac{14}{4}=\frac{7}{2} \text{ or } x=\frac{-8}{4}=-2$$

81. $x=-\frac{3}{8}$ **or** $x=-5$

The equation is in standard form, so $a = 8$, $b = 43$, and $c = 15$. Substitute the values into the quadratic formula.

$$\begin{aligned} x&=\frac{-b\pm\sqrt{b^2-4ac}}{2a}\\ &=\frac{-43\pm\sqrt{43^2-4(8)(15)}}{2(8)}\\ &=\frac{-43\pm\sqrt{1849-480}}{16}=\frac{-43\pm\sqrt{1369}}{16}\\ &=\frac{-43\pm 37}{16} \end{aligned}$$

The fact that the value under the radical was a perfect square means that the quadratic can be factored. Continuing with naming the solutions:

$$\frac{-43\pm 37}{16}=\frac{-43+37}{16} \text{ or } \frac{-43-37}{16}$$

$$x=\frac{-6}{16}=-\frac{3}{8} \text{ or } x=\frac{-80}{16}=-5$$

82. $y=-\frac{5}{2}$ **or** $y=-5$

The equation is in standard form, so $a = 2$, $b = 15$, and $c = 25$. Substitute the values into the quadratic formula.

$$\begin{aligned} y&=\frac{-b\pm\sqrt{b^2-4ac}}{2a}\\ &=\frac{-15\pm\sqrt{15^2-4(2)(25)}}{2(2)}\\ &=\frac{-15\pm\sqrt{225-200}}{4}=\frac{-15\pm\sqrt{25}}{4}\\ &=\frac{-15\pm 5}{4} \end{aligned}$$

The fact that the value under the radical was a perfect square means that the quadratic can be factored. Continuing with naming the solutions:

$$\frac{-15\pm 5}{4}=\frac{-15+5}{4} \text{ or } \frac{-15-5}{4}$$

$$y=\frac{-10}{4}=-\frac{5}{2} \text{ or } y=\frac{-20}{4}=-5$$

83. $x = \frac{5 \pm i\sqrt{11}}{6}$

The equation is in standard form, so $a = 3$, $b = -5$, and $c = 3$. Substitute the values into the quadratic formula.

$$\begin{aligned} x &= \frac{-b \pm \sqrt{b^2 - 4ac}}{2a} \\ &= \frac{-(-5) \pm \sqrt{(-5)^2 - 4(3)(3)}}{2(3)} \\ &= \frac{5 \pm \sqrt{25 - 36}}{6} = \frac{5 \pm \sqrt{-11}}{6} \end{aligned}$$

The value under the radical is negative, so there are no real solutions. The complex answers are written using i for $\sqrt{-1}$.

$$= \frac{5 \pm \sqrt{-11}}{6} = \frac{5 \pm \sqrt{-1}\sqrt{11}}{6} = \frac{5 \pm i\sqrt{11}}{6}$$

In the $a + bi$, form, the answer is written

$$x = \frac{5}{6} + \frac{\sqrt{11}}{6}i \text{ or } x = \frac{5}{6} - \frac{\sqrt{11}}{6}i$$

84. $y = \frac{-3 \pm i\sqrt{7}}{8}$

The equation is in standard form, so $a = 4$, $b = 3$, and $c = 1$. Substitute the values into the quadratic formula.

$$\begin{aligned} y &= \frac{-b \pm \sqrt{b^2 - 4ac}}{2a} \\ &= \frac{-3 \pm \sqrt{3^2 - 4(4)(1)}}{2(4)} \\ &= \frac{-3 \pm \sqrt{9 - 16}}{8} = \frac{-3 \pm \sqrt{-7}}{8} \end{aligned}$$

The value under the radical is negative, so there are no real solutions. The complex answers are written using i for $\sqrt{-1}$.

$$= \frac{-3 \pm \sqrt{-7}}{8} = \frac{-3 \pm \sqrt{-1}\sqrt{7}}{8} = \frac{-3 \pm i\sqrt{7}}{8}$$

In the $a + bi$ form, the answer is written as

$$y = -\frac{3}{8} + \frac{\sqrt{7}}{8}i \text{ or } y = -\frac{3}{8} - \frac{\sqrt{7}}{8}i$$

85. $z = \frac{-4 \pm 2i}{5}$

The equation is in standard form, so $a = 5$, $b = -8$, and $c = 4$. Substitute the values into the quadratic formula.

$$\begin{aligned} z &= \frac{-b \pm \sqrt{b^2 - 4ac}}{2a} \\ &= \frac{-(-8) \pm \sqrt{(-8)^2 - 4(5)(4)}}{2(5)} \\ &= \frac{8 \pm \sqrt{64 - 80}}{10} = \frac{8 \pm \sqrt{-16}}{10} \end{aligned}$$

The value under the radical is negative, so there are no real solutions. The complex answers are written using i for $\sqrt{-1}$.

$$= \frac{8 \pm \sqrt{-16}}{10} = \frac{8 \pm \sqrt{-1}\sqrt{16}}{10} = \frac{8 \pm 4i}{10} = \frac{4 \pm 2i}{5}$$

Written in $a + bi$ form, you have

$$z = \frac{4}{5} + \frac{2}{5}i \text{ or } z = \frac{4}{5} - \frac{2}{5}i$$

86. $z = \frac{-1 \pm 2i\sqrt{2}}{9}$

The equation is in standard form, so $a = 9$, $b = 2$, and $c = 1$. Substitute the values into the formula.

$$\begin{aligned} z &= \frac{-b \pm \sqrt{b^2 - 4ac}}{2a} \\ &= \frac{-2 \pm \sqrt{2^2 - 4(9)(1)}}{2(9)} \\ &= \frac{-2 \pm \sqrt{4 - 36}}{18} = \frac{-2 \pm \sqrt{-32}}{18} \end{aligned}$$

The value under the radical is negative, so there are no real solutions. The complex answers are written using i for $\sqrt{-1}$.

$$\begin{aligned} &= \frac{-2 \pm \sqrt{-32}}{18} = \frac{-2 \pm \sqrt{-1}\sqrt{16}\sqrt{2}}{18} = \frac{-2 \pm 4i\sqrt{2}}{18} \\ &= \frac{-1 \pm 2i\sqrt{2}}{9} \end{aligned}$$

Written in $a + bi$ form, you have

$$z = -\frac{1}{9} + \frac{2\sqrt{2}}{9}i \text{ or } z = -\frac{1}{9} - \frac{2\sqrt{2}}{9}i$$

87. $x = \pm\frac{3}{2}$ **or** $x = \pm 1$

The trinomial is of the form $ax^{2n} + bx^n + c = 0$ and can be factored using the exponent of the middle term.

$$4x^4 - 13x^2 + 9 = (4x^2 - 9)(x^2 - 1) = 0$$

Setting the two binomials equal to 0, use the square root rule to find the solutions.

$$4x^2 - 9 = 0$$
$$x^2 = \frac{9}{4}$$
$$x = \pm\frac{3}{2}$$
$$x^2 - 1 = 0$$
$$x^2 = 1$$
$$x = \pm 1$$

88. $\mathbf{y = -\frac{3}{2}}$ **or** $\mathbf{y = 1}$

The trinomial is of the form $ax^{2n} + bx^n + c = 0$ and can be factored using the exponent of the middle term.

$$8y^6 + 19y^3 - 27 = (8y^3 + 27)(y^3 - 1) = 0$$

Setting the two binomials equal to 0 and solving for y, you need to take the cube root of each side of the equations to solve for y.

$$8y^3 + 27 = 0$$
$$8y^3 = -27$$
$$y^3 = -\frac{27}{8}$$
$$y = -\frac{3}{2}$$
$$y^3 - 1 = 0$$
$$y^3 = 1$$
$$y = 1$$

89. $\mathbf{z = \pm\sqrt{3}}$ **or** $\mathbf{z = \pm i\sqrt{3}}$

First, factor the 5 out of each term.

$$5(z^4 - 9) = 0$$

Now factor the binomial as the difference of squares.

$$5(z^2 - 3)(z^2 + 3) = 0$$

Now set the two binomials equal to 0 and solve for z.

$$z^2 - 3 = 0$$
$$z^2 = 3$$
$$z = \pm\sqrt{3}$$
$$z^2 + 3 = 0$$
$$z^2 = -3$$

Taking the square root yields non-real solutions.

$$z = \pm i\sqrt{3}$$

90. $w = 10$ **or** $w = 1$

The trinomial is of the form $ax^{2n} + bx^{n} + c = 0$ and can be factored using the exponent of the middle term.

$$(w^3 - 1000)(w^3 - 1) = 0$$

Setting the two binomials equal to 0 and solving for w, you need to take the cube root of each side of the equation.

$$w^3 - 1{,}000 = 0$$
$$w^3 = 1{,}000$$
$$w = 10$$
$$w^3 - 1 = 0$$
$$w^3 = 1$$
$$w = 1$$

91. $x = \pm\frac{\sqrt{7}}{7}$ **or** $x = \pm i\frac{\sqrt{3}}{3}$

The trinomial is of the form $ax^{2n} + bx^{n} + c = 0$ and can be factored using the exponent of the middle term.

$$\left(x^{-2} - 7\right)\left(x^{-2} + 3\right) = 0$$

Now set each binomial equal to 0 and solve for x.

$$x^{-2} - 7 = 0$$
$$x^{-2} = 7$$
$$\frac{1}{x^2} = 7$$

Cross multiply and use the square root rule to find the solutions.

$$x^2 = \frac{1}{7}$$
$$x = \pm\sqrt{\frac{1}{7}} = \pm\frac{\sqrt{7}}{7}$$
$$x^{-2} + 3 = 0$$
$$x^{-2} = -3$$
$$\frac{1}{x^2} = -3$$

Cross multiply and use the square root rule to find the solutions. In this case, the solutions are not real numbers.

$$x^2 = -\frac{1}{3}$$
$$x = \pm i\sqrt{\frac{1}{3}} = \pm i\frac{\sqrt{3}}{3}$$

92. $z = \frac{2}{3}$ **or** $z = \frac{2}{9}$

The trinomial is of the form $az^{2n} + bz^{n} + c = 0$ and can be factored using the exponent of the middle term.

$$\left(2z^{-1} - 3\right)\left(2z^{-1} - 9\right) = 0$$

Set the two binomials equal to 0 and solve for z.

$$2z^{-1} - 3 = 0$$
$$2z^{-1} = 3$$
$$z^{-1} = \frac{3}{2}$$
$$\frac{1}{z} = \frac{3}{2}$$
$$z = \frac{2}{3}$$
$$2z^{-1} - 9 = 0$$
$$z^{-1} = \frac{9}{2}$$
$$\frac{1}{z} = \frac{9}{2}$$
$$z = \frac{2}{9}$$

93. $x = -\frac{1}{512}$ **or** $x = 512$

The trinomial is of the form $ax^{2n} + bx^{n} + c = 0$ and can be factored using the exponent of the middle term.

$$\left(8x^{1/3} + 1\right)\left(x^{1/3} - 8\right) = 0$$

Set $8x^{1/3} + 1 = 0$ and solve for x.

$$8x^{1/3} = -1$$
$$x^{1/3} = -\frac{1}{8}$$
$$x = \left(-\frac{1}{8}\right)^3 = -\frac{1}{512}$$

Now, set $x^{1/3} - 8 = 0$ and solve for x.

$$x^{1/3} = 8$$
$$x = 8^3 = 512$$

94. $x = \frac{390,625}{256}$ **or** $x = 1$

The trinomial is of the form $ax^{2n} + bx^{n} + c = 0$ and can be factored using the exponent of the middle term.

$$\left(4x^{1/4} - 25\right)\left(x^{1/4} - 1\right) = 0$$

Set the binomials equal to 0 and solve for x.

$$4x^{1/4}-25=0$$
$$4x^{1/4}=25$$
$$x^{1/4}=\frac{25}{4}$$

Now raise each side to the 4th power.

$$x=\frac{25^4}{4^4}=\frac{390{,}625}{256}$$
$$x^{1/4}-1=0$$
$$x^{1/4}=1$$

Raising each side to the 4th power, $x = 1$.

95. $x=1$ **or** $x=\frac{1}{256}$

The trinomial is of the form $ax^{2n}+bx^n+c=0$ and can be factored using the exponent of the middle term.

$$\left(x^{-1/4}-1\right)\left(x^{-1/4}-4\right)=0$$

Set the binomials equal to 0 and solve for x.

$$x^{-1/4}-1=0$$
$$x^{-1/4}=1$$
$$\frac{1}{x^{1/4}}=1$$
$$x^{1/4}=1$$

Now raise each side to the 4th power.

$$x=1$$
$$x^{-1/4}-4=0$$
$$x^{-1/4}=4$$
$$\frac{1}{x^{1/4}}=4$$
$$x^{1/4}=\frac{1}{4}$$

Now raise each side to the 4th power.

$$x=\frac{1}{256}$$

96. $y=4$ **or** $y=1$

The trinomial is of the form $ay^{2n}+by^n+c=0$ and can be factored using the exponent of the middle term.

$$\left(64y^{-3}-1\right)\left(y^{-3}-1\right)=0$$

Set each binomial equal to 0, and solve for y.

$$64y^{-3}-1=0$$
$$64y^{-3}=1$$
$$y^{-3}=\frac{1}{64}$$
$$\frac{1}{y^3}=\frac{1}{64}$$
$$y^3=64$$

Take the cube root of each side to get $y = 4$.

$$y^{-3}-1=0$$
$$y^{-3}=1$$
$$\frac{1}{y^3}=1$$
$$y^3=1$$

Taking the cube root of each side, $y = 1$.

97. $x = 9$ or $x = -1$

Add 9 to each side of the equation.

$$x^2-8x=9$$

Find the square of half the coefficient of x.

$$\left(\frac{-8}{2}\right)^2=(-4)^2=16$$

Add that square to both sides of the equation.

$$x^2-8x+16=9+16$$

Factor the perfect square trinomial on the left and simplify on the right.

$$(x-4)^2=25$$

Use the square root rule.

$$x-4=\pm 5$$

Add 4 to each side.

$$x=4\pm 5$$

So $x = 4 + 5 = 9$ or $x = 4 - 5 = -1$

98. $x = 4$ or $x = \frac{2}{3}$

First, divide each term by 3 (so that the leading coefficient is 1).

$$x^2-\frac{14}{3}x+\frac{8}{3}=0$$

Add $-\frac{8}{3}$ to each side of the equation.

$$x^2-\frac{14}{3}x=-\frac{8}{3}$$

Find the square of half the coefficient of x.

$$\left(\frac{-14}{3}\cdot\frac{1}{2}\right)^2=\left(\frac{-14}{6}\right)^2=\frac{196}{36}$$

Add that square to both sides of the equation.

$$x^2-\frac{14}{3}x+\frac{196}{36}=-\frac{8}{3}+\frac{196}{36}$$

Factor the perfect square trinomial on the left and simplify on the right.

$$\left(x-\frac{14}{6}\right)^2=\frac{100}{36}$$

Use the square root rule.

$$x-\frac{14}{6}=\pm\frac{10}{6}$$

Add $\frac{14}{6}$ to each side.

$$x=\frac{14}{6}\pm\frac{10}{6}$$

So $x=\frac{14}{6}+\frac{10}{6}=\frac{24}{6}=4$

or $x=\frac{14}{6}-\frac{10}{6}=\frac{4}{6}=\frac{2}{3}$

99. $x=-5\pm3\sqrt{2}$

Add –7 to each side of the equation.

$$x^2+10x=-7$$

Find the square of half the coefficient of x.

$$\left(\frac{10}{2}\right)^2=(5)^2=25$$

Add that square to both sides of the equation.

$$x^2+10x+25=-7+25$$

Factor the perfect square trinomial on the left and simplify on the right.

$$(x+5)^2=18$$

Use the square root rule.

$$x+5=\pm\sqrt{18}=\pm3\sqrt{2}$$

Add –5 to each side.

$$x=-5\pm3\sqrt{2}$$

100. $x=4\pm\sqrt{21}$

Add 5 to each side of the equation.

$$x^2-8x=5$$

Find the square of half the coefficient of x.

$$\left(\frac{-8}{2}\right)^2 = (-4)^2 = 16$$

Add that square to both sides of the equation.

$$x^2 - 8x + 16 = 5 + 16$$

Factor the perfect square trinomial on the left and simplify on the right.

$$(x - 4)^2 = 21$$

Use the square root rule.

$$x - 4 = \pm\sqrt{21}$$

Add 4 to each side.

$$x = 4 \pm \sqrt{21}$$

101. $x = \frac{1 \pm \sqrt{73}}{6}$

First, divide each term by 3 (so that the leading coefficient is 1).

$$x^2 - \frac{1}{3}x - 2 = 0$$

Add 2 to each side of the equation.

$$x^2 - \frac{1}{3}x = 2$$

Find the square of half the coefficient of x.

$$\left(\frac{-1}{3} \cdot \frac{1}{2}\right)^2 = \left(\frac{-1}{6}\right)^2 = \frac{1}{36}$$

Add that square to both sides of the equation.

$$x^2 - \frac{1}{3}x + \frac{1}{36} = 2 + \frac{1}{36}$$

Factor the perfect square trinomial on the left and simplify on the right.

$$\left(x - \frac{1}{6}\right)^2 = \frac{73}{36}$$

Use the square root rule.

$$x - \frac{1}{6} = \pm\sqrt{\frac{73}{36}} = \pm\frac{\sqrt{73}}{6}$$

Add $\frac{1}{6}$ to each side.

$$x = \frac{1}{6} \pm \frac{\sqrt{73}}{6} = \frac{1 \pm \sqrt{73}}{6}$$

102. $x = \frac{-5 \pm \sqrt{33}}{4}$

First, divide each term by 2 (so that the leading coefficient is 1).

$$x^2 + \frac{5}{2}x - \frac{1}{2} = 0$$

Add $\frac{1}{2}$ to each side of the equation.

$$x^2+\frac{5}{2}x=\frac{1}{2}$$

Find the square of half the coefficient of x.

$$\left(\frac{5}{2}\cdot\frac{1}{2}\right)^2=\left(\frac{5}{4}\right)^2=\frac{25}{16}$$

Add that square to both sides of the equation.

$$x^2+\frac{5}{2}x+\frac{25}{16}=\frac{1}{2}+\frac{25}{16}$$

Factor the perfect square trinomial on the left and simplify on the right.

$$\left(x+\frac{5}{4}\right)^2=\frac{33}{16}$$

Use the square root rule.

$$x+\frac{5}{4}=\pm\sqrt{\frac{33}{16}}=\pm\frac{\sqrt{33}}{4}$$

Add $-\frac{5}{4}$ to each side.

$$x=-\frac{5}{4}\pm\frac{\sqrt{33}}{4}=\frac{-5\pm\sqrt{33}}{4}$$

103. $\boldsymbol{x=\frac{-b\pm\sqrt{b^2-4c}}{2}}$

Add $-c$ to each side of the equation.

$$x^2+bx=-c$$

Find the square of half the coefficient of x.

$$\left(\frac{b}{2}\right)^2=\frac{b^2}{4}$$

Add that square to both sides of the equation.

$$x^2+bx+\frac{b^2}{4}=-c+\frac{b^2}{4}$$

Factor the perfect square trinomial on the left and simplify on the right.

$$\left(x+\frac{b}{2}\right)^2=\frac{-4c+b^2}{4}$$

Use the square root rule.

$$x+\frac{b}{2}=\pm\sqrt{\frac{-4c+b^2}{4}}=\pm\sqrt{\frac{b^2-4c}{4}}=\pm\frac{\sqrt{b^2-4c}}{2}$$

Add $-\frac{b}{2}$ to each side of the equation.

$$x=-\frac{b}{2}\pm\frac{\sqrt{b^2-4c}}{2}=\frac{-b\pm\sqrt{b^2-4c}}{2}$$

104. $x = \dfrac{-b \pm \sqrt{b^2 - 4ac}}{2a}$

First, divide each term by a (so that the leading coefficient is 1).

$$x^2 + \frac{b}{a}x + \frac{c}{a} = 0$$

Add $-\frac{c}{a}$ to each side of the equation.

$$x^2 + \frac{b}{a}x = -\frac{c}{a}$$

Find the square of half the coefficient of x.

$$\left(\frac{b}{a}\cdot\frac{1}{2}\right)^2 = \left(\frac{b}{2a}\right)^2 = \frac{b^2}{4a^2}$$

Add that square to both sides of the equation.

$$x^2 + \frac{b}{a}x + \frac{b^2}{4a^2} = -\frac{c}{a} + \frac{b^2}{4a^2}$$

Factor the perfect square trinomial on the left and simplify on the right.

$$\left(x + \frac{b}{2a}\right)^2 = \frac{-4ac + b^2}{4a^2}$$

Use the square root rule.

$$x + \frac{b}{2a} = \pm\sqrt{\frac{-4ac + b^2}{4a^2}} = \pm\frac{\sqrt{b^2 - 4ac}}{2a}$$

Add $-\frac{b}{2a}$ to each side of the equation.

$$x = -\frac{b}{2a} \pm \frac{\sqrt{b^2 - 4ac}}{2a} = \frac{-b \pm \sqrt{b^2 - 4ac}}{2a}$$

Does this look familiar? It should. It's the quadratic formula!

105. $x = 3$ **or** $x = -3$

First, factor out the GCF (greatest common factor), 5.

$$5(x^2 - 9) = 0$$

Now factor the binomial.

$$5(x - 3)(x + 3) = 0$$

Set the binomials equal to 0 and solve for x.

$$x - 3 = 0, x = 3$$

$$x + 3 = 0, x = -3$$

106. $y = -\frac{1}{3}$ **or** $y = 1$

First, factor out the GCF (greatest common factor), –3.

$$-3(6y^2 - 2y - 1) = 0$$

Factor the trinomial.

$$-3(3y+1)(y-1)=0$$

Set the binomials equal to 0 and solve for y.

$$3y+1=0,\ y=-\frac{1}{3}$$

$$y-1=0,\ y=1$$

107. $\boldsymbol{x=\pm 1}$**, or** $\boldsymbol{x=\pm i}$

First, factor the binomial as the difference of squares.

$$(x^2-1)(x^2+1)=0$$

The first binomial factors again because it is still the difference of two perfect squares.

$$(x-1)(x+1)(x^2+1)=0$$

Set the binomials equal to 0 and solve for x.

$$x-1=0,\ x=1$$

$$x+1=0,\ x=-1$$

The last binomial doesn't have real roots.

$$x^2+1=0,\ x^2=-1,\ x=\pm i$$

108. $\boldsymbol{y=\frac{2}{3},\ y=-\frac{2}{3}}$

Factor the *quadratic-like* trinomial as the product of two binomials.

$$(4y^2-9)(4y^2-9)=0$$

The trinomial is actually a *perfect square* trinomial, so the factorization can be written $(4y^2-9)^2=0$.

The binomial factors as the difference of squares, so you have

$$[(2y-3)(2y+3)]^2=0 \text{ or } (2y-3)^2(2y+3)^2=0$$

The solution is two pairs of *double roots:*

$$y=\frac{2}{3},\ y=\frac{2}{3},\ y=-\frac{2}{3},\ y=-\frac{2}{3}$$

109. $\boldsymbol{z=\pm 3,\ z=\pm 2}$

Factor the *quadratic-like* trinomial as the product of two binomials.

$$(z^2-9)(z^2-4)=0$$

Both binomials are a difference of squares.

$$(z-3)(z+3)(z-2)(z+2)=0$$

Setting the binomials equal to 0 gives you

$$z=3,\ z=-3,\ z=2,\ z=-2.$$

110. $y = \pm\frac{1}{2}, y = \pm 5$

Factor the *quadratic-like* trinomial as the product of two binomials.

$$(4y^2 - 1)(y^2 - 25) = 0$$

Both binomials are a difference of squares.

$$(2y - 1)(2y + 1)(y - 5)(y + 5) = 0$$

Setting the binomials equal to 0 gives you

$$y = \frac{1}{2}, y = -\frac{1}{2}, y = 5, y = -5.$$

111. $z = \pm 1, z = \pm\frac{2}{3}$

First, factor out the GCF (greatest common factor), 3.

$$3(9z^4 - 13z^2 + 4) = 0$$

Now factor the *quadratic-like* trinomial as the product of two binomials.

$$3(z^2 - 1)(9z^2 - 4) = 0$$

Both binomials are a difference of squares.

$$3(z - 1)(z + 1)(3z - 2)(3z + 2) = 0$$

Setting the binomials equal to 0 gives you

$$z = 1, z = -1, z = \frac{2}{3}, z = -\frac{2}{3}.$$

112. $x = \pm\frac{1}{5}, x = \pm 5$

Factor the *quadratic-like* trinomial as the product of two binomials.

$$(25x^2 - 1)(x^2 - 25) = 0$$

Both binomials are a difference of squares.

$$(5x - 1)(5x + 1)(x - 5)(x + 5) = 0$$

Setting the binomials equal to 0 gives you

$$x = \frac{1}{5}, x = -\frac{1}{5}, x = 5, x = -5.$$

113. $x = \pm 2, x = \pm 1, x = \pm 2i, x = \pm i$

Factor the *quadratic-like* trinomial as the product of two binomials.

$$(x^4 - 16)(x^4 - 1) = 0$$

Now factor the binomials as the difference of squares.

$$(x^2 - 4)(x^2 + 4)(x^2 - 1)(x^2 + 1) = 0$$

The first and third binomials factor again.

$$(x - 2)(x + 2)(x^2 + 4)(x - 1)(x + 1)(x^2 + 1) = 0$$

Setting the binomials equal to 0 gives you four real roots and four imaginary roots.

$$x = 2, x = -2, x = 2i, x = -2i, x = 1, x = -1, x = i, x = -i$$

114. $y = 0, y = \pm 4, y = \pm 4i$

First, factor out the GCF (greatest common factor), y^4.

$$y^4(y^4 - 256) = 0$$

Now factor the binomial as the difference of squares.

$$y^4(y^2 - 16)(y^2 + 16) = 0$$

The first binomial factors.

$$y^4(y - 4)(y + 4)(y^2 + 16) = 0$$

Set the first factor and the binomials equal to 0 and solve for y.

$$y = 0, y = 4, y = -4, y = 4i, y = -4i$$

115. $x \leq -3$ or $x \geq 4$

First, determine the *critical numbers* by setting the factors equal to 0.

$x = 4$ and $x = -3$ are the critical numbers and are positioned on a number line.

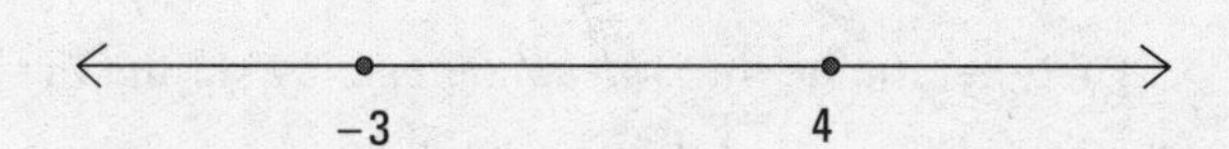

Determine whether each factor is positive or negative in each interval determined by the critical numbers by testing values within each interval.

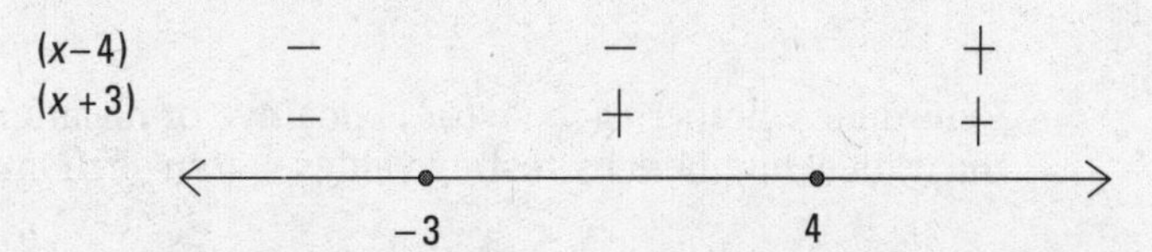

Both factors are negative to the left of –3, so their product is positive.

One factor is negative between –3 and 4, so the product is negative.

Both factors are positive to the right of 4, so the product is positive.

The problem asks for when the product is greater than (positive) or equal to 0, so you include the critical numbers and write:

$x \leq -3$ or $x \geq 4$

In interval notation, this is written $(-\infty, -3] \cup [4, \infty)$.

116. $x < -\frac{1}{2}$ or $x > 5$

First, determine the *critical numbers* by setting the factors equal to 0.

$x = -\frac{1}{2}$ and $x = 5$ are the critical numbers and are positioned on a number line.

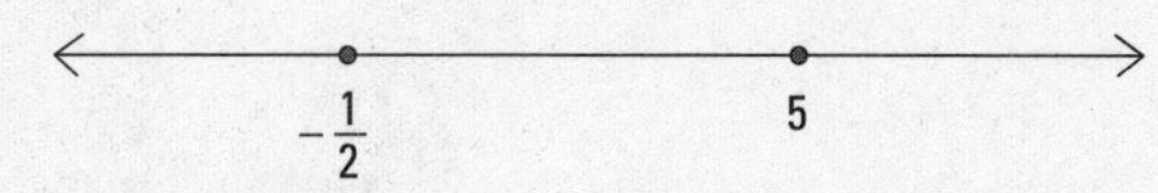

Determine whether each factor is positive or negative in each interval determined by the critical numbers by testing values within each interval.

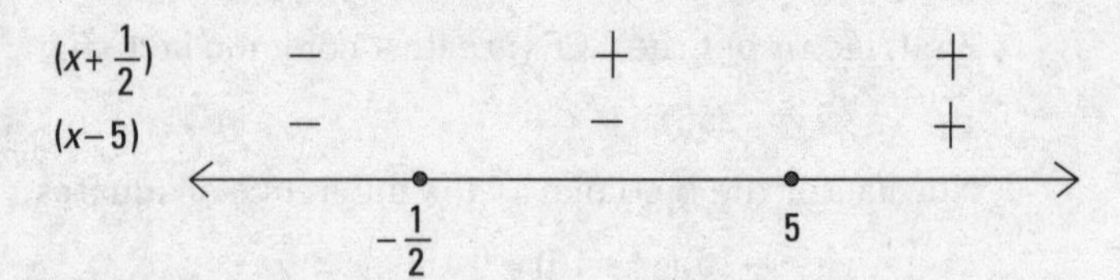

Both factors are negative to the left of $-\frac{1}{2}$, so their product is positive.

One factor is negative between $-\frac{1}{2}$ and 5, so the product is negative.

Both factors are positive to the right of 5, so the product is positive.

The problem asks for when the product is greater than 0 (positive), so:

$$x < -\frac{1}{2} \text{ or } x > 5$$

In interval notation, this is written $\left(-\infty, -\frac{1}{2}\right) \cup (5, \infty)$.

117. $x \leq -6$ or $x \geq 8$

First, determine the *critical numbers* by setting the factors equal to 0.

$x = 8$ and $x = -6$ are the critical numbers and are positioned on a number line.

Determine whether each factor is positive or negative in each interval determined by the critical numbers by testing values within each interval.

One factor is negative to the left of –6, so the product is negative.

Both factors are positive between –6 and 8, so the product is positive.

One factor is negative to the right of 8, so the product is negative.

The problem asks for when the product is less than (negative) or equal to 0, so you include the critical numbers and write:

$$x \leq -6 \text{ or } x \geq 8$$

In interval notation, this is written $(-\infty, -6] \cup [8, \infty)$.

118. $x < -3$ or $x > 0$

First, determine the *critical numbers* by setting the factors equal to 0.

$x = 0$ and $x = -3$ are the critical numbers and are positioned on a number line.

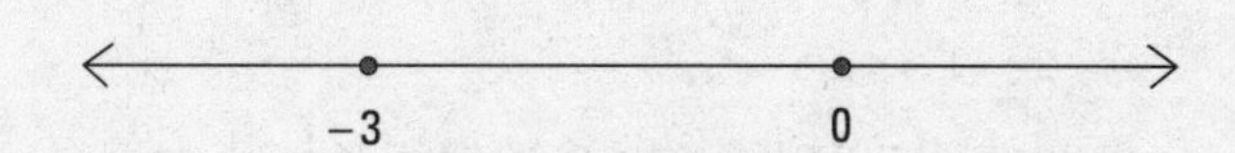

Determine whether each factor is positive or negative in each interval determined by the critical numbers by testing values within each interval.

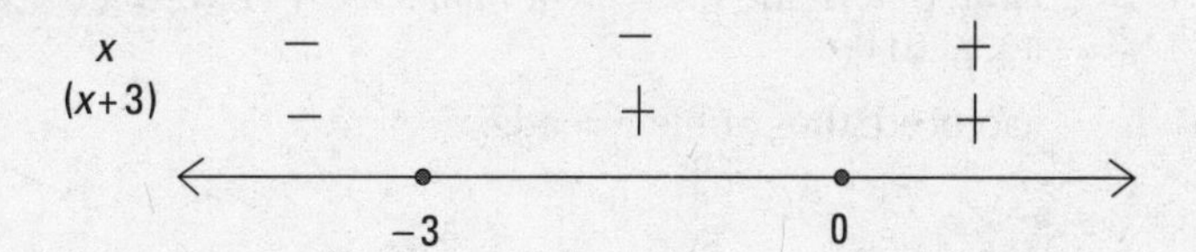

Both factors are negative to the left of –3, so their product is positive.

One factor is negative between –3 and 0, so the product is negative.

Both factors are positive to the right of 0, so the product is positive.

The problem asks for when the product is greater than 0 (positive) so:

$$x < -3 \text{ or } x > 0$$

In interval notation, this is written $(-\infty, -3) \cup (0, \infty)$.

119. $-1 \le x \le 8$

First, determine the *critical numbers* by factoring the quadratic and setting the factors equal to 0.

Factored, the problem reads:

$$(8 - x)(1 + x) \ge 0$$

$x = 8$ and $x = -1$ are the critical numbers and are positioned on a number line.

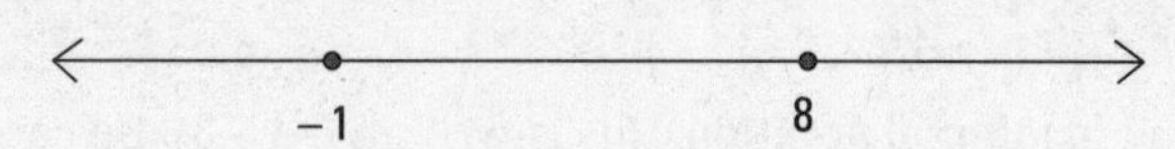

Determine whether each factor is positive or negative in each interval determined by the critical numbers by testing values within each interval.

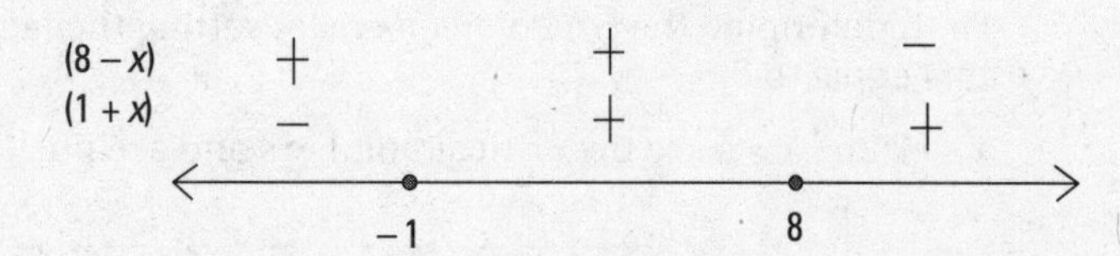

One factor is negative to the left of –1, so the product is negative.

Both factors are positive between –1 and 8, so the product is positive.

One factor is negative to the right of 8, so the product is negative.

The problem asks for when the product is greater than (positive) or equal to 0, so you include the critical numbers and write:

$$-1 \leq x \leq 8$$

In interval notation, this is written [–1, 8].

120. $-4 < x < -3$

First, determine the *critical numbers* by factoring the quadratic and setting the factors equal to 0.

Factored, the problem reads:

$$(x+3)(x+4) < 0$$

$x = -3$ and $x = -4$ are the critical numbers and are positioned on a number line.

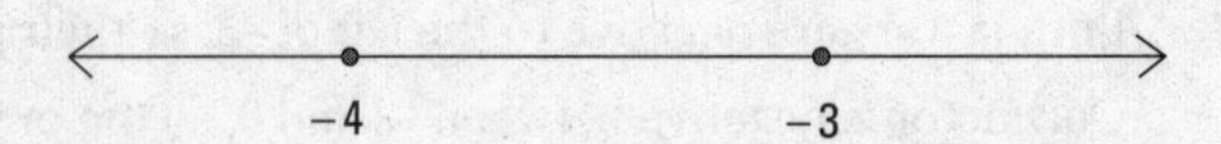

Determine whether each factor is positive or negative in each interval determined by the critical numbers.

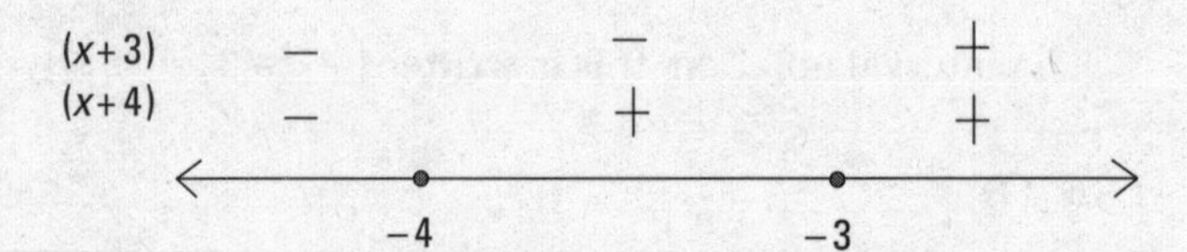

Both factors are negative to the left of –4, so their product is positive.

One factor is negative between –4 and –3, so the product is negative.

Both factors are positive to the right of –3, so the product is positive.

The problem asks for when the product is less than 0 (negative) so:

$$-4 < x < -3$$

In interval notation, this is written (–4, –3). Be careful to interpret this as interval notation and not the coordinates of a point.

121. $x < -2$ or $x > 4$

First, determine the *critical numbers* by setting the factors (numerator and denominator) equal to 0.

$x = -2$ and $x = 4$ are the critical numbers and are positioned on a number line.

Determine whether each factor is positive or negative in each interval determined by the critical numbers.

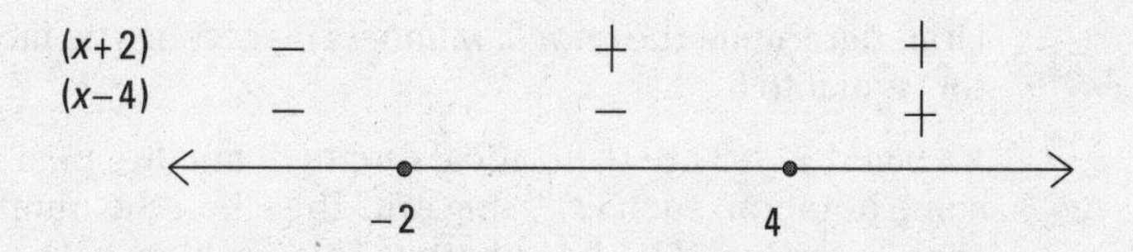

Both factors are negative to the left of –2, so their quotient is positive.

One factor is negative between –2 and 4, so the quotient is negative.

Both factors are positive to the right of 4, so the quotient is positive.

The problem asks for when the quotient is greater than 0 (positive), so:

$x < -2$ or $x > 4$

In interval notation, this is written $\left(-\infty, -2\right)\cup\left(4, \ldots\infty\right)$.

122. $-5 < x < \frac{2}{3}$

First, determine the *critical numbers* by setting the factors (numerator and denominator) equal to 0.

$x = \frac{2}{3}$ and $x = -5$ are the critical numbers and are positioned on a number line.

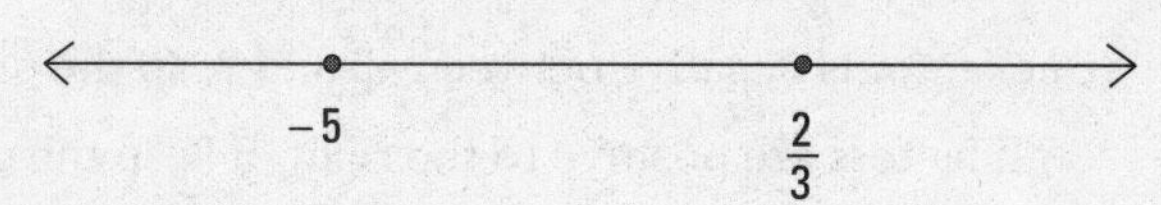

Determine whether each factor is positive or negative in each interval determined by the critical numbers.

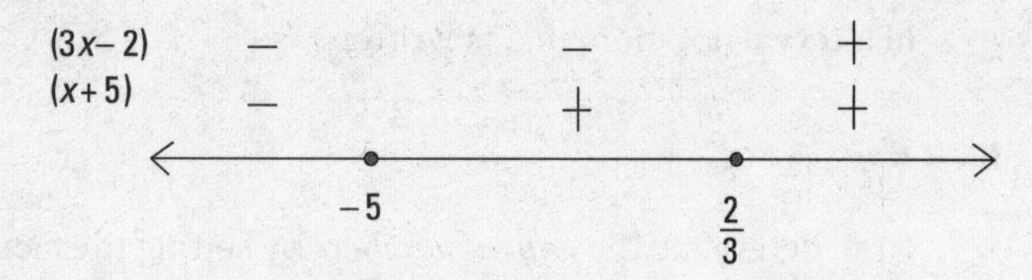

Both factors are negative to the left of –5, so their quotient is positive.

One factor is negative between –5 and $\frac{2}{3}$, so the quotient is negative.

Both factors are positive to the right of $\frac{2}{3}$, so the quotient is positive.

The problem asks for when the quotient is less than 0 (negative), so:

$$-5 < x < \frac{2}{3}$$

In interval notation, this is written $\left(-5, \frac{2}{3}\right)$.

Be careful to interpret this as interval notation and not the coordinates of a point.

123. $x < -3$ or $x \geq 9$

First, determine the *critical numbers* by setting the factors (numerator and denominator) equal to 0.

$x = 9$ and $x = -3$ are the critical numbers and are positioned on a number line. Make some notation, such as "X-ing out" the –3 on the number line, to show that this number can't be included in the solution. This problem asks for greater than *or equal to* 0, so you'll be including the critical number that makes the numerator equal to 0, but not the critical number that makes the denominator equal to 0.

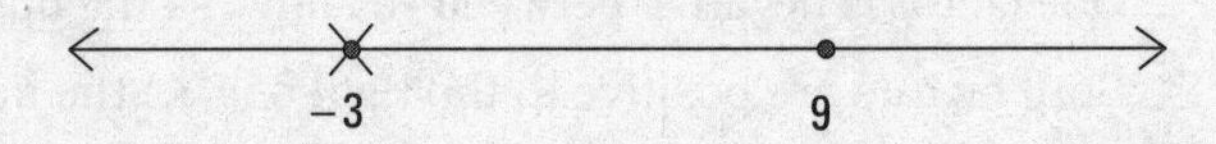

Determine whether each factor is positive or negative in each interval determined by the critical numbers.

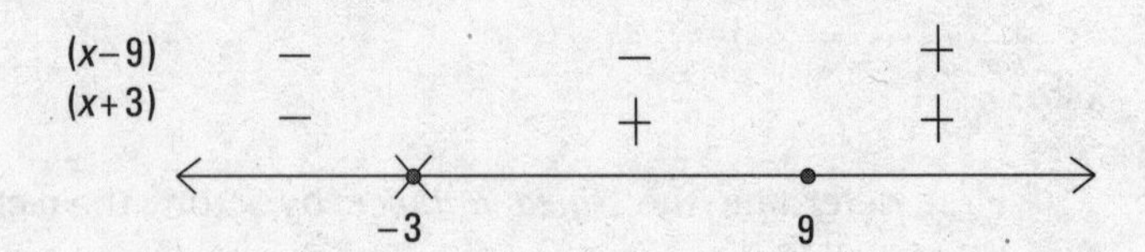

Both factors are negative to the left of –3, so their quotient is positive.

One factor is negative between –3 and 9, so the quotient is negative.

Both factors are positive to the right of 9, so the quotient is positive.

The problem asks for when the quotient is greater than (positive) or equal to 0, so you include the 9 in your answer:

$$x < -3 \text{ or } x \geq 9$$

In interval notation, this is written $(-\infty, -3) \cup [9, \infty)$.

124. $-1 \leq x < \frac{7}{2}$

First, determine the *critical numbers* by setting the factors (numerator and denominator) equal to 0.

$x = -1$ and $x = \frac{7}{2}$ are the critical numbers and are positioned on a number line. Make some notation, such as "X-ing out" the $\frac{7}{2}$ on the number line, to show that this number can't be included in the solution. This problem asks for less than *or equal to* 0, so you'll be including the critical number that makes the numerator equal to 0, but not the critical number that makes the denominator equal to 0.

Determine whether each factor is positive or negative in each interval determined by the critical numbers.

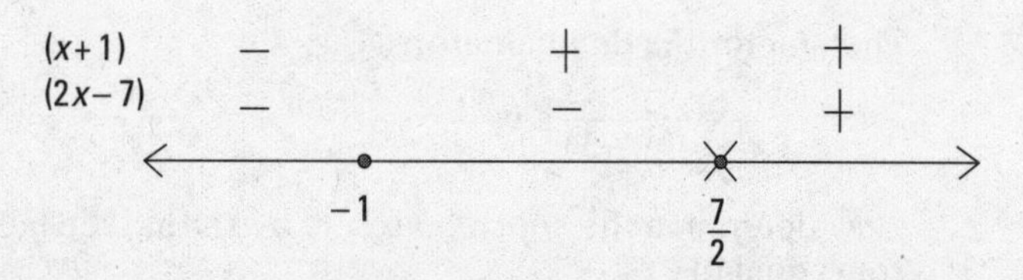

Both factors are negative to the left of –1, so their quotient is positive.

One factor is negative between –1 and $\frac{7}{2}$, so the quotient is negative.

Both factors are positive to the right of $\frac{7}{2}$, so the quotient is positive.

The problem asks for when the quotient is less than (negative) or equal to 0, so you include the –1 in your answer:

$$-1 \le x < \frac{7}{2}$$

In interval notation, this is written $\left[-1, \frac{7}{2}\right)$.

Answers 101–200

125. $-3 < x < -2$ or $x > 2$

First, factor the numerator.

$$\frac{(x-2)(x+2)}{x+3} > 0$$

Now determine the *critical numbers* by setting the factors (numerator and denominator) equal to 0.

$x = 2$, $x = -2$, and $x = -3$ are the critical numbers and are positioned on a number line.

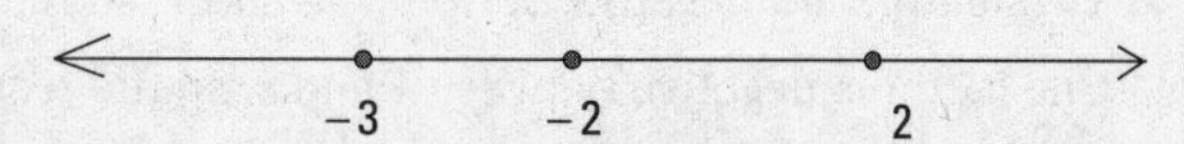

Determine whether each factor is positive or negative in each interval determined by the critical numbers.

$(x-2)$	−	−	−	+
$(x+2)$	−	−	+	+
$(x+3)$	−	+	+	+
		−3	−2	2

All three factors are negative to the left of –3, so the result is negative.

Two factors are negative between –3 and –2, so that result is positive.

One factor is negative between –2 and 2, so the result is negative.

All three factors are positive to the right of 2, so the result is positive.

The problem asks for when the original quotient is greater than 0 (positive), so:

$$-3 < x < -2 \text{ or } x > 2$$

In interval notation, this is written $(-3, -2) \cup (2, \infty)$.

126. $x < -5$ or $1 \le x < 5$

First, factor the denominator.

$$\frac{x-1}{(x-5)(x+5)} \le 0$$

Now determine the *critical numbers* by setting the factors (numerator and denominator) equal to 0.

$x = 1$, $x = 5$, and $x = -5$ are the critical numbers and are positioned on a number line. Make some notation, such as "X-ing out" the 5 and –5 on the number line, to show that these numbers can't be included in the solution. This problem asks for less than *or equal to* 0, so you'll be including the critical number that makes the numerator equal to 0, but not the critical number that makes the denominator equal to 0.

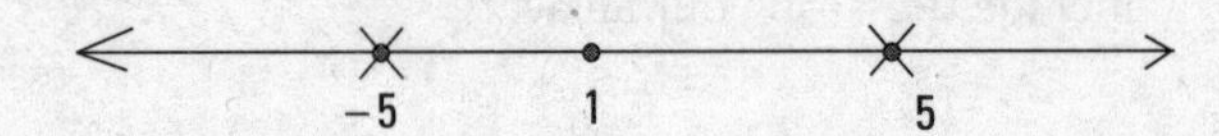

Determine whether each factor is positive or negative in each interval determined by the critical numbers.

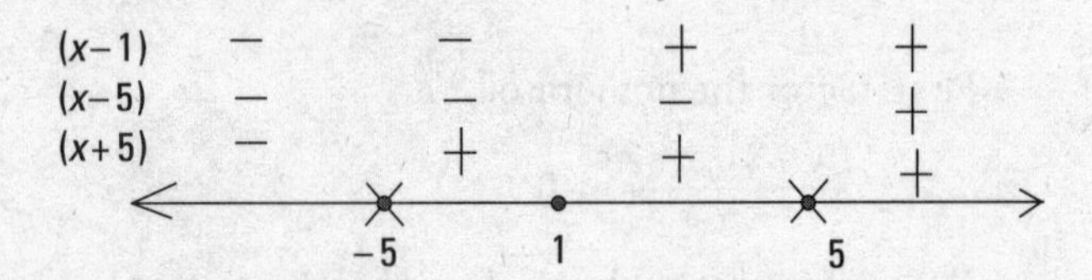

All three factors are negative to the left of –5, so the result is negative.

Two factors are negative between –5 and 1, so that result is positive.

One factor is negative between 1 and 5, so the result is negative.

All three factors are positive to the right of 5, so the result is positive.

The problem asks for when the original quotient is less than (negative) *or equal to* 0, so you include the critical number $x = 1$ and write:

$$x < -5 \text{ or } 1 \le x < 5$$

In interval notation, this is written $(-\infty, -5) \cup [1, 5)$.

127. $x < -4$ or $-3 \le x \le 3$ or $x > 4$

First, factor the numerator and denominator.

$$\frac{(x-3)(x+3)}{(x-4)(x+4)} \ge 0$$

Now determine the *critical numbers* by setting the factors (numerator and denominator) equal to 0.

$x = 3$, $x = -3$, $x = 4$, and $x = -4$ are the critical numbers and are positioned on a number line. Make some notation, such as "X-ing out" the 4 and –4 on the number line, to show that these numbers can't be included in the solution. This problem asks for greater than *or equal to* 0, so you'll be including the critical number that makes the numerator equal to 0, but not the critical number that makes the denominator equal to 0.

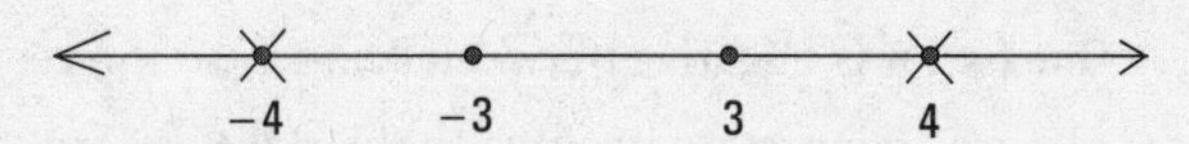

Determine whether each factor is positive or negative in each interval determined by the critical numbers.

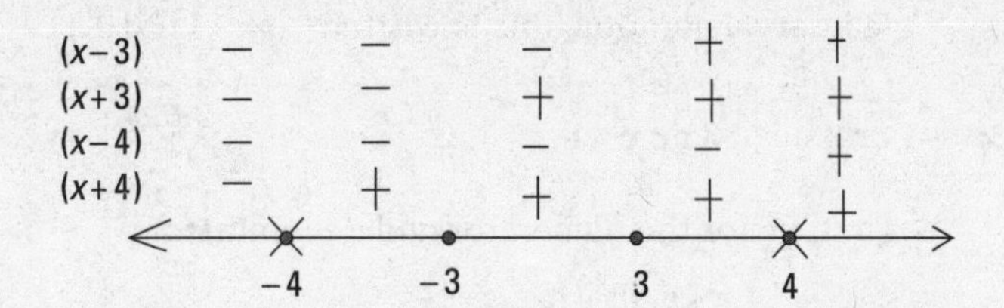

All four factors are negative to the left of –4, so the result is positive.

Three factors are negative between –4 and –3, so the result is negative.

Two factors are negative between –3 and 3, so that result is positive.

One factor is negative between 3 and 4, so the result is negative.

All four factors are positive to the right of 4, so the result is positive.

The problem asks for when the original quotient is greater than (positive) or equal to 0, so you include the critical numbers $x = 3$ and $x = -3$ and write:

$x < -4$ or $-3 \leq x \leq 3$ or $x > 4$

In interval notation, this is written $(-\infty, -4) \cup [-3, 3] \cup (4, \infty)$.

128. $x < -1$ or $0 < x < 1$

First, factor the denominator.

$$\frac{x}{(x-1)(x+1)} < 0$$

Now determine the *critical numbers* by setting the factors (numerator and denominator) equal to 0.

$x = 0$, $x = 1$, and $x = -1$ are the critical numbers and are positioned on a number line.

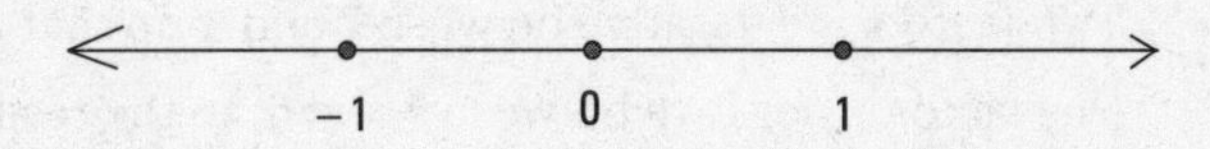

Determine whether each factor is positive or negative in each interval determined by the critical numbers.

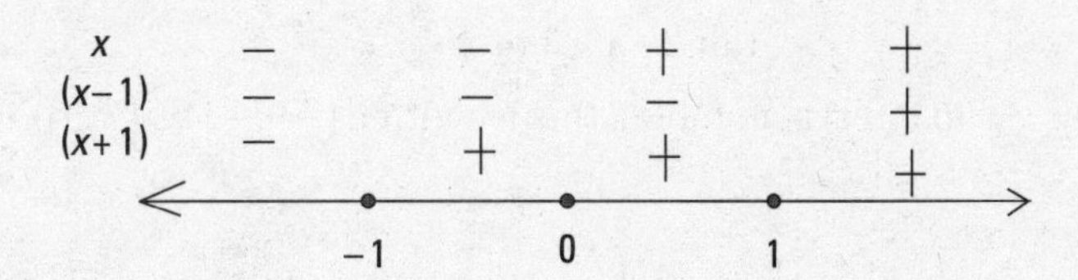

All three factors are negative to the left of –1, so the result is negative.

Two factors are negative between –1 and 0, so that result is positive.

One factor is negative between 0 and 1, so the result is negative.

All three factors are positive to the right of 1, so the result is positive.

The problem asks for when the product is less than 0 (negative), so:

$$x < -1 \text{ or } 0 < x < 1$$

In interval notation, this is written $(-\infty, -1) \cup (0, 1)$.

129. $x < -1$ or $2 < x < 3$ or $x > 6$

First, factor the numerator and denominator.

$$\frac{(x-2)(x-3)}{(x-6)(x+1)} > 0$$

Now determine the *critical numbers* by setting the factors (numerator and denominator) equal to 0.

$x = 2$, $x = 3$, $x = 6$, and $x = -1$ are the critical numbers and are positioned on a number line.

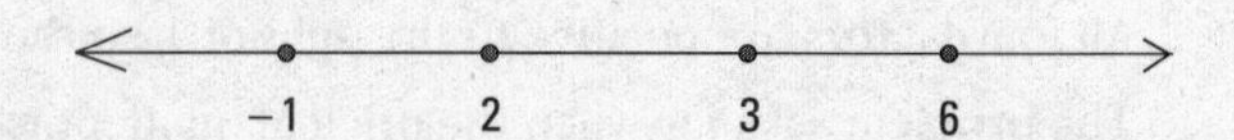

Determine whether each factor is positive or negative in each interval determined by the critical numbers.

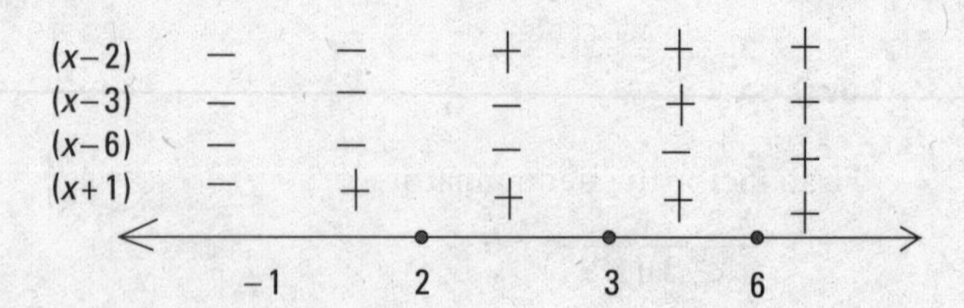

All four factors are negative to the left of –1, so the result is positive.

Three factors are negative between –1 and 2, so the result is negative.

Two factors are negative between 2 and 3, so that result is positive.

One factor is negative between 3 and 6, so the result is negative.

All four factors are positive to the right of 6, so the result is positive.

The problem asks for when the original quotient is greater than 0 (positive), so you write:

$$x < -1 \text{ or } 2 < x < 3 \text{ or } x > 6$$

In interval notation, this is written $(-\infty, -1) \cup (2, 3) \cup (6, \infty)$.

130. $-\frac{1}{2} \leq x < -\frac{1}{5}$ **or** $\frac{1}{4} < x \leq \frac{1}{3}$

First, factor the numerator and denominator.

$$\frac{(2x+1)(3x-1)}{(4x-1)(5x+1)} \leq 0$$

Now determine the *critical numbers* by setting the factors (numerator and denominator) equal to 0.

$x = -\frac{1}{2}, x = \frac{1}{3}, x = \frac{1}{4}$ and $x = -\frac{1}{5}$ are the critical numbers and are positioned on a number line. Make some notation, such as "X-ing out" the $x = \frac{1}{4}$ and $x = -\frac{1}{5}$ on the number line, to show that these numbers can't be included in the solution. This problem asks for less than *or equal to* 0, so you'll be including the critical number that makes the numerator equal to 0, but not the critical number that makes the denominator equal to 0.

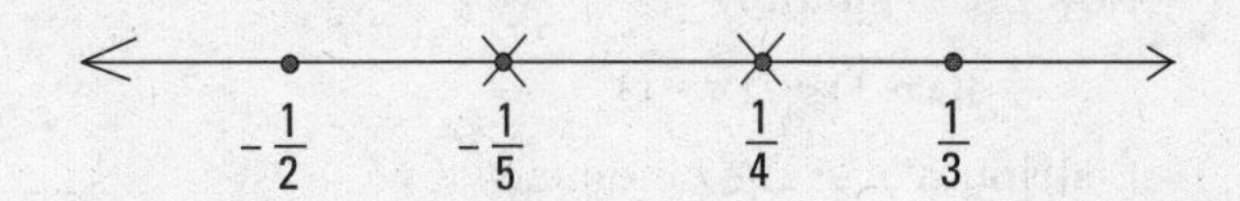

Determine whether each factor is positive or negative in each interval determined by the critical numbers.

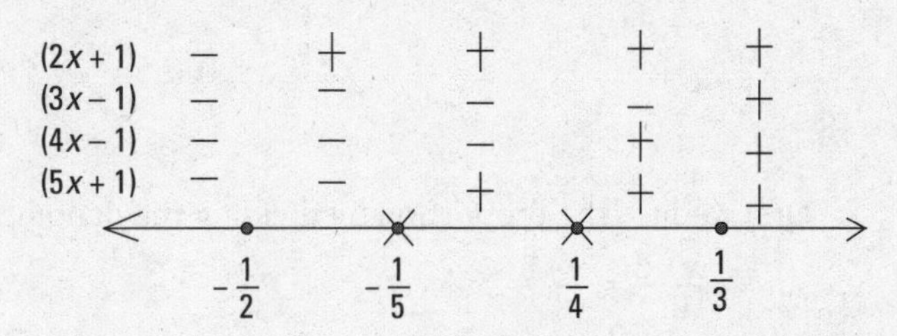

All four factors are negative to the left of $-\frac{1}{2}$, so the result is positive.

Three factors are negative between $-\frac{1}{2}$ and $-\frac{1}{5}$, so the result is negative.

Two factors are negative between $-\frac{1}{5}$ and $\frac{1}{4}$, so that result is positive.

One factor is negative between $\frac{1}{4}$ and $\frac{1}{3}$, so the result is negative.

All four factors are positive to the right of $\frac{1}{3}$, so the result is positive.

The problem asks for when the product is less than (negative) or equal to 0, so you include the critical numbers $-\frac{1}{2}$ and $\frac{1}{3}$ and write:

$$-\frac{1}{2} \leq x < -\frac{1}{5} \text{ or } \frac{1}{4} < x \leq \frac{1}{3}$$

In interval notation, this is written $\left[-\frac{1}{2}, -\frac{1}{5}\right) \cup \left(\frac{1}{4}, \frac{1}{3}\right]$.

131. $x = 30$

First, reduce the proportion by dividing the denominators by 8.

$$\frac{x}{\cancel{40}_5} = \frac{18}{\cancel{24}_3}$$

Now cross-multiply.

$$3x = 90$$

Divide by 3.

$$x = 30$$

After dividing by 8, the new proportion could have been reduced by dividing the right fraction by 3. You get the same answer, of course.

132. $x = 5$

First, reduce the proportion by dividing the numerators by 4.

$$\frac{\cancel{12}^{3}}{x+1} = \frac{\cancel{8}^{2}}{x-1}$$

Now cross-multiply.

$$3(x - 1) = 2(x + 1)$$

Distribute over the binomials.

$$3x - 3 = 2x + 2$$

Solve for x.

$$x = 5$$

133. $x = -8$

First, reduce the proportion by dividing the denominators by 2.

$$\frac{x+3}{\cancel{10}_{5}} = \frac{x+5}{\cancel{6}_{3}}$$

Now cross-multiply.

$$3(x + 3) = 5(x + 5)$$

Distribute over the binomials.

$$3x + 9 = 5x + 25$$

Solve for x.

$$-2x = 16$$

$$x = -8$$

134. $x = \frac{5}{3}$ or $x = -3$

As tempting as it is, you can't reduce the proportion. So cross-multiply.

$$(3x + 1)(x + 1) = 16$$

Multiply the binomials.

$$3x^2 + 4x + 1 = 16$$

Subtract 16 from each side. Then solve the quadratic equation by factoring.

$$3x^2 + 4x - 15 = 0$$

$$(3x - 5)(x + 3) = 0$$

Setting the binomials equal to 0, you get

$$x = \frac{5}{3} \text{ or } x = -3.$$

135. $x = 5$

Take advantage of having two fractions with the same denominator, and subtract $\frac{7}{x+2}$ from each side.

$$\frac{5}{x} = \frac{14}{x+2} - \frac{7}{x+2}$$

$$\frac{5}{x} = \frac{7}{x+2}$$

Now cross-multiply.

$$5(x + 2) = 7x$$

Distribute the 5. Then solve for x.

$$5x + 10 = 7x$$

$$10 = 2x$$

$$x = 5$$

136. $x = -8$ or $x = -12$

The common denominator for the three fractions is $12x$. Rewrite each fraction with that denominator.

$$\frac{x}{4} \cdot \frac{3x}{3x} - \frac{8}{x} \cdot \frac{12}{12} = \frac{x+5}{3} \cdot \frac{4x}{4x}$$

$$\frac{3x^2}{12x} - \frac{96}{12x} = \frac{4x^2 + 20x}{12x}$$

Now multiply each term in the equation by $12x$. Be sure to check for extraneous solutions when using this method.

$$\cancel{12x}\left(\frac{3x^2}{\cancel{12x}}\right) - \cancel{12x}\left(\frac{96}{\cancel{12x}}\right) = \cancel{12x}\left(\frac{4x^2 + 20x}{\cancel{12x}}\right)$$

$$3x^2 - 96 = 4x^2 + 20x$$

Add $-3x^2$ and $+96$ to each side; then solve the quadratic equation by factoring.

$$0 = x^2 + 20x + 96$$

$$0 = (x + 8)(x + 12)$$

Set the binomials equal to 0 and solve for x.

$x = -8$ or $x = -12$

Both solutions work when you check them in the original equation.

Looking back at the original equation, you can save time by just multiplying each term by $12x$. When you can easily identify a common denominator for all the terms in an equation, this is a nice way to approach the solution.

137. $x = -2$

The common denominator of the two fractions on the left is $x^2 - 1$. Add those two left-hand equations together.

$$\frac{5}{x+1}\cdot\frac{x-1}{x-1}+\frac{3}{x-1}\cdot\frac{x+1}{x+1}=\frac{x-16}{x^2-1}$$

$$\frac{5x-5}{x^2-1}+\frac{3x+3}{x^2-1}=\frac{x-16}{x^2-1}$$

$$\frac{8x-2}{x^2-1}=\frac{x-16}{x^2-1}$$

Multiplying each side of the equation by $x^2 - 1$ or reducing the proportion by dividing the denominators by $x^2 - 1$ produces the same result. And both methods may introduce an extraneous solution.

Multiplying, you get $8x - 2 = x - 16$.

Solving for x, you have $7x = -14$, or $x = -2$.

This solution does work when checked.

138. $x = 2$

The common denominator of the two fractions on the left is $x^2 + x - 12$. Add those two left-hand equations together.

$$\frac{6}{x+4}\cdot\frac{x-3}{x-3}+\frac{2}{x-3}\cdot\frac{x+4}{x+4}=\frac{6}{x^2+x-12}$$

$$\frac{6x-18}{x^2+x-12}+\frac{2x+8}{x^2+x-12}=\frac{6}{x^2+x-12}$$

$$\frac{8x-10}{x^2+x-12}=\frac{6}{x^2+x-12}$$

Multiplying each side of the equation by $x^2 + x - 12$ or reducing the proportion by dividing the denominators by $x^2 + x - 12$ produces the same result. And both methods may introduce an extraneous solution.

Multiplying, you get $8x - 10 = 6$.

Solving for x:

$8x = 16$

$x = 2$

This solution does work when checked.

139. $x = 3$

The common denominator of the two fractions on the left is $x^2 - 2x$. Subtract those two left-hand equations.

$$\frac{9}{x}\cdot\frac{x-2}{x-2}-\frac{2}{x-2}\cdot\frac{x}{x}=\frac{3}{x^2-2x}$$

$$\frac{9x-18}{x^2-2x}-\frac{2x}{x^2-2x}=\frac{3}{x^2-2x}$$

$$\frac{7x-18}{x^2-2x}=\frac{3}{x^2-2x}$$

Multiplying each side of the equation by $x^2 - 2x$ or reducing the proportion by dividing the denominators by $x^2 - 2x$ produces the same result. And both methods may introduce an extraneous solution.

Multiplying, you get $7x - 18 = 3$.

Solving for x:

$$7x = 21$$

$$x = 3$$

This solution does work when checked.

140. $x = \pm 2$

The common denominator for the three fractions is $6x$. Rewrite each fraction with that denominator, and add the fractions on the left.

$$\frac{x}{6}\cdot\frac{x}{x}+\frac{1}{x}\cdot\frac{6}{6}=\frac{5}{3x}\cdot\frac{2}{2}$$

$$\frac{x^2}{6x}+\frac{6}{6x}=\frac{10}{6x}$$

$$\frac{x^2+6}{6x}=\frac{10}{6x}$$

Multiplying each side of the equation by $6x$ or reducing the proportion by dividing the denominators by $6x$ produces the same result. And both methods may introduce an extraneous solution.

Multiplying, you get $x^2 + 6 = 10$

Solving for x:

$$x^2 = 4$$

$$x = \pm 2$$

Both solutions work when checked.

141. $x = -4$

The common denominator of the two fractions on the left is $2x^2 + 5x - 3$. Subtract those two left-hand equations.

$$\frac{5}{2x-1}\cdot\frac{x+3}{x+3}-\frac{1}{x+3}\cdot\frac{2x-1}{2x-1}=\frac{4}{2x^2+5x-3}$$

$$\frac{5x+15}{2x^2+5x-3}-\frac{2x-1}{2x^2+5x-3}=\frac{4}{2x^2+5x-3}$$

$$\frac{3x+16}{2x^2+5x-3}=\frac{4}{2x^2+5x-3}$$

Multiplying each side of the equation by $2x^2 + 5x - 3$ or reducing the proportion by dividing the denominators by $2x^2 + 5x - 3$ produces the same result. And both methods may introduce an extraneous solution.

Multiplying, you get $3x + 16 = 4$

Solving for x, $3x = -12$, so $x = -4$.

This solution does work when checked.

142. $x = \frac{1}{3}$

The common denominator of the two fractions on the left is $x^2 + x$. Subtract those two left-hand equations.

$$\frac{4}{x}\cdot\frac{x+1}{x+1} - \frac{1}{x+1}\cdot\frac{x}{x} = \frac{5}{x^2+x}$$

$$\frac{4x+4}{x^2+x} - \frac{x}{x^2+x} = \frac{5}{x^2+x}$$

$$\frac{3x+4}{x^2+x} = \frac{5}{x^2+x}$$

Multiplying each side of the equation by $x^2 + x$ or reducing the proportion by dividing the denominators by $x^2 + x$ produces the same result. And both methods may introduce an extraneous solution.

Multiplying and solving for x, you get

$$3x + 4 = 5$$

$$3x = 1$$

$$x = \frac{1}{3}$$

This solution does work when checked.

143. $x = -\frac{4}{3}$ or $x = 3$

Add $\frac{4}{x+1}$ to both sides of the equation. Then add the two fractions on the right.

$$3 = \frac{10}{x+2} + \frac{4}{x+1}$$

$$3 = \frac{10}{x+2}\cdot\frac{x+1}{x+1} + \frac{4}{x+1}\cdot\frac{x+2}{x+2}$$

$$3 = \frac{10x+10}{x^2+3x+2} + \frac{4x+8}{x^2+3x+2}$$

$$3 = \frac{14x+18}{x^2+3x+2}$$

Multiply each side of the equation by the denominator of the fraction on the right. If you think of the 3 on the left as being $\frac{3}{1}$, then this is like cross-multiplying with a proportion.

$$3(x^2 + 3x + 2) = 14x + 18$$

Distribute on the left, and then solve the quadratic equation by factoring.

$$3x^2 + 9x + 6 = 14x + 18$$

$$3x^2 - 5x - 12 = 0$$

$$(3x + 4)(x - 3) = 0$$

Set the binomials equal to 0 and solve for x.

$$x = -\frac{4}{3} \text{ or } x = 3$$

144. $x = \frac{9}{2}$ **or** $x = 5$

Subtract $\frac{3}{x-6}$ from both sides of the equation. Then subtract the two fractions on the right.

$$4 = \frac{1}{x-4} - \frac{3}{x-6}$$
$$4 = \frac{1}{x-4} \cdot \frac{x-6}{x-6} - \frac{3}{x-6} \cdot \frac{x-4}{x-4}$$
$$4 = \frac{x-6}{x^2-10x+24} - \frac{3x-12}{x^2-10x+24}$$
$$4 = \frac{-2x+6}{x^2-10x+24}$$

Multiply each side of the equation by the denominator of the fraction on the right. If you think of the 4 on the left as being $\frac{4}{1}$, then this is like cross-multiplying with a proportion.

$$4(x^2 - 10x + 24) = -2x + 6$$

Distribute on the left, and then solve the quadratic equation by factoring.

$$4x^2 - 40x + 96 = -2x + 6$$
$$4x^2 - 38x + 90 = 0$$
$$2x^2 - 19x + 45 = 0$$
$$(2x - 9)(x - 5) = 0$$

Set the binomials equal to 0 and solve for x.

$$x = \frac{9}{2} \text{ or } x = 5$$

145. $x = -\frac{11}{3}$ **or** $x = -4$

Add $\frac{x}{x+3}$ to both sides of the equation. Then add the two fractions on the right.

$$10 = \frac{6}{x+5} + \frac{x}{x+3}$$
$$10 = \frac{6}{x+5} \cdot \frac{x+3}{x+3} + \frac{x}{x+3} \cdot \frac{x+5}{x+5}$$
$$10 = \frac{6x+18}{x^2+8x+15} + \frac{x^2+5x}{x^2+8x+15}$$
$$10 = \frac{x^2+11x+18}{x^2+8x+15}$$

Multiply each side of the equation by the denominator of the fraction on the right. If you think of the 10 on the left as being $\frac{10}{1}$, then this is like cross-multiplying with a proportion.

$$10(x^2 + 8x + 15) = x^2 + 11x + 18$$

Distribute on the left, and then solve the quadratic equation by factoring.

$10x^2 + 80x + 150 = x^2 + 11x + 18$

$9x^2 + 69x + 132 = 0$

$3x^2 + 23x + 44 = 0$

$(3x + 11)(x + 4) = 0$

Set the binomials equal to 0 and solve for x.

$x = -\frac{11}{3}$ or $x = -4$

146. $x = -\frac{4}{5}$ **or** $x = -3$

Subtract $\frac{x+6}{x}$ from both sides of the equation. Then subtract the two fractions on the right.

$$9 = \frac{8}{x+4} - \frac{x+6}{x}$$

$$9 = \frac{8}{x+4} \cdot \frac{x}{x} - \frac{x+6}{x} \cdot \frac{x+4}{x+4}$$

$$9 = \frac{8x}{x^2+4x} - \frac{x^2+10x+24}{x^2+4x}$$

$$9 = \frac{-x^2-2x-24}{x^2+4x}$$

Multiply each side of the equation by the denominator of the fraction on the right. If you think of the 9 on the left as being $\frac{9}{1}$, then this is like cross-multiplying with a proportion.

$9(x^2 + 4x) = -x^2 - 2x - 24$

Distribute on the left, and then solve the quadratic equation by factoring.

$9x^2 + 36x = -x^2 - 2x - 24$

$10x^2 + 38x + 24 = 0$

$5x^2 + 19x + 12 = 0$

$(5x + 4)(x + 3) = 0$

Set the binomials equal to 0 and solve for x.

$x = -\frac{4}{5}$ or $x = -3$

147. $x = 0$

The common denominator for the three fractions is $x^2 - 9$. Rewrite the first two fractions with that denominator.

$$\frac{x}{x-3} \cdot \frac{x+3}{x+3} = \frac{3}{x+3} \cdot \frac{x-3}{x-3} + \frac{9}{x^2-9}$$

$$\frac{x^2+3x}{x^2-9} = \frac{3x-9}{x^2-9} + \frac{9}{x^2-9}$$

Add the fractions on the right.

$$\frac{x^2+3x}{x^2-9} = \frac{3x}{x^2-9}$$

Now multiply each side of the equation by $x^2 - 9$.

$$x^2 + 3x = 3x$$

Solve for x.

$$x^2 = 0$$

$$x = 0$$

148. $x = -\frac{1}{2}$ or $x = 6$

The common denominator of the three fractions is $x^2 - 2x - 3$. Rewrite the first two fractions with that denominator.

$$\frac{2x}{x+1} \cdot \frac{x-3}{x-3} = \frac{5}{x-3} \cdot \frac{x+1}{x+1} + \frac{1}{x^2-2x-3}$$

$$\frac{2x^2-6x}{x^2-2x-3} = \frac{5x+5}{x^2-2x-3} + \frac{1}{x^2-2x-3}$$

Now multiply each term by that common denominator. All the fractions will reduce, leaving just the numerators.

$$2x^2 - 6x = 5x + 5 + 1$$

Solve the quadratic equation by factoring.

$$2x^2 - 11x - 6 = 0$$

$$(2x + 1)(x - 6) = 0$$

Set the two binomials equal to 0 and solve for x.

$$x = -\frac{1}{2} \text{ or } x = 6$$

149. $x = -1$ or $x = -4$

The common denominator of the three fractions is x2 – 2x – 8. Rewrite the first two fractions with that denominator.

$$\frac{3x}{x-4} \cdot \frac{x+2}{x+2} = \frac{x}{x+2} \cdot \frac{x-4}{x-4} - \frac{8}{x^2-2x-8}$$

$$\frac{3x^2+6x}{x^2-2x-8} = \frac{x^2-4x}{x^2-2x-8} - \frac{8}{x^2-2x-8}$$

Now multiply each term by that common denominator. All the fractions will reduce, leaving just the numerators.

$$3x^2 + 6x = x^2 - 4x - 8$$

Solve the quadratic equation by factoring.

$$2x^2 + 10x + 8 = 0$$

$$x^2 + 5x + 4 = 0$$

$$(x + 1)(x + 4) = 0$$

Set the two binomials equal to 0 and solve for x.

$$x = -1 \text{ or } x = -4$$

150. $x = 0$ or $x = 9$

The common denominator of the three fractions is $x^2 - 2x - 15$. Rewrite the first two fractions with that denominator.

$$\frac{x}{x-5}\cdot\frac{x+3}{x+3}=\frac{2x}{x+3}\cdot\frac{x-5}{x-5}+\frac{4x}{x^2-2x-15}$$

$$\frac{x^2+3x}{x^2-2x-15}=\frac{2x^2-10x}{x^2-2x-15}+\frac{4x}{x^2-2x-15}$$

Now multiply each term by that common denominator. All the fractions will reduce, leaving just the numerators.

$$x^2 + 3x = 2x^2 - 10x + 4x$$

Solve the quadratic equation by factoring.

$$0 = x^2 - 9x$$

$$0 = x(x - 9)$$

Set the two factors equal to 0 and solve for x.

$$x = 0 \text{ or } x = 9$$

151. $x = 15$

Square both sides of the equation.

$$x + 1 = 16$$

Solve for x.

$$x = 15$$

Check the solution.

$$\sqrt{15+1} = \sqrt{16} = 4 \checkmark$$

152. c. $x = 9$

Square both sides of the equation.

$$3x - 2 = 25$$

Solve for x.

$$3x = 27$$

$$x = 9$$

Check the solution.

$$\sqrt{3(9)-2} = \sqrt{27-2} = \sqrt{25} = 5 \checkmark$$

153. $x = 12$

First, add 4 to both sides of the equation.

$$\sqrt{2x+1} = 5$$

Now square both sides of the equation.

$2x + 1 = 25$

Solve for x.

$2x = 24$

$x = 12$

Check the solution (using the original equation).

$\sqrt{2(12)+1} - 4 = \sqrt{25} - 4 = 5 - 4 = 1$ ✓

154. $x = -5$

First, add –5 to both sides of the equation.

$\sqrt{1-3x} = 4$

Now square both sides of the equation.

$1 - 3x = 16$

Solve for x.

$-3x = 15$

$x = -5$

Check the solution (using the original equation).

$5 + \sqrt{1-3(-5)} = 5 + \sqrt{16} = 5 + 4 = 9$ ✓

155. No solution

First, add –3 to both sides of the equation.

$\sqrt{x-1} = -1$

Now square both sides of the equation (even though you're a bit suspicious at this point because the square root value is equal to a negative number).

$x - 1 = 1$

Solve for x.

$x = 2$

Check the solution (using the original equation).

$\sqrt{2-1} + 3 = \sqrt{1} + 3 = 4 \neq 2$

It didn't check. There is no solution. You may have caught this earlier when you saw the radical equal to a negative number.

156. No solution

First, add –1 to both sides of the equation.

$-\sqrt{3x-2} = 2$

Now square both sides of the equation.

$$3x - 2 = 4$$

Solve for x.

$$3x = 6$$

$$x = 2$$

Check the solution (using the original equation).

$$1 - \sqrt{3(2) - 2} = 1 - \sqrt{4} = 1 - 2 = -1 \neq 3$$

It didn't check. There is no solution. The first clue to this situation was in the step where the opposite (negative) of the radical was equal to a positive number.

157. $x = 7$

First, add –2 to both sides of the equation.

$$\sqrt{3x + 4} = x - 2$$

Now square both sides of the equation.

$$3x + 4 = x^2 - 4x + 4$$

Set the quadratic equation equal to 0 and solve for x.

$$0 = x^2 - 7x$$

$$0 = x(x - 7)$$

So $x = 0$ or $x = 7$

Check $x = 0$ (using the original equation).

$$\sqrt{3(0) + 4} + 2 \stackrel{?}{=} 0$$

$$\sqrt{4} + 2 \stackrel{?}{=} 0$$

$$2 + 2 \neq 0$$

This doesn't check; $x = 0$ is extraneous.

Now check $x = 7$ (using the original equation).

$$\sqrt{3(7) + 4} + 2 \stackrel{?}{=} 7$$

$$\sqrt{25} + 2 \stackrel{?}{=} 7$$

$$5 + 2 = 7 \checkmark$$

The solution $x = 7$ works.

158. $x = 10$

First, add –4 to both sides of the equation.

$$\sqrt{4x - 4} = x - 4$$

Now square both sides of the equation.

$$4x - 4 = x^2 - 8x + 16$$

Set the quadratic equation equal to 0 and solve for x.

$$0 = x^2 - 12x + 20$$

$$0 = (x - 2)(x - 10)$$

So $x = 2$ or $x = 10$

Check $x = 2$ (using the original equation).

$$\sqrt{4(2)-4} + 4 \stackrel{?}{=} 2$$

$$\sqrt{4} + 4 \stackrel{?}{=} 2$$

$$2 + 4 \neq 2$$

This doesn't check; $x = 2$ is extraneous.

Now check $x = 10$ (using the original equation).

$$\sqrt{4(10)-4} + 4 \stackrel{?}{=} 10$$

$$\sqrt{36} + 4 \stackrel{?}{=} 10$$

$$6 + 4 = 10 \checkmark$$

The solution $x = 10$ works.

159. $x = -2$

First, add 6 to both sides of the equation.

$$\sqrt{6-5x} = x + 6$$

Now square both sides of the equation.

$$6 - 5x = x^2 + 12x + 36$$

Set the quadratic equation equal to 0 and solve for x.

$$0 = x^2 + 17x + 30$$

$$0 = (x + 2)(x + 15)$$

So $x = -2$ or $x = -15$

Check $x = -2$ (using the original equation).

$$\sqrt{6-5(-2)} - 6 \stackrel{?}{=} -2$$

$$\sqrt{16} - 6 \stackrel{?}{=} -2$$

$$4 - 6 = -2 \checkmark$$

The solution $x = -2$ works.

Now check $x = -15$ (using the original equation).

$$\sqrt{6-5(-15)} - 6 \stackrel{?}{=} -15$$

$$\sqrt{81} - 6 \stackrel{?}{=} -15$$

$$9 - 6 \neq -15$$

The solution $x = -15$ is extraneous.

160. $x = -4$

First, add 7 to both sides of the equation.

$$\sqrt{1-2x} = x+7$$

Now square both sides of the equation.

$$1 - 2x = x^2 + 14x + 49$$

Set the quadratic equation equal to 0 and solve for x.

$$0 = x^2 + 16x + 48$$

$$0 = (x + 4)(x + 12)$$

So $x = -4$ or $x = -12$

Check $x = -4$ (using the original equation).

$$\sqrt{1-2(-4)} - 7 \stackrel{?}{=} -4$$

$$\sqrt{9} - 7 \stackrel{?}{=} -4$$

$$3 - 7 = -4 \checkmark$$

The solution $x = -4$ works.

Now check $x = -12$ (using the original equation).

$$\sqrt{1-2(-12)} - 7 \stackrel{?}{=} -12$$

$$\sqrt{25} - 7 \stackrel{?}{=} -12$$

$$5 - 7 \neq -12$$

The solution $x = -12$ is extraneous.

161. $x = 10$

First, add $-x$ to both sides of the equation.

$$-\sqrt{5x-1} = 3 - x$$

Now square both sides of the equation.

$$5x - 1 = 9 - 6x + x^2$$

Set the quadratic equation equal to 0 and solve for x.

$$0 = x^2 - 11x + 10$$

$$0 = (x - 1)(x - 10)$$

So $x = 1$ or $x = 10$

Check $x = 1$ (using the original equation).

$$1 - \sqrt{5(1)-1} \stackrel{?}{=} 3$$

$$1 - \sqrt{4} \stackrel{?}{=} 3$$

$$1 - 2 \neq 3$$

The solution $x = 1$ is extraneous.

Now check $x = 10$ (using the original equation).

$$10-\sqrt{5(10)-1}\stackrel{?}{=}3$$

$$10-\sqrt{49}\stackrel{?}{=}3$$

$$10-7=3\ \checkmark$$

The solution $x = 10$ works.

162. $x = 3$

First, add $-x$ to both sides of the equation.

$$\sqrt{7x+4}=8-x$$

Now square both sides of the equation.

$$7x + 4 = 64 - 16x + x^2$$

Set the quadratic equation equal to 0 and solve for x.

$$0 = x^2 - 23x + 60$$

$$0 = (x - 3)(x - 20)$$

So $x = 3$ or $x = 20$

Check $x = 3$ (using the original equation).

$$3+\sqrt{7(3)+4}\stackrel{?}{=}8$$

$$3+\sqrt{25}\stackrel{?}{=}8$$

$$3+5=8\ \checkmark$$

The solution $x = 3$ works.

Now check $x = 20$ (using the original equation).

$$20+\sqrt{7(20)+4}\stackrel{?}{=}8$$

$$20+\sqrt{144}\stackrel{?}{=}8$$

$$20+12\neq 8$$

The solution $x = 20$ is extraneous.

163. $x = 2$ or $x = 5$

First, add $-x$ to both sides of the equation.

$$-\sqrt{3x-6}=2-x$$

Now square both sides of the equation.

$$3x - 6 = 4 - 4x + x^2$$

Set the quadratic equation equal to 0 and solve for x.

$$0 = x^2 - 7x + 10$$

$$0 = (x - 2)(x - 5)$$

So $x = 2$ or $x = 5$

Check $x = 2$ (using the original equation).

$$2-\sqrt{3(2)-6}\stackrel{?}{=}2$$
$$2-\sqrt{0}\stackrel{?}{=}2$$
$$2-0=2\checkmark$$

The solution $x = 2$ works.

Now check $x = 5$ (using the original equation).

$$5-\sqrt{3(5)-6}\stackrel{?}{=}2$$
$$5-\sqrt{9}\stackrel{?}{=}2$$
$$5-3=2\checkmark$$

The solution $x = 5$ also works.

164. $x = 7$

First, add $-x$ to both sides of the equation.

$$\sqrt{2x+11}=2-x$$

Now square both sides of the equation.

$$2x + 11 = 4 - 4x + x^2$$

Set the quadratic equation equal to 0 and solve for x.

$$0 = x^2 - 6x - 7$$
$$0 = (x - 7)(x + 1)$$

So $x = 7$ or $x = -1$

Check $x = 7$ (using the original equation).

$$7-\sqrt{2(7)+11}\stackrel{?}{=}2$$
$$7-\sqrt{25}\stackrel{?}{=}2$$
$$7-5=2\checkmark$$

The solution $x = 7$ works.

Now check $x = -1$ (using the original equation).

$$-1-\sqrt{2(-1)+11}\stackrel{?}{=}2$$
$$-1-\sqrt{9}\stackrel{?}{=}2$$
$$-1-3\neq 2$$

The solution $x = -1$ is extraneous.

165. $x = 2$ or $x = -2$

First, add $-x$ to both sides of the equation.

$$\sqrt{8-4x}=2-x$$

Now square both sides of the equation.

$$8 - 4x = 4 - 4x + x^2$$

Set the quadratic equation equal to 0 and solve for x.

$$0 = x^2 - 4$$

$$0 = (x - 2)(x + 2)$$

So $x = 2$ or $x = -2$

Check $x = 2$ (using the original equation).

$$2 + \sqrt{8 - 4(2)} \stackrel{?}{=} 2$$

$$2 + \sqrt{0} \stackrel{?}{=} 2$$

$$2 + 0 = 2 \checkmark$$

The solution $x = 2$ works.

Now check $x = -2$ (using the original equation).

$$-2 + \sqrt{8 - 4(-2)} \stackrel{?}{=} 2$$

$$-2 + \sqrt{16} \stackrel{?}{=} 2$$

$$-2 + 4 = 2 \checkmark$$

The solution $x = -2$ also works.

166. $x = 6$

First, add $-x$ to both sides of the equation.

$$-\sqrt{13 - 2x} = 5 - x$$

Now square both sides of the equation.

$$13 - 2x = 25 - 10x + x^2$$

Set the quadratic equation equal to 0 and solve for x.

$$0 = x^2 - 8x + 12$$

$$0 = (x - 2)(x - 6)$$

So $x = 2$ or $x = 6$

Check $x = 2$ (using the original equation).

$$2 - \sqrt{13 - 2(2)} \stackrel{?}{=} 5$$

$$2 - \sqrt{9} \stackrel{?}{=} 5$$

$$2 - 3 \neq 5$$

The solution $x = 2$ is extraneous.

Now check $x = 6$ (using the original equation).

$$6 - \sqrt{13 - 2(6)} \stackrel{?}{=} 5$$

$$6 - \sqrt{1} \stackrel{?}{=} 5$$

$$6 - 1 = 5 \checkmark$$

The solution $x = 6$ works.

167. $x = -3$ or $x = -4$

First, add x to both sides of the equation.

$$\sqrt{x+4} = 4 + x$$

Now square both sides of the equation.

$$x + 4 = 16 + 8x + x^2$$

Set the quadratic equation equal to 0 and solve for x.

$$0 = x^2 + 7x + 12$$

$$0 = (x + 3)(x + 4)$$

So $x = -3$ or $x = -4$

Check $x = -3$ (using the original equation).

$$\sqrt{-3+4} - (-3) \stackrel{?}{=} 4$$

$$\sqrt{1} + 3 \stackrel{?}{=} 4$$

$$1 + 3 = 4 \checkmark$$

The solution $x = -3$ works.

Now check $x = -4$ (using the original equation).

$$\sqrt{-4+4} - (-4) \stackrel{?}{=} 4$$

$$\sqrt{0} + 4 \stackrel{?}{=} 4$$

$$0 + 4 = 4 \checkmark$$

The solution $x = -4$ works, also.

168. $x = -6$ or $x = -7$

First, add x to both sides of the equation.

$$\sqrt{x+7} = 7 + x$$

Now square both sides of the equation.

$$x + 7 = 49 + 14x + x^2$$

Set the quadratic equation equal to 0 and solve for x.

$$0 = x^2 + 13x + 42$$

$$0 = (x + 6)(x + 7)$$

So $x = -6$ or $x = -7$

Check $x = -6$ (using the original equation).

$$\sqrt{-6+7} - (-6) \stackrel{?}{=} 7$$

$$\sqrt{1} + 6 \stackrel{?}{=} 7$$

$$1 + 6 = 7 \checkmark$$

The solution $x = -6$ works.

Now check $x = -7$ (using the original equation).

$$\sqrt{-7+7}-(-7)\stackrel{?}{=}7$$

$$\sqrt{0}+7\stackrel{?}{=}7$$

$$0+7=7 ✓$$

The solution $x = -7$ works, also.

169. $\mathbf{x = 11}$

Raise each side of the equation to the third power.

$$x-3=8$$

Solve for x.

$$x=11$$

170. $\mathbf{x = 40}$

Raise each side of the equation to the fourth power.

$$2x+1=81$$

Solve for x.

$$2x=80$$

$$x=40$$

171. $\mathbf{x = 40}$

Raise each side of the equation to the fifth power.

$$x-8=32$$

Solve for x.

$$x=40$$

172. $\mathbf{x = 4}$

Raise each side of the equation to the fourth power.

$$5-x=1$$

Solve for x.

$$-x=-4$$

$$x=4$$

173. $\mathbf{x = -3}$ **or** $\mathbf{x = -4}$ **or** $\mathbf{x = -5}$

First, add x to each side of the equation.

$$\sqrt[3]{x+4}=4+x$$

Now raise each side of the equation to the third power.

$$x + 4 = 64 + 48x + 12x^2 + x^3$$

Set the equation equal to 0 and solve for x.

$$0 = x^3 + 12x^2 + 47x + 60$$

You can use synthetic division to find a solution/factor; try $x = -3$.

$$\begin{array}{r|rrrr} -3 & 1 & 12 & 47 & 60 \\ & & -3 & -27 & -60 \\ \hline & 1 & 9 & 20 & 0 \end{array}$$

$x = -3$ is a solution, so $x + 3$ is a factor of the polynomial. You can write $x^3 + 12x^2 + 47x + 60$ as $(x + 3)(x^2 + 9x + 20)$, using the coefficients in the bottom row of the synthetic division.

Factoring the quadratic, you have $(x + 3)(x + 4)(x + 5) = 0$ with solutions $x = -3$, $x = -4$, and $x = -5$. Now check the solutions in the original equation.

Check $x = -3$.

$$\sqrt[3]{-3+4} - (-3) \stackrel{?}{=} 4$$
$$\sqrt[3]{1} + 3 \stackrel{?}{=} 4$$
$$1 + 3 = 4 \checkmark$$

So the solution $x = -3$ works.

Now check $x = -4$.

$$\sqrt[3]{-4+4} - (-4) \stackrel{?}{=} 4$$
$$\sqrt[3]{0} + 4 \stackrel{?}{=} 4$$
$$0 + 4 = 4 \checkmark$$

So the solution $x = -4$ also works.

Finally, check $x = -5$.

$$\sqrt[3]{-5+4} - (-5) \stackrel{?}{=} 4$$
$$\sqrt[3]{-1} + 5 \stackrel{?}{=} 4$$
$$-1 + 5 = 4 \checkmark$$

So the solution $x = -5$ works, as well.

174. $\boldsymbol{x = -1}$ **or** $\boldsymbol{x = -4}$

First, add x to each side of the equation.

$$\sqrt[3]{3x+11} = 3 + x$$

Now raise each side of the equation to the third power.

$$3x + 11 = 27 + 27x + 9x^2 + x^3$$

Set the equation equal to 0 and solve for x.

$$0 = x^3 + 9x^2 + 24x + 16$$

You can use synthetic division to find a solution/factor; try $x = -1$.

$$\begin{array}{r|rrrr} -1 & 1 & 9 & 24 & 16 \\ & & -1 & -8 & -16 \\ \hline & 1 & 8 & 16 & 0 \end{array}$$

$x = -1$ is a solution, so $x + 1$ is a factor of the polynomial. You can write $x^3 + 9x^2 + 24x + 16$ as $(x + 1)(x^2 + 8x + 16)$, using the coefficients in the bottom row of the synthetic division.

Factoring the quadratic, you have $(x + 1)(x + 4)(x + 4) = 0$ with solutions $x = -1$, $x = -4$, and $x = -1$. Now check the solutions in the original equation.

Check $x = -1$.

$$\sqrt[3]{3(-1)+11}-(-1)\stackrel{?}{=}3$$
$$\sqrt[3]{8}+1\stackrel{?}{=}3$$
$$2+1=3\ \checkmark$$

So the solution $x = -1$ works.

Now check $x = -4$.

$$\sqrt[3]{3(-4)+11}-(-4)\stackrel{?}{=}3$$
$$\sqrt[3]{-1}+4\stackrel{?}{=}3$$
$$-1+4=3\ \checkmark$$

So the solution $x = -4$ also works.

175. $x = 10$

First, add –7 to each side of the equation.

$$\sqrt[3]{2x+7} = x-7$$

Now raise each side of the equation to the third power.

$$2x + 7 = x^3 - 21x^2 + 147x - 343$$

Set the equation equal to 0 and solve for x.

$$0 = x^3 - 21x^2 + 145x - 350$$

You can use synthetic division to find a solution/factor; try $x = 10$.

$$\begin{array}{r|rrrr} 10 & 1 & -21 & 145 & -350 \\ & & 10 & -110 & 350 \\ \hline & 1 & -11 & 35 & 0 \end{array}$$

$x = 10$ is a solution, so $x - 10$ is a factor of the polynomial. You can write $x^3 - 21x^2 + 145x - 350$ as $(x - 10)(x^2 - 11x + 35)$, using the coefficients in the bottom row of the synthetic division.

The quadratic doesn't factor, and, using the quadratic formula, you get imaginary numbers. The imaginary/complex answers are $x = \frac{11 \pm i\sqrt{19}}{2}$. So the only possible real answer is $x = 10$.

Check $x = 10$ using the original equation.

$$\sqrt[3]{2(10)+7}+7\stackrel{?}{=}10$$

$$\sqrt[3]{27}+7\stackrel{?}{=}10$$

$$3+7=10 \checkmark$$

So the solution $x = 10$ works.

176. $x = -1$

First, add 3 to each side of the equation.

$$\sqrt[3]{4-4x}=x+3$$

Now raise each side of the equation to the third power.

$$4 - 4x = x^3 + 9x^2 + 27x + 27$$

Set the equation equal to 0 and solve for x.

$$0 = x^3 + 9x^2 + 31x + 23$$

You can use synthetic division to find a solution/factor; try $x = -1$.

$$\begin{array}{r|rrrr} -1 & 1 & 9 & 31 & 23 \\ & & -1 & -8 & -23 \\ \hline & 1 & 8 & 23 & 0 \end{array}$$

$x = -1$ is a solution, so $x + 1$ is a factor of the polynomial. You can write $x^3 + 9x^2 + 31x + 23$ as $(x + 1)(x^2 + 8x + 23)$, using the coefficients in the bottom row of the synthetic division.

The quadratic doesn't factor, and, using the quadratic formula, you get imaginary numbers. The imaginary/complex answers are $x=\frac{-8\pm 2i\sqrt{7}}{2}=-4\pm i\sqrt{7}$.

So the only possible real answer is $x = -1$.

Check $x = -1$ using the original equation.

$$\sqrt[3]{4-4(-1)}-3\stackrel{?}{=}-1$$

$$\sqrt[3]{8}-3\stackrel{?}{=}-1$$

$$2-3=-1 \checkmark$$

So the solution $x = -1$ works.

177. $x = 13$

Square both sides of the equation.

$$5+\sqrt{x+3}=9$$

Now add –5 to both sides.

$$\sqrt{x+3}=4$$

Square both sides again.

$$x + 3 = 16$$

Solve for x.

$$x = 13$$

Check $x = 13$ in the original equation.

$$\sqrt{5+\sqrt{13+3}} \stackrel{?}{=} 3$$
$$\sqrt{5+\sqrt{16}} \stackrel{?}{=} 3$$
$$\sqrt{5+4} \stackrel{?}{=} 3$$
$$\sqrt{9} \stackrel{?}{=} 3$$
$$3 = 3 \checkmark$$

So the solution $x = 13$ works.

178. $x = 90$

Square both sides of the equation.

$$\sqrt{x-9} - 8 = 1$$

Now add 8 to both sides.

$$\sqrt{x-9} = 9$$

Square both sides again.

$$x - 9 = 81$$

Solve for x.

$$x = 90$$

Check $x = 90$ in the original equation.

$$\sqrt{\sqrt{90-9} - 8} \stackrel{?}{=} 1$$
$$\sqrt{\sqrt{81} - 8} \stackrel{?}{=} 1$$
$$\sqrt{9-8} \stackrel{?}{=} 1$$
$$\sqrt{1} \stackrel{?}{=} 1$$
$$1 = 1 \checkmark$$

So the solution $x = 90$ works.

179. $x = 7$

Square both sides of the equation.

$$1 + \sqrt{x+2} = 4$$

Now add –1 to both sides.

$$\sqrt{x+2} = 3$$

Square both sides again.

$$x + 2 = 9$$

Solve for x.

$$x = 7$$

Check $x = 7$ in the original equation.

$$\sqrt{1+\sqrt{7+2}} \stackrel{?}{=} 2$$
$$\sqrt{1+\sqrt{9}} \stackrel{?}{=} 2$$
$$\sqrt{1+3} \stackrel{?}{=} 2$$
$$\sqrt{4} \stackrel{?}{=} 2$$
$$2 = 2 \checkmark$$

So the solution $x = 7$ works.

180. $x = 7$

Square both sides of the equation.

$$x - \sqrt{x+2} = 4$$

Now add $-x$ to both sides.

$$-\sqrt{x+2} = 4 - x$$

Square both sides again.

$$x + 2 = 16 - 8x + x^2$$

Set the quadratic equal to 0 and factor.

$$0 = x^2 - 9x + 14 = (x-2)(x-7)$$

Setting the binomials equal to 0, you have $x = 2$ or $x = 7$.

Check $x = 2$ in the original equation.

$$\sqrt{2-\sqrt{2+2}} \stackrel{?}{=} 2$$
$$\sqrt{2-\sqrt{4}} \stackrel{?}{=} 2$$
$$\sqrt{2-2} \stackrel{?}{=} 2$$
$$\sqrt{0} \stackrel{?}{=} 2$$
$$2 \neq 2$$

The solution $x = 2$ is extraneous.

Now check $x = 7$ in the original equation.

$$\sqrt{7-\sqrt{7+2}} \stackrel{?}{=} 2$$
$$\sqrt{7-\sqrt{9}} \stackrel{?}{=} 2$$
$$\sqrt{7-3} \stackrel{?}{=} 2$$
$$\sqrt{4} \stackrel{?}{=} 2$$
$$2 = 2 \checkmark$$

The solution $x = 7$ works.

181. $x = 8$ or $x = 16$

Square both sides of the equation.

$$3x+1-8\sqrt{3x+1}+16=x-7$$

Notice that the radical term is twice the product of the two original terms on the left.

Now subtract terms to create an equation with the radical on one side and the other terms on the other side.

$$-8\sqrt{3x+1}=-2x-24$$

Before squaring both sides again, divide each term by –2.

$$4\sqrt{3x+1}=x+12$$

Now square both sides of the equation.

$$16(3x+1) = x^2 + 24x + 144$$

Distribute on the left.

$$48x + 16 = x^2 + 24x + 144$$

Now set the quadratic equation equal to 0 and solve for x.

$$0 = x^2 - 24x + 128 = (x - 8)(x - 16)$$

The two solutions are $x = 8$ or $x = 16$.

Check $x = 8$ in the original equation.

$$\sqrt{3(8)+1}-4\stackrel{?}{=}\sqrt{8-7}$$

$$\sqrt{24+1}-4\stackrel{?}{=}\sqrt{1}$$

$$\sqrt{25}-4\stackrel{?}{=}1$$

$$5-4=1\checkmark$$

The solution $x = 8$ works.

Now check $x = 16$ in the original equation.

$$\sqrt{3(16)+1}-4\stackrel{?}{=}\sqrt{16-7}$$

$$\sqrt{48+1}-4\stackrel{?}{=}\sqrt{9}$$

$$\sqrt{49}-4\stackrel{?}{=}3$$

$$7-4=3\checkmark$$

The solution $x = 16$ also works.

182. $x = 5$

Square both sides of the equation.

$$x+4+2\sqrt{x+4}+1=2x+6$$

Notice that the radical term is twice the product of the two original terms on the left.

Now subtract terms to create an equation with the radical on one side and the other terms on the other side.

$$2\sqrt{x+4}=x+1$$

Now square both sides of the equation.

$$4(x+4)=x^2+2x+1$$

Distribute on the left.

$$4x+16=x^2+2x+1$$

Now set the quadratic equation equal to 0 and solve for x.

$$0=x^2-2x-15=(x-5)(x+3)$$

The two solutions are $x=5$ or $x=-3$.

Check $x=5$ in the original equation.

$$\sqrt{5+4}+1\overset{?}{=}\sqrt{2(5)+6}$$

$$\sqrt{9}+1\overset{?}{=}\sqrt{16}$$

$$3+1=4\checkmark$$

The solution $x=5$ works.

Now check $x=-3$ in the original equation.

$$\sqrt{-3+4}+1\overset{?}{=}\sqrt{2(-3)+6}$$

$$\sqrt{1}+1\overset{?}{=}\sqrt{0}$$

$$1+1\neq 0$$

The solution $x=-3$ is extraneous.

183. $x=-5$

Square both sides of the equation.

$$5-4x-6\sqrt{5-4x}+9=x+9$$

Notice that the radical term is twice the product of the two original terms on the left.

Now subtract terms to create an equation with the radical on one side and the other terms on the other side.

$$-6\sqrt{5-4x}=5x-5$$

Now square both sides of the equation.

$$36(5-4x)=25x^2-50x+25$$

Distribute on the left.

$$180-144x=25x^2-50x+25$$

Now set the quadratic equation equal to 0 and solve for x.

$$0=25x^2+94x-155=(25x-31)(x+5)$$

The two solutions are $x = \frac{31}{25}$ or $x = -5$.

Check $x = \frac{31}{25}$ in the original equation.

$$\sqrt{5-4\left(\frac{31}{25}\right)}-3\overset{?}{=}\sqrt{\frac{31}{25}+9}$$

$$\sqrt{\frac{125}{25}-\frac{124}{25}}-3\overset{?}{=}\sqrt{\frac{31}{25}+\frac{225}{25}}$$

$$\sqrt{\frac{1}{25}}-3\overset{?}{=}\sqrt{\frac{256}{25}}$$

$$\frac{1}{5}-3\overset{?}{=}\frac{16}{5}$$

$$\frac{1}{5}-\frac{15}{5}\overset{?}{=}\frac{16}{5}$$

$$-\frac{14}{5}\neq\frac{16}{5}$$

The solution $x = \frac{31}{25}$ is extraneous.

Now check $x = -5$ in the original equation.

$$\sqrt{5-4(-5)}-3\overset{?}{=}\sqrt{-5+9}$$

$$\sqrt{5+20}-3\overset{?}{=}\sqrt{4}$$

$$\sqrt{25}-3\overset{?}{=}2$$

$$5-3=2\ \checkmark$$

The solution $x = -5$ works.

184. $x = 1$ or $x = -7$

Square both sides of the equation.

$$2-2x+2\sqrt{2-2x}+1=4-3x$$

Notice that the radical term is twice the product of the two original terms on the left.

Now subtract terms to create an equation with the radical on one side and the other terms on the other side.

$$2\sqrt{2-2x}=1-x$$

Now square both sides of the equation.

$$4(2-2x) = 1-2x+x^2$$

Distribute on the left.

$$8-8x = 1-2x+x^2$$

Now set the quadratic equation equal to 0 and solve for x.

$$0 = x^2+6x-7 = (x-1)(x+7)$$

The two solutions are $x = 1$ or $x = -7$.

Check $x = 1$ in the original equation.

$$\sqrt{2-2(1)}+1\stackrel{?}{=}\sqrt{4-3(1)}$$

$$\sqrt{2-2}+1\stackrel{?}{=}\sqrt{4-3}$$

$$\sqrt{0}+1\stackrel{?}{=}\sqrt{1}$$

$$0+1=1\checkmark$$

The solution $x = 1$ works.

Now check $x = -7$ in the original equation.

$$\sqrt{2-2(-7)}+1\stackrel{?}{=}\sqrt{4-3(-7)}$$

$$\sqrt{2+14}+1\stackrel{?}{=}\sqrt{4+21}$$

$$\sqrt{16}+1\stackrel{?}{=}\sqrt{25}$$

$$4+1=5\checkmark$$

The solution $x = -7$ also works.

185. $x = 4$ or $x = -4$

First, subtract $\sqrt{5-x}$ from each side of the equation so that there's a radical on each side.

$$\sqrt{x+5}=4-\sqrt{5-x}$$

Now square both sides of the equation.

$$x+5=16-8\sqrt{5-x}+5-x$$

Notice that the radical term is twice the product of the two terms on the right.

Now subtract terms to create an equation with the radical on one side and the other terms on the other side.

$$2x-16=-8\sqrt{5-x}$$

Before squaring both sides again, divide each term by 2.

$$x-8=-4\sqrt{5-x}$$

Now square both sides of the equation.

$$x^2-16x+64=16(5-x)$$

Distribute on the right.

$$x^2-16x+64=80-16x$$

Now set the quadratic equation equal to 0 and solve for x.

$$x^2-16=(x-4)(x+4)=0$$

The two solutions are $x = 4$ or $x = -4$.

Check $x = 4$ in the original equation.

$$\sqrt{4+5}+\sqrt{5-4}\stackrel{?}{=}4$$

$$\sqrt{9}+\sqrt{1}\stackrel{?}{=}4$$

$$3+1=4\checkmark$$

The solution $x = 4$ works.

Now check $x = -4$ in the original equation.

$$\sqrt{-4+5}+\sqrt{5-(-4)}\stackrel{?}{=}4$$
$$\sqrt{1}+\sqrt{9}\stackrel{?}{=}4$$
$$1+3=4 ✓$$

The solution $x = -4$ also works.

186. $x = 5$

First, subtract $\sqrt{2x-1}$ from each side of the equation so that there's a radical on each side.

$$\sqrt{3x+1}=7-\sqrt{2x-1}$$

Now square both sides of the equation.

$$3x+1=49-14\sqrt{2x-1}+2x-1$$

Notice that the radical term is twice the product of the two terms on the right.

Now subtract terms to create an equation with the radical on one side and the other terms on the other side.

$$x-47=-14\sqrt{2x-1}$$

Now square both sides of the equation.

$$x^2-94x+2209=196(2x-1)$$

Distribute on the right.

$$x^2-94x+2209=392x-196$$

Now set the quadratic equation equal to 0 and solve for x.

$$x^2-486x+2405=(x-5)(x-481)=0$$

The two solutions are $x = 5$ or $x = 481$.

Check $x = 5$ in the original equation.

$$\sqrt{3(5)+1}+\sqrt{2(5)-1}\stackrel{?}{=}7$$
$$\sqrt{16}+\sqrt{9}\stackrel{?}{=}7$$
$$4+3=7 ✓$$

The solution $x = 5$ works.

Now check $x = 481$ in the original equation.

$$\sqrt{3(481)+1}+\sqrt{2(481)-1}\stackrel{?}{=}7$$
$$\sqrt{1,443+1}+\sqrt{962-1}\stackrel{?}{=}7$$
$$\sqrt{1,444}+\sqrt{961}\stackrel{?}{=}7$$
$$38+31\neq 7$$

The solution $x = 481$ is extraneous.

187. $x = -2$

First, add $\sqrt{x+6}$ to each side of the equation so that there's a radical on each side.

$$\sqrt{8-4x} = 2 + \sqrt{x+6}$$

Now square both sides of the equation.

$$8 - 4x = 4 + 4\sqrt{x+6} + x + 6$$

Notice that the radical term is twice the product of the two terms on the right.

Now subtract terms to create an equation with the radical on one side and the other terms on the other side.

$$-2 - 5x = 4\sqrt{x+6}$$

Now square both sides of the equation.

$$4 + 20x + 25x^2 = 16(x + 6)$$

Distribute on the right.

$$4 + 20x + 25x^2 = 16x + 96$$

Now set the quadratic equation equal to 0 and solve for x.

$$25x^2 + 4x - 92 = (25x - 46)(x + 2) = 0$$

The two solutions are $x = \frac{46}{25}$ or $x = -2$.

Check $x = \frac{46}{25}$ in the original equation.

$$\sqrt{8 - 4\left(\frac{46}{25}\right)} - \sqrt{\frac{46}{25} + 6} \stackrel{?}{=} 2$$

$$\sqrt{\frac{200}{25} - \frac{184}{25}} - \sqrt{\frac{46}{25} + \frac{150}{25}} \stackrel{?}{=} 2$$

$$\sqrt{\frac{16}{25}} - \sqrt{\frac{196}{25}} \stackrel{?}{=} 2$$

$$\frac{4}{5} - \frac{14}{5} \stackrel{?}{=} 2$$

$$\frac{-10}{5} \neq 2$$

The solution $x = \frac{46}{25}$ is extraneous.

Now check $x = -2$ in the original equation.

$$\sqrt{8 - 4(-2)} - \sqrt{-2 + 6} \stackrel{?}{=} 2$$

$$\sqrt{8+8} - \sqrt{4} \stackrel{?}{=} 2$$

$$\sqrt{16} - \sqrt{4} \stackrel{?}{=} 2$$

$$4 - 2 = 2 \checkmark$$

The solution $x = -2$ works.

188. $x = 12$

First, add $\sqrt{2x+1}$ to each side of the equation so that there's a radical on each side.

$$\sqrt{5x+4} = 3 + \sqrt{2x+1}$$

Now square both sides of the equation.

$$5x+4=9+6\sqrt{2x+1}+2x+1$$

Notice that the radical term is twice the product of the two terms on the right.

Now subtract terms to create an equation with the radical on one side and the other terms on the other side.

$$3x-6=6\sqrt{2x+1}$$

Before squaring both sides, divide each term by 3.

$$x-2=2\sqrt{2x+1}$$

Now square both sides of the equation.

$$x^2-4x+4=4(2x+1)$$

Distribute on the right.

$$x^2-4x+4=8x+4$$

Now set the quadratic equation equal to 0 and solve for x.

$$x^2-12x=x(x-12)=0$$

The two solutions are $x=0$ or $x=12$.

Check $x=0$ in the original equation.

$$\sqrt{5(0)+4}-\sqrt{2(0)+1}\stackrel{?}{=}3$$

$$\sqrt{0+4}-\sqrt{0+1}\stackrel{?}{=}3$$

$$\sqrt{4}-\sqrt{1}\stackrel{?}{=}3$$

$$2-1\neq 3$$

The solution $x=0$ is extraneous.

Now check $x=12$ in the original equation.

$$\sqrt{5(12)+4}-\sqrt{2(12)+1}\stackrel{?}{=}3$$

$$\sqrt{60+4}-\sqrt{24+1}\stackrel{?}{=}3$$

$$\sqrt{64}-\sqrt{25}\stackrel{?}{=}3$$

$$8-5=3\checkmark$$

The solution $x=12$ works.

189. $x=-1$

Square both sides of the equation. Remember that you'll have twice the product of the two radicals as one of the terms on the left.

$$8-x-2\sqrt{8-x}\sqrt{x+2}+x+2=x+5$$

Now isolate the radical term on the left by simplifying and subtracting terms.

$$-2\sqrt{8-x}\sqrt{x+2}=x-5$$

Square both sides again.

$$4(8-x)(x+2)=x^2-10x+25$$

Multiply on the left.

$$64 + 24x - 4x^2 = x^2 - 10x + 25$$

Set the equation equal to 0 and solve for x.

$$0 = 5x^2 - 34x - 39 = (5x - 39)(x + 1)$$

The two solutions are $x = \frac{39}{5}$ or $x = -1$.

Check the solution $x = \frac{39}{5}$ in the original equation.

$$\sqrt{8 - \frac{39}{5}} - \sqrt{\frac{39}{5} + 2} \stackrel{?}{=} \sqrt{\frac{39}{5} + 5}$$

$$\sqrt{\frac{40}{5} - \frac{39}{5}} - \sqrt{\frac{39}{5} + \frac{10}{5}} \stackrel{?}{=} \sqrt{\frac{39}{5} + \frac{25}{5}}$$

$$\sqrt{\frac{1}{5}} - \sqrt{\frac{49}{5}} \stackrel{?}{=} \sqrt{\frac{64}{5}}$$

$$\frac{1}{\sqrt{5}} - \frac{7}{\sqrt{5}} \stackrel{?}{=} \frac{8}{\sqrt{5}}$$

$$-\frac{6}{\sqrt{5}} \neq \frac{8}{\sqrt{5}}$$

The solution $x = \frac{39}{5}$ is extraneous.

Now check $x = -1$ in the original equation.

$$\sqrt{8 - (-1)} - \sqrt{(-1) + 2} \stackrel{?}{=} \sqrt{(-1) + 5}$$

$$\sqrt{9} - \sqrt{1} \stackrel{?}{=} \sqrt{4}$$

$$3 - 1 = 2 \checkmark$$

The solution $x = -1$ works.

190. $x = 2$

First, square both sides of the equation. Remember that you'll have twice the product of the two radicals as one of the terms on the left.

$$11x + 3 - 2\sqrt{11x + 3}\sqrt{x + 7} + x + 7 = x + 2$$

Now isolate the radical term on the left by simplifying and subtracting terms.

$$-2\sqrt{11x + 3}\sqrt{x + 7} = -11x - 8$$

Square both sides again.

$$4(11x + 3)(x + 7) = 121x^2 + 176x + 64$$

Multiply on the left.

$$44x^2 + 320x + 84 = 121x^2 + 176x + 64$$

Set the equation equal to 0 and solve for x.

$$0 = 77x^2 - 144x - 20 = (77x + 10)(x - 2)$$

The two solutions are $x = -\frac{10}{77}$ or $x = 2$.

Check $x=-\frac{10}{77}$ in the original equation.

$$\sqrt{11\left(-\frac{10}{77}\right)+3}-\sqrt{-\frac{10}{77}+7}\stackrel{?}{=}\sqrt{-\frac{10}{77}+2}$$

$$\sqrt{-\frac{110}{77}+\frac{231}{77}}-\sqrt{-\frac{10}{77}+\frac{539}{77}}\stackrel{?}{=}\sqrt{-\frac{10}{77}+\frac{154}{77}}$$

$$\sqrt{\frac{121}{77}}-\sqrt{\frac{529}{77}}\stackrel{?}{=}\sqrt{\frac{144}{77}}$$

$$\frac{11}{\sqrt{77}}-\frac{23}{\sqrt{77}}\stackrel{?}{=}\frac{12}{\sqrt{77}}$$

$$-\frac{12}{\sqrt{77}}\neq\frac{12}{\sqrt{77}}$$

The solution $x=-\frac{10}{77}$ is extraneous.

Now check $x = 2$.

$$\sqrt{11(2)+3}-\sqrt{2+7}\stackrel{?}{=}\sqrt{2+2}$$

$$\sqrt{25}-\sqrt{9}\stackrel{?}{=}\sqrt{4}$$

$$5-3=2\ \checkmark$$

The solution $x = 2$ works.

191. Moves upward from left to right; also through (4, 4)

First mark the point at (3, 2). Then think of the slope as the fraction $\frac{2}{1}=\frac{\text{change in } y}{\text{change in } x}$. Starting at the point (3, 2), move 1 unit to the right and then 2 units up, ending at the point (4, 4). Draw the line through those two points.

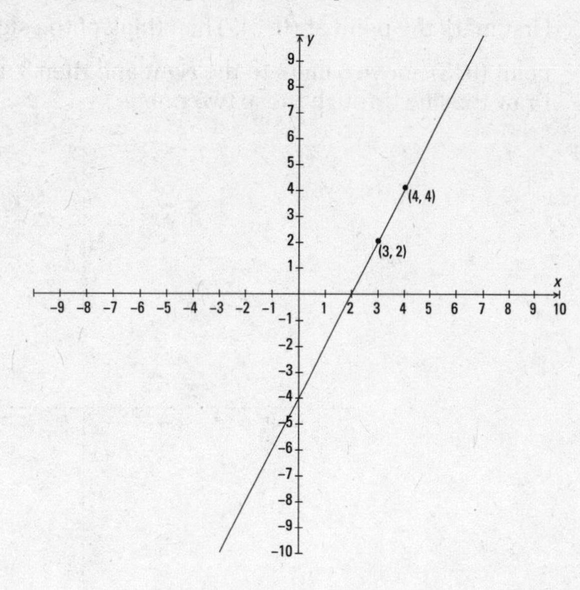

Answers 101–200

192. Moves downward from left to right; also through (–1, –6)

First mark the point at (–2, –3). Then think of the slope as the fraction $\frac{-3}{1} = \frac{\text{change in } y}{\text{change in } x}$. Starting at the point (–2, –3), move 1 unit to the right and then 3 units down, ending at the point (–1, –6). Draw the line through those two points.

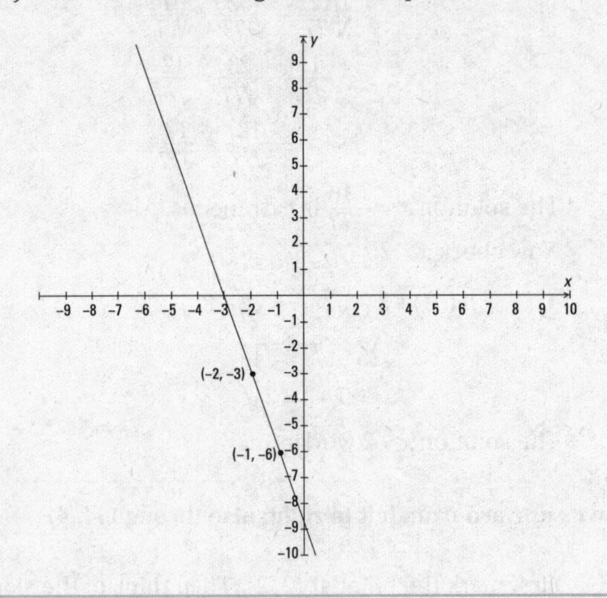

193. Moves upward from left to right; also through (5, 6)

First mark the point at (0, 3). Then think of the slope as $\frac{3}{5} = \frac{\text{change in } y}{\text{change in } x}$. Starting at the point (0, 3), move 5 units to the right and then 3 units up, ending at the point (5, 6). Draw the line through those two points.

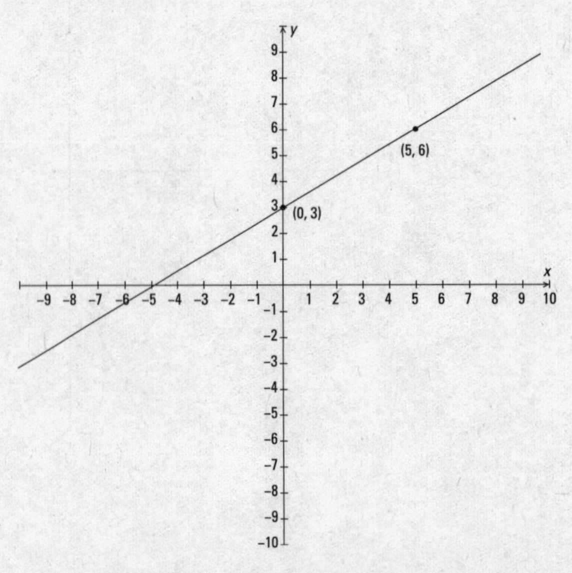

194. Moves downward from left to right; also through (5, –7)

First mark the point at (3, –2). Then think of the slope as $\frac{-5}{2} = \frac{\text{change in } y}{\text{change in } x}$. Starting at the point (3, –2), move 2 units to the right and then 5 units down, ending at the point (5, –7). Draw the line through those two points.

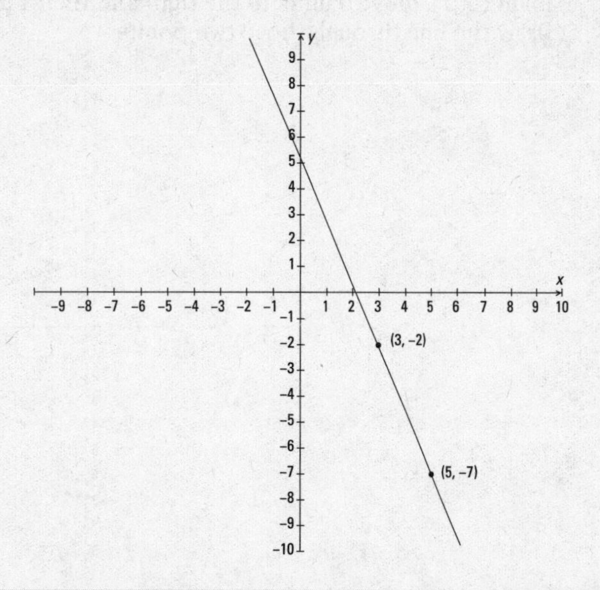

195. Moves upward from left to right; also through (2, 6)

The line is in slope-intercept form, so you can read the slope value directly: $m = 3$. So mark the point at (1, 3). Then think of the slope as the fraction $\frac{3}{1} = \frac{\text{change in } y}{\text{change in } x}$. Starting at the point (1, 3), move 1 unit to the right and then 3 units up, ending at the point (2, 6). Draw the line through those two points.

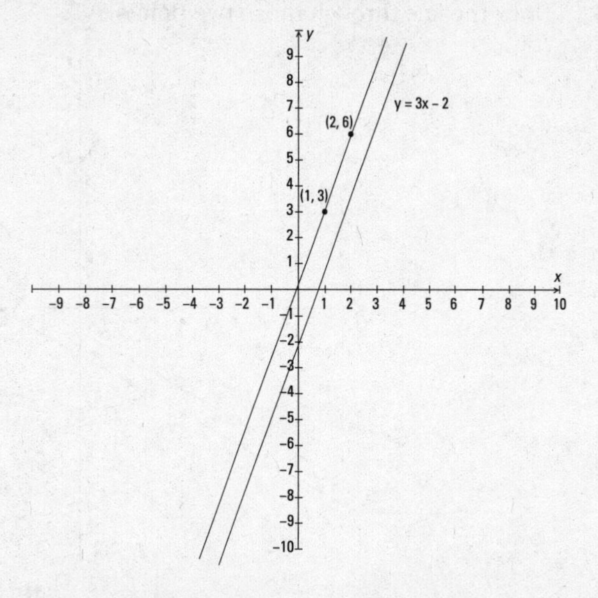

196. **Moves downward from left to right; also through (8, 0)**

The line is in slope-intercept form, so you can read the slope value directly: $m = -\frac{1}{3}$. So mark the point at (5, 1). Then think of the slope as $\frac{-1}{3} = \frac{\text{change in } y}{\text{change in } x}$. Starting at the point (5, 1), move 3 units to the right and then 1 unit down, ending at the point (8, 0). Draw the line through those two points.

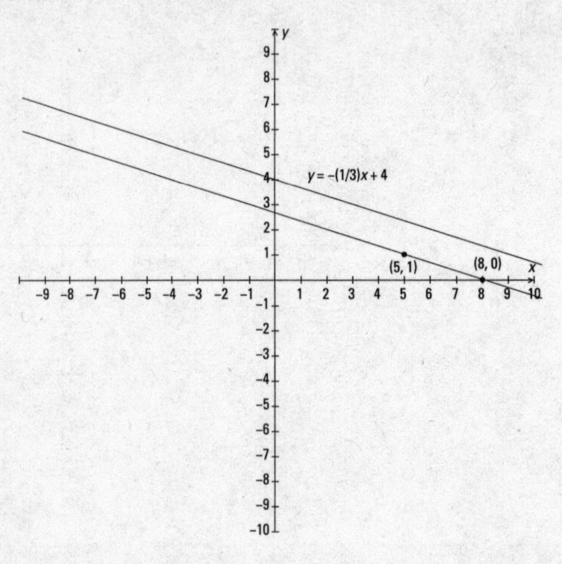

197. **Moves upward from left to right; also through (5, 1)**

The line is in slope-intercept form, so you can read the slope value directly: $m = \frac{4}{7}$. So mark the point at (–2, –3). Then think of the slope as $\frac{4}{7} = \frac{\text{change in } y}{\text{change in } x}$. Starting at the point (–2, –3), move 7 units to the right and then 4 units up, ending at the point (5, 1). Draw the line through those two points.

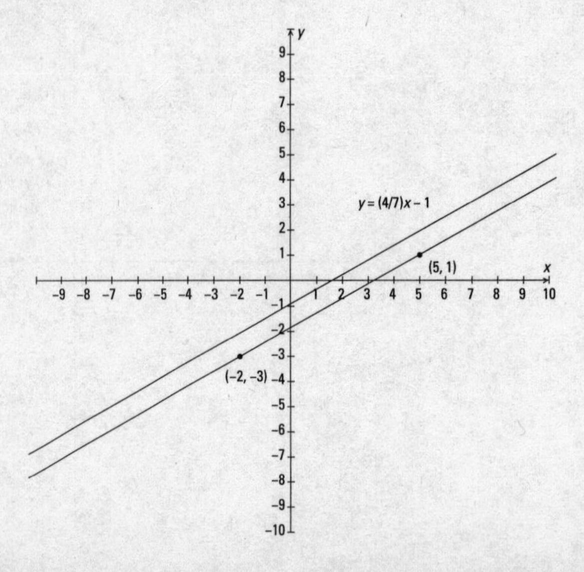

198. Moves downward from left to right; also through (–2, –3)

The line is in slope-intercept form, so you can read the slope value directly: $m = -2$. So mark the point at (–3, –1). Then think of the slope as the fraction $\frac{-2}{1} = \frac{\text{change in } y}{\text{change in } x}$. Starting at the point (–3, –1), move 1 unit to the right and then 2 units down, ending at the point (–2, –3). Draw the line through those two points.

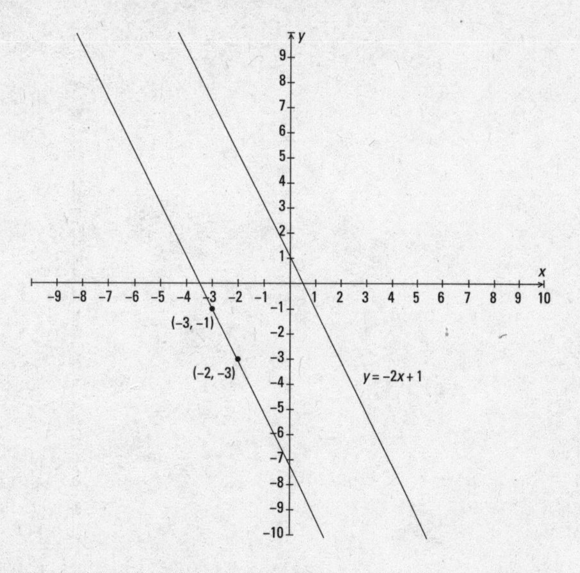

199. Moves upward from left to right; also through (6, 2)

The line is in slope-intercept form, so you can read the slope value directly: $m = \frac{1}{2}$. So mark the point at (4, 1). Then think of the slope as $\frac{1}{2} = \frac{\text{change in } y}{\text{change in } x}$. Starting at the point (4, 1), move 2 units to the right and then 1 unit up, ending at the point (6, 2). Draw the line through those two points.

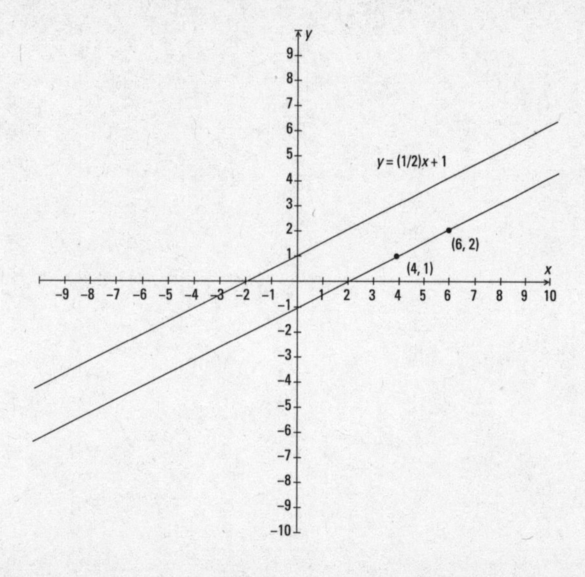

200. Moves downward from left to right; also through (1, –7)

The line is in slope-intercept form, so you can read the slope value directly: $m = -7$. So mark the point at (0, 0). Then think of the slope as the fraction $\frac{-7}{1} = \frac{\text{change in } y}{\text{change in } x}$. Starting at the point (0, 0), move 1 unit to the right and then 7 units down, ending at the point (1, –7). Draw the line through those two points.

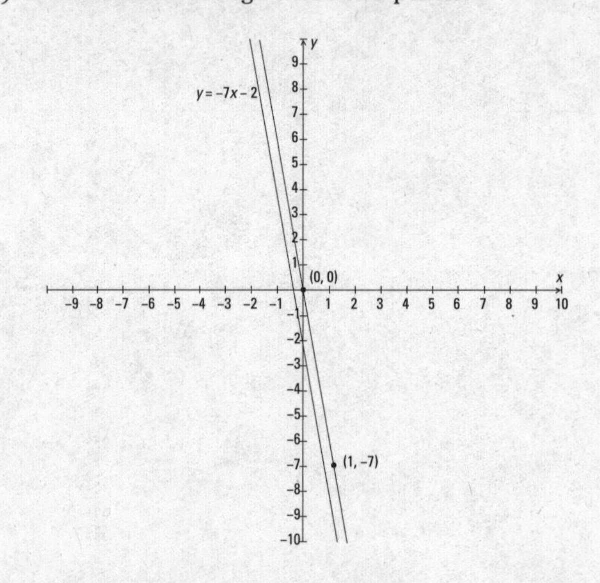

201. Moves downward from left to right; also through (4, 3)

The line is in slope-intercept form, so you can read its slope value directly and find it to be $m = 2$. A line perpendicular to this line has $m = -\frac{1}{2}$. To draw that line, first mark the point (2, 4). Then move 2 units to the right and 1 unit down to the point (4, 3). Draw a line through these points.

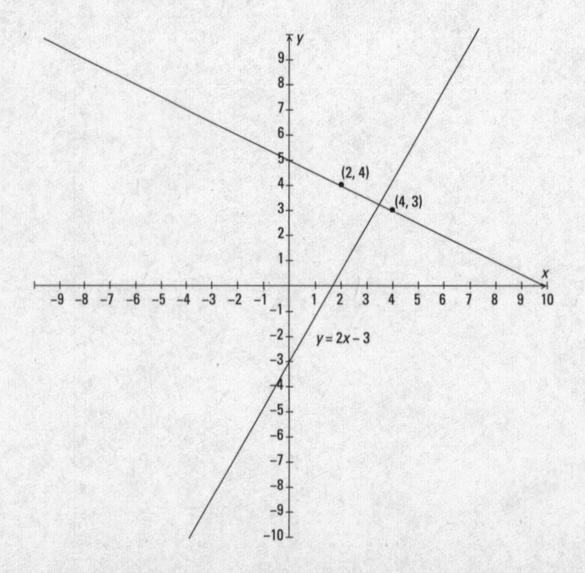

202. Moves upward from left to right; also through (–2, 0)

The line is in slope-intercept form, so you can read its slope value directly and find it to be $m = -\frac{1}{3}$. A line perpendicular to this line has $m = 3$. To draw that line, first mark the point (–3, –3). Then move 1 unit to the right and 3 units up to the point (–2, 0). Draw a line through these points.

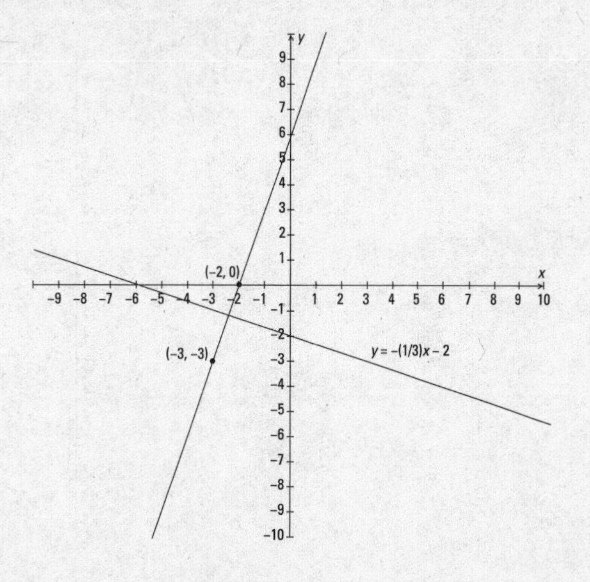

203. Moves upward from left to right; also through (1, 8)

The line is in slope-intercept form, so you can read its slope value directly and find it to be $m = -3$. A line perpendicular to this line has $m = \frac{1}{3}$. To draw that line, first mark the point (–2, 7). Then move 3 units to the right and 1 unit up to the point (1, 8). Draw a line through these points.

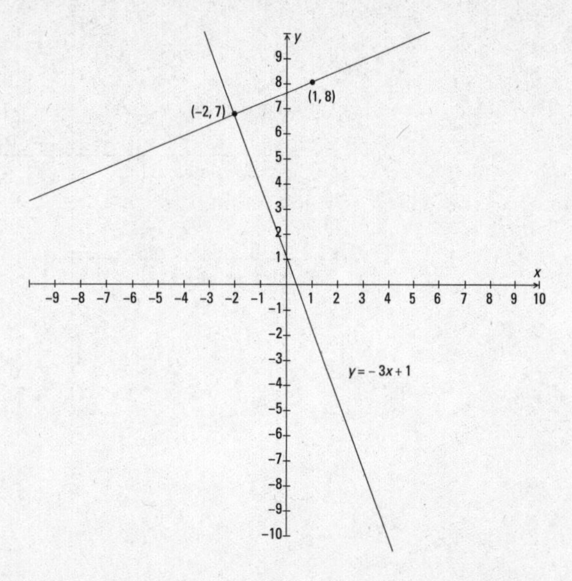

204. **Moves downward from left to right; also through (5, –5)**

The line is in slope-intercept form, so you can read its slope value directly and find it to be $m = \frac{1}{4}$. A line perpendicular to this line has $m = -4$. To draw that line, first mark the point (4, –1). Then move 1 unit to the right and 4 units down to the point (5, –5). Draw a line through these points.

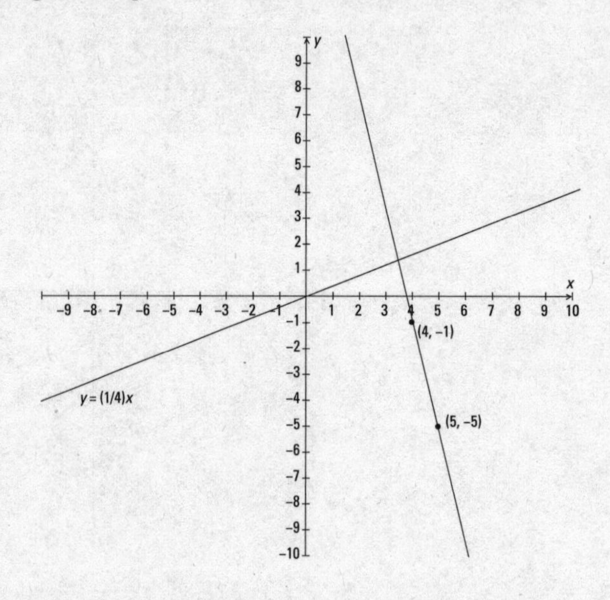

205. **Vertical through (2, 3) and (2, 4)**

The line $y = 4$ is horizontal, so its slope is $m = 0$. A line perpendicular to this line has an undefined slope — it's vertical. So draw a vertical line through the point (2, 3); it will cross $y = 4$ at (2, 4).

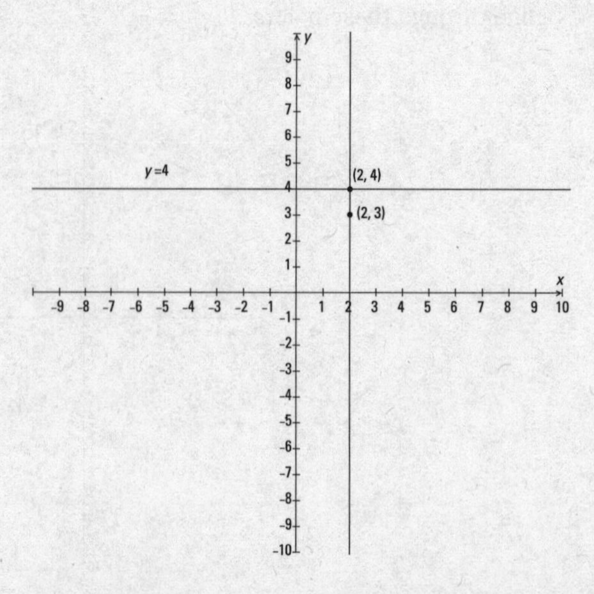

206. Vertical through (–2, 4) and (–2, 2)

The line $y = -2$ is horizontal, so its slope is $m = 0$. A line perpendicular to this line has an undefined slope — it's vertical. So draw a vertical line through the point (–2, 4); it will cross $y = -2$ at (–2, –2).

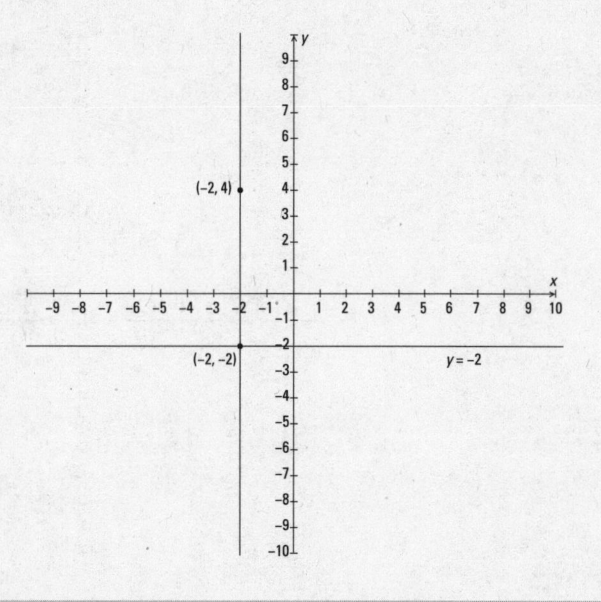

Answers 201–300

207. Horizontal through (2, 5) and (1, 5)

The line $x = 1$ is vertical, so its slope is undefined. A line perpendicular to this line has a slope of 0 — it's horizontal. So draw a horizontal line through the point (2, 5); it will cross $x = 1$ at (1, 5).

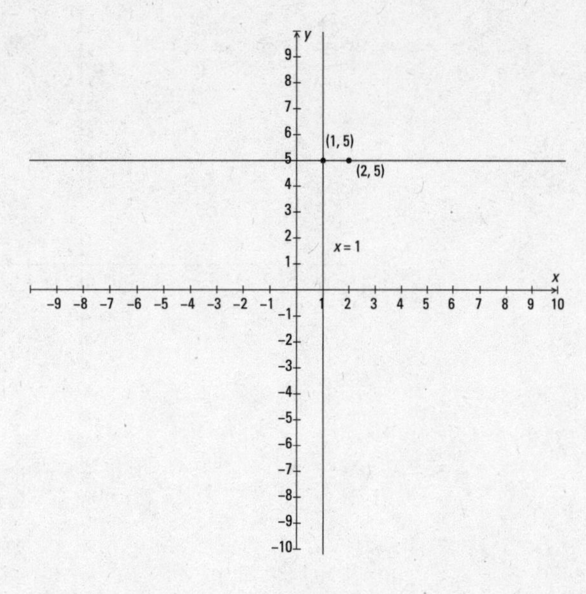

208. Horizontal through (1, –1) and (3, –1)

The line $x = 3$ is vertical, so its slope is undefined. A line perpendicular to this line has a slope of 0 — it's horizontal. So draw a horizontal line through the point (1, –1); it will cross $x = 3$ at (3, –1).

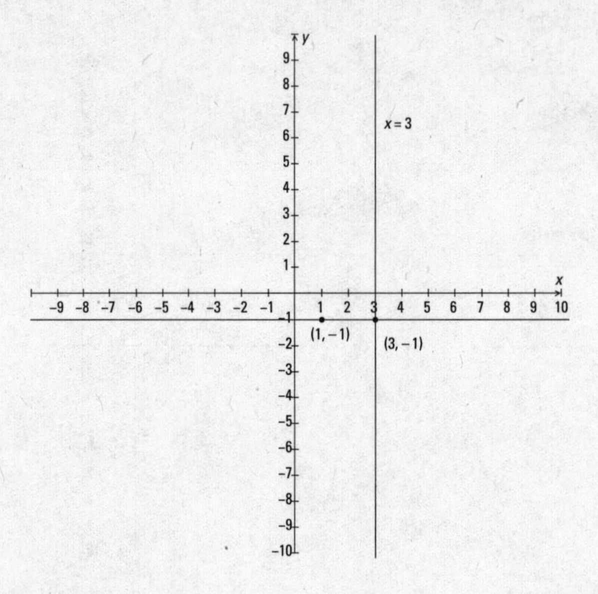

209. Vertical through (4, 3) and (4, 0)

The x-axis is horizontal with a slope of 0, so the line must be vertical through the point (4, 3) and passes through the x-axis at (4, 0).

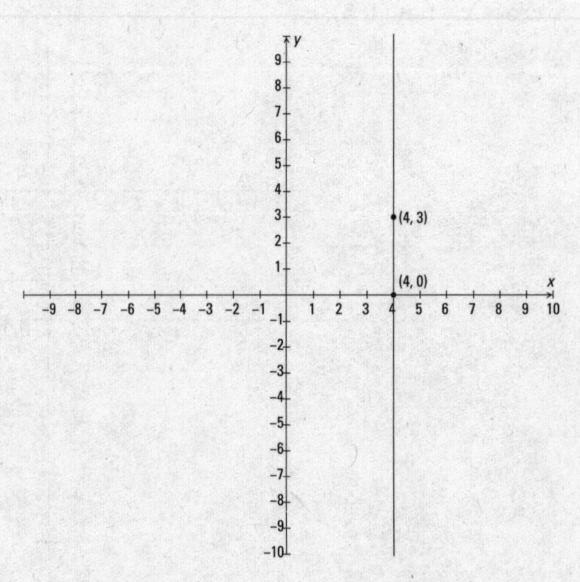

210. Horizontal through (3, –5) and (0, –5)

The *y*-axis is vertical with an undefined slope, so the line must be horizontal through the point (3, –5) and passes through the *y*-axis at (0, –5).

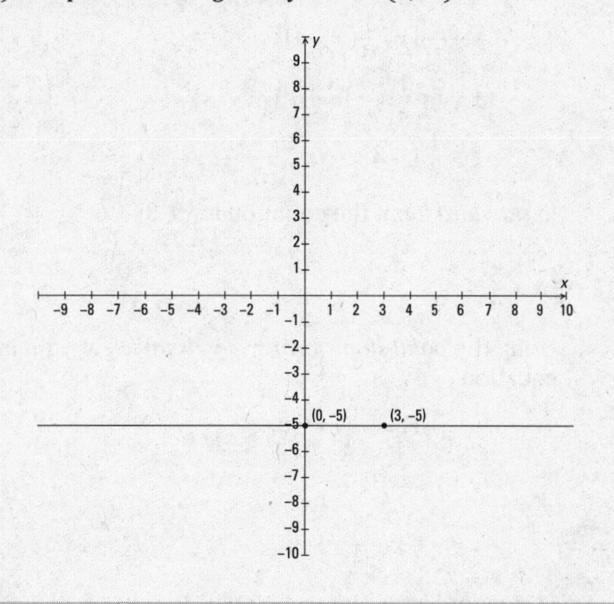

211. $y = 2x - 7$

Using the *point-slope form*, $y - y_1 = m(x - x_1)$, fill in the given values and simplify the equation.

$$y - (-1) = 2(x - 3)$$

$$y + 1 = 2x - 6$$

$$y = 2x - 7$$

In *standard form*, the equation is $2x - y = 7$

212. $y = -2x + 5$

Using the *point-slope form*, $y - y_1 = m(x - x_1)$, fill in the given values and simplify the equation.

$$y - (-1) = -2(x - 3)$$

$$y + 1 = -2x + 6$$

$$y = -2x + 5$$

In *standard form*, the equation is $2x + y = 5$

213. $y = \frac{1}{3}x - 2$

Using the *point-slope form*, $y - y_1 = m(x - x_1)$, fill in the given values and simplify the equation.

$$y - (-1) = \frac{1}{3}(x - 3)$$

$$y + 1 = \frac{1}{3}x - 1$$

$$y = \frac{1}{3}x - 2$$

In *standard form*, the equation is $x - 3y = 6$.

214. $y = -\frac{3}{4}x + \frac{5}{4}$

Using the *point-slope form*, $y - y_1 = m(x - x_1)$, fill in the given values and simplify the equation.

$$y - (-1) = -\frac{3}{4}(x - 3)$$

$$y + 1 = -\frac{3}{4}x + \frac{9}{4}$$

$$y = -\frac{3}{4}x + \frac{5}{4}$$

In *standard form*, the equation is $3x + 4y = 5$.

215. $y = \frac{4}{5}x - \frac{17}{5}$

Using the *point-slope form*, $y - y_1 = m(x - x_1)$, fill in the given values and simplify the equation.

$$y - (-1) = \frac{4}{5}(x - 3)$$

$$y + 1 = \frac{4}{5}x - \frac{12}{5}$$

$$y = \frac{4}{5}x - \frac{17}{5}$$

In *standard form*, the equation is $4x - 5y = 17$

216. $y = -5x + 14$

Using the *point-slope form*, $y - y_1 = m(x - x_1)$, fill in the given values and simplify the equation.

$$y - (-1) = -5(x - 3)$$

$$y + 1 = -5x + 15$$

$$y = -5x + 14$$

In *standard form*, the equation is $5x + y = 14$

217. $y = \frac{2}{7}x - \frac{13}{7}$

Using the *point-slope form,* $y - y_1 = m(x - x_1)$, fill in the given values and simplify the equation.

$$y-(-1)=\frac{2}{7}(x-3)$$
$$y+1=\frac{2}{7}x-\frac{6}{7}$$
$$y=\frac{2}{7}x-\frac{13}{7}$$

In *standard form*, the equation is $2x - 7y = 13$

218. $y = -\frac{5}{3}x + 4$

Using the *point-slope form,* $y - y_1 = m(x - x_1)$, fill in the given values and simplify the equation.

$$y-(-1)=-\frac{5}{3}(x-3)$$
$$y+1=-\frac{5}{3}x+5$$
$$y=-\frac{5}{3}x+4$$

In *standard form*, the equation is $5x + 3y = 12$

219. $y = -1$

Using the *point-slope form,* $y - y_1 = m(x - x_1)$, fill in the given values and simplify the equation.

$$y - (-1) = 0(x - 3)$$
$$y + 1 = 0$$
$$y = -1$$

220. $x = 3$

It's not appropriate to use the *point-slope form* when there is no slope. So just remember that vertical lines of the form $x = h$ are in this category. Because the line has to go through (3, –1), look at the *x*-value. $x = 3$

221. $y = 4x - 14$

First find the slope of the line by filling in the slope formula, $m = \frac{y_2 - y_1}{x_2 - x_1} = \frac{2-(-2)}{4-3} = \frac{4}{1} = 4$.

Now use the *point-slope* formula, inserting this slope and the coordinates of one of the points.

Using the second point:

$$y - 2 = 4(x - 4)$$

$$y - 2 = 4x - 16$$

$$y = 4x - 14$$

In *standard form*, the equation is written $4x - y = 14$.

222. $y = 3x + 16$

First find the slope of the line by filling in the slope formula, $m = \frac{y_2 - y_1}{x_2 - x_1} = \frac{7-1}{-3-(-5)} = \frac{6}{2} = 3$.

Now use the *point-slope* formula, inserting this slope and the coordinates of one of the points.

Using the second point:

$$y - 7 = 3(x - (-3))$$

$$y - 7 = 3x + 9$$

$$y = 3x + 16$$

In *standard form*, the equation is written $3x - y = -16$.

223. $y = -3x + 9$

First find the slope of the line by filling in the slope formula, $m = \frac{y_2 - y_1}{x_2 - x_1} = \frac{6-(-3)}{1-4} = \frac{9}{-3} = -3$.

Now use the *point-slope* formula, inserting this slope and the coordinates of one of the points.

Using the second point:

$$y - 6 = -3(x - 1)$$

$$y - 6 = -3x + 3$$

$$y = -3x + 9$$

In *standard form*, the equation is written $3x + y = 9$.

224. $y = \frac{7}{5}x - \frac{1}{5}$

First find the slope of the line by filling in the slope formula, $m = \frac{y_2 - y_1}{x_2 - x_1} = \frac{4-(-3)}{3-(-2)} = \frac{7}{5}$.

Now use the *point-slope* formula, inserting this slope and the coordinates of one of the points.

Using the second point:

$$y - 4 = \frac{7}{5}(x - 3)$$

$$y - 4 = \frac{7}{5}x - \frac{21}{5}$$

$$y = \frac{7}{5}x - \frac{1}{5}$$

In *standard form*, the equation is written $7x - 5y = 1$.

225. $y = -x + 11$

First find the slope of the line by filling in the slope formula, $m = \frac{y_2 - y_1}{x_2 - x_1} = \frac{12-6}{-1-5} = \frac{6}{-6} = -1$. Now use the *point-slope* formula, inserting this slope and the coordinates of one of the points.

Using the first point:

$$y - 6 = -1(x - 5)$$

$$y - 6 = -x + 5$$

$$y = -x + 11$$

In *standard form*, the equation is written $x + y = 11$.

226. $y = -\frac{9}{7}x + \frac{8}{7}$

First find the slope of the line by filling in the slope formula, $m = \frac{y_2 - y_1}{x_2 - x_1} = \frac{-4-5}{4-(-3)} = \frac{-9}{7} = -\frac{9}{7}$. Now use the *point-slope* formula, inserting this slope and the coordinates of one of the points. Using the first point:

$$y - 5 = -\frac{9}{7}(x + 3)$$

$$y - 5 = -\frac{9}{7}x - \frac{27}{7}$$

$$y = -\frac{9}{7}x + \frac{8}{7}$$

In *standard form*, the equation is written $9x + 7y = 8$.

227. $x = 6$

First find the slope of the line by filling in the slope formula, $m = \frac{y_2 - y_1}{x_2 - x_1} = \frac{-8-3}{6-6} = \frac{-11}{0}$. The slope doesn't exist. The line must be vertical, so its equation is $x = 6$.

228. $y = -2$

First find the slope of the line by filling in the slope formula, $m = \frac{y_2 - y_1}{x_2 - x_1} = \frac{-2-(-2)}{5-4} = \frac{0}{1} = 0$. The slope is 0, so this is a horizontal line, and its equation is $y = -2$.

229. $y = 3x - 2$

You want the same slope as that of the given line, so $m = 3$. Using this slope and the given point in the *point-slope* formula:

$$y - 1 = 3(x - 1)$$

$$y - 1 = 3x - 3$$

$$y = 3x - 2$$

In *standard form*, the equation is $3x - y = 2$.

230. $y = -\frac{4}{3}x + \frac{17}{3}$

You want the same slope as that of the given line, so $m = -\frac{4}{3}$. Using this slope and the given point in the *point-slope* formula:

$$y - 3 = -\frac{4}{3}(x - 2)$$

$$y - 3 = -\frac{4}{3}x + \frac{8}{3}$$

$$y = -\frac{4}{3}x + \frac{17}{3}$$

In standard form, the equation is $4x + 3y = 17$.

231. $y = -\frac{1}{3}x + 1$

You want the same slope as that of the given line, so first change the equation of the line to slope-intercept form.

$y = -\frac{1}{3}x + \frac{4}{3}$ and $m = -\frac{1}{3}$

Using this slope and the given point in the *point-slope* formula:

$$y - 0 = -\frac{1}{3}(x - 3)$$

$$y = -\frac{1}{3}x + 1$$

In *standard form,* the equation is $x + 3y = 3$.

232. $y = 2x + 5$

You want the same slope as that of the given line, so first change the equation of the line to slope-intercept form.

$y = 2x - \frac{7}{2}$ and $m = 2$

Using this slope and the given point in the *point-slope* formula:

$$y - 5 = 2(x - 0)$$

$$y = 2x + 5$$

In *standard form,* the equation is $2x - y = -5$.

233. $y = 8$

The line $y = 6$ is horizontal, so the slope is 0.

Using this slope and the given point in the *point-slope* formula:

$$y - 8 = 0(x - 4)$$

$$y - 8 = 0$$

$$y = 8$$

234. $x = 4$

The line $x = -3$ is vertical and has no slope. Another vertical line through (4, –7) has the equation $x = 4$.

235. $y = \frac{1}{2}x + 2$

The line $y = -2x + 1$ has slope $m = -2$, so a line perpendicular to this line has $m = \frac{1}{2}$.

Using this slope and the given point in the *point-slope* formula:

$$y - 3 = \frac{1}{2}(x - 2)$$

$$y - 3 = \frac{1}{2}x - 1$$

$$y = \frac{1}{2}x + 2$$

The standard equation of this line is $x - 2y = -4$.

236. $y = -\frac{8}{5}x + 9$

The line $y = \frac{5}{8}x - 2$ has slope $m = \frac{5}{8}$, so a line perpendicular to this line has $m = -\frac{8}{5}$.

Using this slope and the given point in the *point-slope* formula:

$$y - (-7) = -\frac{8}{5}(x - 10)$$

$$y + 7 = -\frac{8}{5}x + 16$$

$$y = -\frac{8}{5}x + 9$$

The standard equation of this line is $8x + 5y = 45$.

237. $y = -5x - 7$

To find the slope of $x - 5y = 10$, put it in slope-intercept form.

$y = \frac{1}{5}x - 2$ and $m = \frac{1}{5}$

A line perpendicular to this line has slope $m = -5$.

Using this slope and the given point in the *point-slope* formula:

$$y - 3 = -5(x + 2)$$

$$y - 3 = -5x - 10$$

$$y = -5x - 7$$

The standard equation of this line is $5x + y = -7$.

238. $y = \frac{2}{3}x + 5$

To find the slope of $3x + 2y = 4$, put it in slope-intercept form.

$y = -\frac{3}{2}x + 2$ and $m = -\frac{3}{2}$

A line perpendicular to this line has slope $m = \frac{2}{3}$.

Using this slope and the given point in the *point-slope* formula:

$$y - (-1) = \frac{2}{3}(x - (-9))$$
$$y + 1 = \frac{2}{3}x + 6$$
$$y = \frac{2}{3}x + 5$$

The standard equation of this line is $2x - 3y = -15$.

239. $y = 2$

The line $x = 4$ is vertical, so a line perpendicular to it will be horizontal with $m = 0$.

Using the *point-slope* formula:

$$y - 2 = 0(x - (-6))$$
$$y - 2 = 0$$
$$y = 2$$

240. $x = 5$

The line $y = -8$ is horizontal, so a line perpendicular to it will be vertical. The vertical line through $(5, -3)$ is $x = 5$.

241. **D: $(-\infty, \infty)$, R: $[5, \infty)$**

The input value, x, can be any real number, because it's being squared.

D: $(-\infty, \infty)$

The output value, $f(x)$, is the result of adding a non-negative number to 5, so it can't be less than 5.

R: $[5, \infty)$

242. **D: $(-\infty, \infty)$, R: $(-\infty, 7]$**

The input value, x, can be any real number, because it's being squared.

D: $(-\infty, \infty)$

The output value, $g(x)$, is the result of adding a negative number or 0 to 7, so it can't be greater than 7.

R: $(-\infty, 7]$

243. **D: $(-\infty, 3) \cup (3, \infty)$, R: $(-\infty, 1) \cup (1, \infty)$**

The input value, x, can be any real number except for $x = 3$, because you can't divide by 0.

D: $(-\infty, 3) \cup (3, \infty)$

The output value, $h(x)$, cannot be equal to 1, because the numerator and denominator can never be exactly the same.

R: $(-\infty, 1) \cup (1, \infty)$

244. **D: $(-\infty, -4) \cup (-4, \infty)$, R: $(-\infty, 2) \cup (2, \infty)$**

The input value, x, can be any real number except for $x = -4$, because you can't divide by 0.

D: $(-\infty, -4) \cup (-4, \infty)$

The output value, $k(x)$, cannot be equal to 2 because the ratio between the numerator and denominator is always smaller than 2 (up to slightly smaller than 2), or it is larger than 2 (down to just slightly larger than 2).

R: $(-\infty, 2) \cup (2, \infty)$

245. **D: $[3, \infty)$, R: $[0, \infty)$**

The input value, x, has to be 3 or greater so that either 0 or a positive number will be under the radical.

D: $[3, \infty)$

The output value, $f(x)$, will always be either 0 or positive.

R: $[0, \infty)$

246. **D: $(-\infty, 5]$, R: $[0, \infty)$**

The input value, x, has to be 5 or smaller so that either 0 or a positive number will be under the radical.

D: $(-\infty, 5]$

The output value, $g(x)$, will always be either 0 or positive.

R: $[0, \infty)$

247. **D: $[-4, 4]$, R: $[0, 4]$**

The input value, x, has to be between –4 and 4, including those two numbers, so that either 0 or a positive number will be under the radical.

D: $[-4, 4]$

The output value, $h(x)$, will always lie between 0 and 4, including 0 and 4.

R: $[0, 4]$

248. **D: $[-4, 1) \cup (1, \infty)$, R: $(-\infty, \infty)$**

The input value, x, has to be –4 or larger so that either 0 or a positive number will be under the radical. One exception is $x = 1$, because that would put a 0 in the denominator.

D: $[-4, 1) \cup (1, \infty)$

The output value, $k(x)$, goes from negative infinity to positive infinity. Even though the value stays just above 0 moving to the right, it is exactly 0 when $x = -4$.

R: $(-\infty, \infty)$

249. **D: $(-\infty, \infty)$, R: $[0, \infty)$**

The input value, x, can be any real number, because you can take the absolute value of all reals.

D: $(-\infty, \infty)$

The output value, $f(x)$, is the result of finding the absolute of any real number, so it will always be either 0 or a positive number.

R: $[0, \infty)$

250. **D: $(-\infty, \infty)$, R: $[8, \infty)$**

The input value, x, can be any real number, because you can take the absolute value of all reals.

D: $(-\infty, \infty)$

The output value, $f(x)$, is the result of finding the absolute of any real number, doubling it, and then adding 8; so $f(x)$ will always be either 8 or a number greater than 8.

R: $[8, \infty)$

251. $f^{-1}(x) = x + 3$

Rewrite the equation as $y = x - 3$.

Now change y to x and x to y.

$$x = y - 3$$

Solve for y.

$$y = x + 3$$

Rewrite using inverse function notation.

$$f^{-1}(x) = x + 3$$

252. $f^{-1}(x)=\frac{x-4}{3}$

Rewrite the equation as $y = 3x + 4$.

Now change y to x and x to y.

$$x = 3y + 4$$

Solve for y.

$$3y = x - 4$$

$$y=\frac{x-4}{3}$$

Rewrite using inverse function notation.

$$f^{-1}(x)=\frac{x-4}{3}$$

253. $f^{-1}(x)=\sqrt[3]{x-7}$

Rewrite the equation as $y = x^3 + 7$.

Now change y to x and x to y.

$$x = y^3 + 7$$

Solve for y.

$$y^3 = x - 7$$

$$y=\sqrt[3]{x-7}$$

Rewrite using inverse function notation.

$$f^{-1}(x)=\sqrt[3]{x-7}$$

254. $f^{-1}(x)=\sqrt[3]{\frac{x+5}{3}}$

Rewrite the equation as $y = 3x^3 - 5$.

Now change y to x and x to y.

$$x = 3y^3 - 5$$

Solve for y.

$$3y^3 = x + 5$$

$$y^3=\frac{x+5}{3}$$

$$y=\sqrt[3]{\frac{x+5}{3}}$$

Rewrite using inverse function notation.

$$f^{-1}(x)=\sqrt[3]{\frac{x+5}{3}}$$

255. $f^{-1}(x) = \sqrt[5]{2x+4}$

Rewrite the equation as $y = \frac{x^5 - 4}{2}$.

Now change y to x and x to y.

$$x = \frac{y^5 - 4}{2}$$

Solve for y.

$$2x = y^5 - 4$$

$$y^5 = 2x + 4$$

$$y = \sqrt[5]{2x+4}$$

Rewrite using inverse function notation.

$$f^{-1}(x) = \sqrt[5]{2x+4}$$

256. $f^{-1}(x) = \sqrt[3]{\frac{2+x}{x}}$

Rewrite the equation as $y = \frac{2}{x^3 - 1}$.

Now change y to x and x to y.

$$x = \frac{2}{y^3 - 1}$$

Solve for y.

$$x(y^3 - 1) = 2$$

$$y^3 - 1 = \frac{2}{x}$$

$$y^3 = \frac{2}{x} + 1 = \frac{2+x}{x}$$

$$y = \sqrt[3]{\frac{2+x}{x}}$$

Rewrite using inverse function notation.

$$f^{-1}(x) = \sqrt[3]{\frac{2+x}{x}}$$

257. $f^{-1}(x) = \log_2(x)$

Rewrite the equation as $y = 2^x$.

Now change y to x and x to y.

$$x = 2^y$$

Solve for y. To do this, you need to take a log of both sides. Use $\log_2$.

$$\log_2(x) = \log_2(2^y)$$

Apply the laws of logarithms.

$\log_2(x) = y\log_2(2)$, using the log of a power.

$\log_2(x) = y(1)$, using the log of the base.

$$y = \log_2(x)$$

Rewrite using inverse function notation.

$$f^{-1}(x) = \log_2(x)$$

258. $f^{-1}(x) = \ln(x)$

Rewrite the equation as $y = e^x$.

Now change y to x and x to y.

$$x = e^y$$

Solve for y. To do this, you need to take a log of both sides. Use $\log_e = \ln$.

$$\ln(x) = \ln(e^y)$$

Apply the laws of logarithms.

$\ln(x) = y\ln(e)$, using the log of a power.

$\ln(x) = y(1)$, using the log of the base.

$$y = \ln(x)$$

Rewrite using inverse function notation.

$$f^{-1}(x) = \ln(x)$$

259. $f^{-1}(x) = \dfrac{3-x}{2x-1}$

Rewrite the equation as $y = \dfrac{x+3}{2x+1}$.

Now change y to x and both x's to y's.

$$x = \frac{y+3}{2y+1}$$

Solve for y.

$$x(2y+1) = y+3$$

$$2xy + x = y + 3$$

$$2xy - y = 3 - x$$

$$y(2x-1) = 3 - x$$

$$y = \frac{3-x}{2x-1}$$

Rewrite using inverse function notation.

$$f^{-1}(x) = \frac{3-x}{2x-1}$$

260. $f^{-1}(x)=\frac{x}{x-1}$

Rewrite the equation as $y=\frac{x}{x-1}$.

Now change y to x and both x's to y's.

$$x=\frac{y}{y-1}$$

Solve for y.

$$x(y-1)=y$$

$$xy-x=y$$

$$xy-y=x$$

$$y(x-1)=x$$

$$y=\frac{x}{x-1}$$

Rewrite using inverse function notation.

$$f^{-1}(x)=\frac{x}{x-1}$$

The function is its own inverse.

261. **Odd and one-to-one**

The function rule has an odd exponent on the variable and is therefore an odd function.

The function has an inverse that is also a function, $f^{-1}(x)=\sqrt[3]{x}$, so it's one-to-one.

262. **Even only**

The function rule has two even exponents on the variables; the constant 2 represents $2x^0$ and is therefore an even function.

263. **One-to-one only**

The function has one odd and one even exponent on the variables; the constant 5 represents $5x^0$. So it's neither even nor odd.

The function has an inverse that is also a function, $h^{-1}(x)=\sqrt[3]{x-5}$, so it's one-to-one.

264. **Even only**

The function rule has three even exponents on the variables; the constant 3 represents $3x^0$ and is therefore an even function. The function has an inverse that can be found using the quadratic formula; the inverse involves taking two square roots with ±. The function has an inverse, but the inverse is not a function.

265. **One-to-one only**

The function has one odd and one even exponent on the variables; the constant 1 represents $1x^0$. So it's neither even nor odd.

The function has an inverse that is also a function, $f^{-1}(x)=x^3-1$, so it's one-to-one.

266. **One-to-one only**

The function has two odd and two even exponents on the variables; the constants have x^0 multipliers. So it's neither even nor odd.

The function has an inverse that is also a function, $g^{-1}(x)=\frac{2x+3}{x-1}$, so it's one-to-one.

267. **Odd and one-to-one**

The function rule has an odd exponent; think of it as being written x^{-1}. The function has an inverse, $h^{-1}(x)=\frac{1}{x}$; as you see, the function is its own inverse, so it's one-to-one.

268. **Even only**

The function rule has two even exponents on the variables; the constant 9 represents $9x^0$, so it's an even function.

The inverse $k^{-1}(x)=\pm\sqrt{9-x^2}$ is not a function, so k is not one-to-one.

269. **Odd and one-to-one**

The function rule has an odd exponent, so it's an odd function. The function has an inverse that is also a function, $f^{-1}(x)=\sqrt[5]{3x}$, so it's one-to-one.

270. **Even only**

The function rule has an even exponent on the variable; the constant 6 represents $6x^0$, so it's an even function. And since $g(x)$ is a horizontal line, its inverse would have to be a vertical line, which is not a function. So g is not one-to-one.

271. $(f\circ g)(x)=x^2+2x+4$

The composition $(f\circ g)(x)$ means to find $f(g(x))$.

Applying the function rule for f to the input $g(x)$,

$$
\begin{aligned}
f(g(x)) &= (g(x))^2+3 \\
f(x+1) &= (x+1)^2+3 \\
&= x^2+2x+1+3 \\
&= x^2+2x+4
\end{aligned}
$$

272. $(f \circ g)(x) = 2x^2 - 3x - 2$

The composition $(f \circ g)(x)$ means to find $f(g(x))$.

Applying the function rule for f to the input $g(x)$,

$$\begin{aligned} f(g(x)) &= 2(g(x))^2 + 5(g(x)) \\ f(x-2) &= 2(x-2)^2 + 5(x-2) \\ &= 2x^2 - 8x + 8 + 5x - 10 \\ &= 2x^2 - 3x - 2 \end{aligned}$$

273. $(f \circ g)(x) = -x^2 + 5$

The composition $(f \circ g)(x)$ means to find $f(g(x))$.

Applying the function rule for f to the input $g(x)$,

$$\begin{aligned} f(g(x)) &= -(g(x)) + 3 \\ f(x^2 - 2) &= -(x^2 - 2) + 3 \\ &= -x^2 + 2 + 3 \\ &= -x^2 + 5 \end{aligned}$$

274. $(f \circ g)(x) = -5x^2 + 30x - 41$

The composition $(f \circ g)(x)$ means to find $f(g(x))$.

Applying the function rule for f to the input $g(x)$,

$$\begin{aligned} f(g(x)) &= 4 - 5(g(x))^2 \\ f(3-x) &= 4 - 5(3-x)^2 \\ &= 4 - 45 + 30x - 5x^2 \\ &= -5x^2 + 30x - 41 \end{aligned}$$

275. $(f \circ g)(x) = \frac{6-x}{x-3}$

The composition $(f \circ g)(x)$ means to find $f(g(x))$.

Applying the function rule for f to the input $g(x)$,

$$\begin{aligned} f(g(x)) &= \frac{1-g(x)}{2+g(x)} \\ f(x-5) &= \frac{1-(x-5)}{2+(x-5)} \\ &= \frac{1-x+5}{2+x-5} \\ &= \frac{6-x}{x-3} \end{aligned}$$

276. $(f \circ g)(x) = \dfrac{x^2+8x+7}{x^2+8x+15}$

The composition $(f \circ g)(x)$ means to find $f(g(x))$.

Applying the function rule for *f* to the input *g*(*x*),

$$\begin{aligned} f(g(x)) &= \frac{(g(x))^2-9}{(g(x))^2-1} \\ f(x+4) &= \frac{(x+4)^2-9}{(x+4)^2-1} \\ &= \frac{x^2+8x+16-9}{x^2+8x+16-1} \\ &= \frac{x^2+8x+7}{x^2+8x+15} \end{aligned}$$

277. $(f \circ g)(x) = \sqrt{x^2-3}$

The composition $(f \circ g)(x)$ means to find $f(g(x))$.

Applying the function rule for *f* to the input *g*(*x*),

$$\begin{aligned} f(g(x)) &= \sqrt{g(x)-4} \\ f(x^2+1) &= \sqrt{x^2+1-4} \\ &= \sqrt{x^2-3} \end{aligned}$$

278. $(f \circ g)(x) = \sqrt{5+4x-x^2}$

The composition $(f \circ g)(x)$ means to find $f(g(x))$.

Applying the function rule for *f* to the input *g*(*x*),

$$\begin{aligned} f(g(x)) &= \sqrt{9-(g(x))^2} \\ f(x-2) &= \sqrt{9-(x-2)^2} \\ &= \sqrt{9-(x^2-4x+4)} \\ &= \sqrt{9-x^2+4x-4} \\ &= \sqrt{5+4x-x^2} \end{aligned}$$

279. $(f \circ g)(x) = x^3+3x^2+5x+4$

The composition $(f \circ g)(x)$ means to find $f(g(x))$.

Applying the function rule for *f* to the input *g*(*x*),

$$\begin{aligned} f(g(x)) &= (g(x))^3+2(g(x))+1 \\ f(x+1) &= (x+1)^3+2(x+1)+1 \\ &= x^3+3x^2+3x+1+2x+2+1 \\ &= x^3+3x^2+5x+4 \end{aligned}$$

280. $(f \circ g)(x) = x^3 + 6x^2 + 12x$

The composition $(f \circ g)(x)$ means to find $f(g(x))$.

Applying the function rule for f to the input $g(x)$,

$$\begin{aligned} f(g(x)) &= (g(x))^3 - 8 \\ f(x+2) &= (x+2)^3 - 8 \\ &= x^3 + 6x^2 + 12x + 8 - 8 \\ &= x^3 + 6x^2 + 12x \end{aligned}$$

281. 1

To find the difference quotient, you replace each variable x in the function rule with the input $(x + h)$, subtract the original function, and then divide by h.

$$\begin{aligned} \frac{f(x+h)-f(x)}{h} &= \frac{(x+h)+3-(x+3)}{h} \\ &= \frac{x+h+3-x-3}{h} \\ &= \frac{h}{h} = 1 \end{aligned}$$

282. 4

To find the difference quotient, you replace each variable x in the function rule with the input $(x + h)$, subtract the original function, and then divide by h.

$$\begin{aligned} \frac{f(x+h)-f(x)}{h} &= \frac{4(x+h)-2-(4x-2)}{h} \\ &= \frac{4x+4h-2-4x+2}{h} \\ &= \frac{4h}{h} = 4 \end{aligned}$$

283. $2x + h - 2$

To find the difference quotient, you replace each variable x in the function rule with the input $(x + h)$, subtract the original function, and then divide by h.

$$\begin{aligned} &\frac{f(x+h)-f(x)}{h} \\ &= \frac{(x+h)^2 - 2(x+h)+3-\left(x^2-2x+3\right)}{h} \\ &= \frac{x^2+2xh+h^2-2x-2h+3-x^2+2x-3}{h} \\ &= \frac{2xh+h^2-2h}{h} \end{aligned}$$

Notice that the only terms remaining in the numerator are those with a factor of h. The other terms were all added to their opposite with a sum of 0.

$$= \frac{h(2x+h-2)}{h} = 2x+h-2$$

284. $\mathbf{10x + 5h + 4}$

To find the difference quotient, you replace each variable x in the function rule with the input $(x + h)$, subtract the original function, and then divide by h.

$$\frac{f(x+h)-f(x)}{h}$$

$$=\frac{5(x+h)^2+4(x+h)-5-(5x^2+4x-5)}{h}$$

$$=\frac{5x^2+10xh+5h^2+4x+4h-5-5x^2-4x+5}{h}$$

$$=\frac{10xh+5h^2+4h}{h}$$

Notice that the only terms remaining in the numerator are those with a factor of h. The other terms were all added to their opposite with a sum of 0.

$$=\frac{h(10x+5h+4)}{h}=10x+5h+4$$

285. $\mathbf{3x^2 + 3xh + h^2 - 3}$

To find the difference quotient, you replace each variable x in the function rule with the input $(x + h)$, subtract the original function, and then divide by h.

$$\frac{f(x+h)-f(x)}{h}$$

$$=\frac{(x+h)^3-3(x+h)+4-(x^3-3x+4)}{h}$$

$$=\frac{x^3+3x^2h+3xh^2+h^3-3x-3h+4-x^3+3x-4}{h}$$

$$=\frac{3x^2h+3xh^2+h^3-3h}{h}$$

Notice that the only terms remaining in the numerator are those with a factor of h. The other terms were all added to their opposite with a sum of 0.

$$=\frac{h(3x^2+3xh+h^2-3)}{h}=3x^2+3xh+h^2-3$$

286. $\mathbf{6x^2 + 6xh + 2h^2 + 8x + 4h}$

To find the difference quotient, you replace each variable x in the function rule with the input $(x + h)$, subtract the original function, and then divide by h.

$$\frac{f(x+h)-f(x)}{h}$$

$$=\frac{2(x+h)^3+4(x+h)^2-(2x^3+4x^2)}{h}$$

$$=\frac{2x^3+6x^2h+6xh^2+2h^3+4x^2+8xh+4h^2-2x^3-4x^2}{h}$$

$$=\frac{6x^2h+6xh^2+2h^3+8xh+4h^2}{h}$$

Notice that the only terms remaining in the numerator are those with a factor of h. The other terms were all added to their opposite with a sum of 0.

$$=\frac{h\left(6x^2+6xh+2h^2+8x+4h\right)}{h}$$
$$=6x^2+6xh+2h^2+8x+4h$$

287. $\dfrac{-1}{(x+h+3)(x+3)}$

To find the difference quotient, you replace each variable x in the function rule with the input $(x+h)$, subtract the original function, and then divide by h.

$$\frac{f(x+h)-f(x)}{h}=\frac{\frac{1}{(x+h)+3}-\frac{1}{x+3}}{h}$$

To simplify the complex fraction, find a common denominator for the two fractions and subtract.

$$=\frac{\frac{1}{x+h+3}\cdot\frac{x+3}{x+3}-\frac{1}{x+3}\cdot\frac{x+h+3}{x+h+3}}{h}$$
$$=\frac{\frac{x+3}{(x+h+3)(x+3)}-\frac{x+h+3}{(x+h+3)(x+3)}}{h}$$
$$=\frac{\frac{x+3-(x+h+3)}{(x+h+3)(x+3)}}{h}=\frac{\frac{x+3-x-h-3}{(x+h+3)(x+3)}}{h}$$
$$=\frac{\frac{-h}{(x+h+3)(x+3)}}{h}$$

Now multiply the numerator times the reciprocal of the denominator.

$$=\frac{-\cancel{h}}{(x+h+3)(x+3)}\cdot\frac{1}{\cancel{h}}$$
$$=\frac{-1}{(x+h+3)(x+3)}$$

288. $\dfrac{-3}{(x+h-2)(x-2)}$

To find the difference quotient, you replace each variable x in the function rule with the input $(x+h)$, subtract the original function, and then divide by h.

$$\frac{f(x+h)-f(x)}{h}=\frac{\frac{3}{(x+h)-2}-\frac{3}{x-2}}{h}$$

To simplify the complex fraction, find a common denominator for the two fractions and subtract.

$$=\frac{\frac{3}{x+h-2}\cdot\frac{x-2}{x-2}-\frac{3}{x-2}\cdot\frac{x+h-2}{x+h-2}}{h}$$

$$=\frac{\frac{3x-6}{(x+h-2)(x-2)}-\frac{3x+3h-6}{(x+h-2)(x-2)}}{h}$$

$$=\frac{\frac{3x-6-(3x+3h-6)}{(x+h-2)(x-2)}}{h}=\frac{\frac{3x-6-3x-3h+6}{(x+h-2)(x-2)}}{h}$$

$$=\frac{\frac{-3h}{(x+h-2)(x-2)}}{h}$$

Now multiply the numerator times the reciprocal of the denominator.

$$=\frac{-3\cancel{h}}{(x+h-2)(x-2)}\cdot\frac{1}{\cancel{h}}=\frac{-3}{(x+h-2)(x-2)}$$

289. $\mathbf{\frac{1}{\sqrt{x+h+4}+\sqrt{x+4}}}$

To find the difference quotient, you replace each variable x in the function rule with the input $(x + h)$, subtract the original function, and then divide by h.

$$\frac{f(x+h)-f(x)}{h}=\frac{\sqrt{(x+h)+4}-\sqrt{x+4}}{h}$$

To simplify the numerator, multiply both the numerator and denominator by the conjugate of the numerator.

$$=\frac{\sqrt{x+h+4}-\sqrt{x+4}}{h}\cdot\frac{\sqrt{x+h+4}+\sqrt{x+4}}{\sqrt{x+h+4}+\sqrt{x+4}}$$

$$=\frac{\left(\sqrt{x+h+4}-\sqrt{x+4}\right)\left(\sqrt{x+h+4}+\sqrt{x+4}\right)}{h\left(\sqrt{x+h+4}+\sqrt{x+4}\right)}$$

The two binomials in the numerator are the difference and sum of the same two expressions. Their product will be the first term squared minus the last term squared.

$$=\frac{\left(\sqrt{x+h+4}\right)^2-\left(\sqrt{x+4}\right)^2}{h\left(\sqrt{x+h+4}+\sqrt{x+4}\right)}$$

$$=\frac{(x+h+4)-(x+4)}{h\left(\sqrt{x+h+4}+\sqrt{x+4}\right)}=\frac{x+h+4-x-4}{h\left(\sqrt{x+h+4}+\sqrt{x+4}\right)}$$

$$=\frac{h}{h\left(\sqrt{x+h+4}+\sqrt{x+4}\right)}$$

Now just reduce the fraction.

$$=\frac{1}{\sqrt{x+h+4}+\sqrt{x+4}}$$

290. $\frac{-1}{\sqrt{1-x-h}+\sqrt{1-x}}$

To find the difference quotient, you replace each variable x in the function rule with the input $(x + h)$, subtract the original function, and then divide by h.

$$\frac{f(x+h)-f(x)}{h}=\frac{\sqrt{1-(x+h)}-\sqrt{1-x}}{h}$$

To simplify the numerator, multiply both the numerator and denominator by the conjugate of the numerator.

$$=\frac{\sqrt{1-(x+h)}-\sqrt{1-x}}{h}\cdot\frac{\sqrt{1-(x+h)}+\sqrt{1-x}}{\sqrt{1-(x+h)}+\sqrt{1-x}}$$

$$=\frac{\left(\sqrt{1-(x+h)}-\sqrt{1-x}\right)\left(\sqrt{1-(x+h)}+\sqrt{1-x}\right)}{h\left(\sqrt{1-(x+h)}+\sqrt{1-x}\right)}$$

The two binomials in the numerator are the difference and sum of the same two expressions. Their product will be the first term squared minus the last term squared.

$$=\frac{\left(\sqrt{1-(x+h)}\right)^2-\left(\sqrt{1-x}\right)^2}{h\left(\sqrt{1-(x+h)}+\sqrt{1-x}\right)}$$

$$=\frac{1-(x+h)-(1-x)}{h\left(\sqrt{1-(x+h)}+\sqrt{1-x}\right)}$$

$$=\frac{1-x-h-1+x}{h\left(\sqrt{1-x-h}+\sqrt{1-x}\right)}$$

$$=\frac{-h}{h\left(\sqrt{1-x-h}+\sqrt{1-x}\right)}$$

Now just reduce the fraction.

$$=\frac{-1}{\sqrt{1-x-h}+\sqrt{1-x}}$$

291. 4

To evaluate $f(1)$, you choose the part of the function rule that applies to an input of $x = 1$. Because $1 \geq 0$, use $f(x) = x + 3$, giving you $f(1) = 1 + 3 = 4$.

292. –5

To evaluate $f(-1)$, you choose the part of the function rule that applies to an input of $x = -1$. Because $-1 \geq -1$, use $f(x) = x - 4$, giving you $f(-1) = -1 - 4 = -5$.

293. 8

To evaluate $f(2)$, you choose the part of the function rule that applies to an input of $x = 2$. Because $2 \leq 3$, use $f(x) = 4x$, giving you $f(2) = 4(2) = 8$.

294. **3**

To evaluate $f(0)$, you choose the part of the function rule that applies to an input of $x = 0$. Because $0 \le 0$, use $f(x) = 3 - x$, giving you $f(0) = 3 - 0 = 3$.

295. **-2**

To evaluate $f(-1)$, you choose the part of the function rule that applies to an input of $x = -1$. Because $-1 > -3$, use $f(x) = x - 1$, giving you $f(-1) = -1 - 1 = -2$.

296. **9**

To evaluate $f(-4)$, you choose the part of the function rule that applies to an input of $x = -4$. Because $-4 \ge -5$, use $f(x) = 5 - x$, giving you $f(-4) = 5 - (-4) = 9$.

297. **-25**

To evaluate $f(-5)$, you choose the part of the function rule that applies to an input of $x = -5$. Because -5 is an odd number, use $f(x) = -x^2$, giving you $f(-5) = -(-5)^2 = -25$.

298. **-51**

To evaluate $f(51)$, you choose the part of the function rule that applies to an input of $x = 51$. Because 51 is a composite number, use $f(x) = -x$, giving you $f(51) = -51$.

299. **1**

To evaluate $f(0)$, you choose the part of the function rule that applies to an input of $x = 0$. Because 0 falls within the bounds of $0 \le x < 9$, use $f(x) = x!$, giving you $f(0) = 0! = 1$.

Recall that, by special definition, $0! = 1$.

300. **1**

To evaluate $f(4)$, you choose the part of the function rule that applies to an input of $x = 4$. Because $4 = 4$, use $f(x) = 1$, giving you $f(4) = 1$.

301. ***x*-intercepts: (1, 0), (–1, 0); *y*-intercept: (0, –1); vertex: (0, –1)**

To find the *y*-intercept, set $x = 0$.

$$y = 0^2 - 1 = -1$$

So the *y*-intercept is $(0, -1)$.

To find the *x*-intercepts, set $y = 0$ and solve for x.

$$0 = x^2 - 1$$

$$0 = (x - 1)(x + 1)$$

$x = 1$ or -1, so the *x*-intercepts are $(1, 0)$ and $(-1, 0)$.

Using the formula for the x-coordinate of the vertex, $x = -\frac{b}{2a} = -\frac{0}{2(1)} = 0$.

And, when $x = 0$, $y = 0^2 - 1 = -1$, so the vertex is at $(0, -1)$.

302. ***x*-intercepts: (3, 0), (–3, 0); *y*-intercept: (0, 9); vertex: (0, 9)**

To find the y-intercept, set $x = 0$.

$$y = 9 - 0^2 = 9$$

So the y-intercept is $(0, 9)$.

To find the x-intercepts, set $y = 0$ and solve for x.

$$0 = 9 - x^2$$

$$0 = (3 - x)(3 + x)$$

$x = 3$ or -3, so the x-intercepts are $(3, 0)$ and $(-3, 0)$.

Rewrite the equation as $y = -x^2 + 9$, and then use the formula for the x-coordinate of the vertex, $x = -\frac{b}{2a} = -\frac{0}{2(-1)} = 0$.

And, when $x = 0$, $y = 9 - 0^2 = 9$, so the vertex is at $(0, 9)$.

303. ***x*-intercepts: (0, 0), (10, 0); *y*-intercept: (0, 0); vertex: (5, 25)**

To find the y-intercept, set $x = 0$.

$$y = 10(0) - 0^2 = 0$$

So the y-intercept is $(0, 0)$.

To find the x-intercepts, set $y = 0$ and solve for x.

$$0 = 10x - x^2$$

$$0 = x(10 - x)$$

$x = 0$ or 10, so the x-intercepts are $(0, 0)$ and $(10, 0)$.

Rewrite the equation as $y = -x^2 + 10x$, and then use the formula for the x-coordinate of the vertex, $x = -\frac{b}{2a} = -\frac{10}{2(-1)} = 5$.

And, when $x = 5$, $y = 10(5) - (5)^2 = 25$, so the vertex is at $(5, 25)$.

304. ***x*-intercepts: (5, 0), (1, 0); *y*-intercept: (0, 5); vertex: (3, –4)**

To find the y-intercept, set $x = 0$.

$$y = 0^2 - 6(0) + 5 = 5$$

So the y-intercept is $(0, 5)$.

To find the x-intercepts, set $y = 0$ and solve for x.

$$0 = x^2 - 6x + 5$$

$$0 = (x - 5)(x - 1)$$

$x = 5$ or 1, so the x-intercepts are $(5, 0)$ and $(1, 0)$.

Using the formula for the x-coordinate of the vertex, $x = -\frac{b}{2a} = -\frac{-6}{2(1)} = 3$.
And, when $x = 3$, $y = (3)^2 - 6(3) + 5 = -4$, so the vertex is at (3, –4).

305. **x-intercepts: (4, 0), (–8, 0); y-intercept: (0, –32); vertex: (–2, –36)**

To find the y-intercept, set $x = 0$.

$$y = 0^2 + 4(0) - 32 = -32$$

So the y-intercept is (0, –32).

To find the x-intercepts, set $y = 0$ and solve for x.

$$0 = x^2 + 4x - 32$$

$$0 = (x - 4)(x + 8)$$

$x = 4$ or -8, so the x-intercepts are (4, 0) and (–8, 0).

Using the formula for the x-coordinate of the vertex, $x = -\frac{b}{2a} = -\frac{4}{2(1)} = -2$.

And, when $x = -2$, $y = (-2)^2 + 4(-2) - 32 = -36$, so the vertex is at (–2, –36).

306. **x-intercept: (3, 0); y-intercept: (0, 9); vertex: (3, 0)**

To find the y-intercept, set $x = 0$.

$$y = 0^2 - 6(0) + 9 = 9$$

So the y-intercept is (0, 9).

To find the x-intercepts, set $y = 0$ and solve for x.

$$0 = x^2 - 6x + 9$$

$$0 = (x - 3)(x - 3) = (x - 3)^2$$

$x = 3$, so the only x-intercept is (3, 0).

Using the formula for the x-coordinate of the vertex, $x = -\frac{b}{2a} = -\frac{-6}{2(1)} = 3$.

And, when $x = 3$, $y = (3)^2 - 6(3) + 9 = 0$, so the vertex is at (3, 0).

307. **No x-intercept; y-intercept: (0, 7); vertex: (0, 7)**

To find the y-intercept, set $x = 0$.

$$y = 0^2 + 7 = 7$$

So the y-intercept is (0, 7).

To find the x-intercepts, set $y = 0$ and solve for x.

$$0 = x^2 + 7$$

$$x^2 = -7$$

There are no real roots, so there are no x-intercepts.

Using the formula for the x-coordinate of the vertex, $x = -\frac{b}{2a} = -\frac{0}{2(1)} = 0$.

And, when $x = 0$, $y = (0)^2 + 7 = 7$, so the vertex is at (0, 7).

308. **x-intercepts: $\left(-2+\sqrt{10}, 0\right), \left(-2-\sqrt{10}, 0\right)$; y-intercept: (0, –6); vertex: (–2, –10)**

To find the y-intercept, set $x = 0$.

$$y = 0^2 + 4(0) - 6 = -6$$

So the y-intercept is (0, –6).

To find the x-intercepts, set $y = 0$ and solve for x.

$$0 = x^2 + 4x - 6$$

The trinomial doesn't factor, so use the quadratic formula.

$$x = \frac{-4 \pm \sqrt{4^2 - 4(1)(-6)}}{2(1)}$$

$$= \frac{-4 \pm \sqrt{40}}{2} = \frac{-4 \pm 2\sqrt{10}}{2} = -2 \pm \sqrt{10}$$

So the x-intercepts are $\left(-2+\sqrt{10}, 0\right)$ and $\left(-2-\sqrt{10}, 0\right)$.

Using the formula for the x-coordinate of the vertex, $x = -\frac{b}{2a} = -\frac{4}{2(1)} = -2$.

And, when $x = -2$, $y = (-2)^2 + 4(-2) - 6 = -10$, so the vertex is at (–2, –10).

309. **x-intercepts: $\left(\frac{1}{2}, 0\right)$, (–3, 0); y-intercept: (0, –3); vertex: $\left(-\frac{5}{4}, -\frac{49}{8}\right)$**

To find the y-intercept, set $x = 0$.

$$y = 2(0)^2 + 5(0) - 3 = -3$$

So the y-intercept is (0, –3).

To find the x-intercepts, set $y = 0$ and solve for x.

$$0 = 2x^2 + 5x - 3$$

$$0 = (2x - 1)(x + 3)$$

$x = \frac{1}{2}$ or $x = -3$, so the x-intercepts are $\left(\frac{1}{2}, 0\right)$ and (–3, 0).

Using the formula for the x-coordinate of the vertex, $x = -\frac{b}{2a} = -\frac{5}{2(2)} = -\frac{5}{4}$.

And, when $x = -\frac{5}{4}$, $y = 2\left(-\frac{5}{4}\right)^2 + 5\left(-\frac{5}{4}\right) - 3 = -\frac{49}{8}$, so the vertex is at $\left(-\frac{5}{4}, -\frac{49}{8}\right)$.

310. **x-intercepts: $\left(\frac{4}{3}, 0\right)$, (–2, 0); y-intercept: (0, –8); vertex: $\left(-\frac{1}{3}, -\frac{25}{3}\right)$**

To find the y-intercept, set $x = 0$.

$$y = 3(0)^2 + 2(0) - 8 = -8$$

So the y-intercept is (0, –8).

To find the x-intercepts, set $y = 0$ and solve for x.

$$0 = 3x^2 + 2x - 8$$

$$0 = (3x - 4)(x + 2)$$

$x = \frac{4}{3}$ or $x = -2$, so the x-intercepts are $\left(\frac{4}{3}, 0\right)$ and $(-2, 0)$.

Using the formula for the x-coordinate of the vertex, $x = -\frac{b}{2a} = -\frac{2}{2(3)} = -\frac{1}{3}$.

And, when $x = -\frac{1}{3}$, $y = 3\left(-\frac{1}{3}\right)^2 + 2\left(-\frac{1}{3}\right) - 8 = -\frac{25}{3}$, so the vertex is at $\left(-\frac{1}{3}, -\frac{25}{3}\right)$.

311. **(–2, –5)**

First, add 1 to both sides of the equation.

$$y + 1 = x^2 + 4x$$

Now *complete the square* on the right by adding 4 to both sides.

$$y + 5 = x^2 + 4x + 4$$

The right side can be written as a perfect square.

$$y + 5 = (x + 2)^2$$

The vertex is (–2, –5).

312. **(3, –16)**

First, add 7 to both sides of the equation.

$$y + 7 = x^2 - 6x$$

Now *complete the square* on the right by adding 9 to both sides.

$$y + 16 = x^2 - 6x + 9$$

The right side can be written as a perfect square.

$$y + 16 = (x - 3)^2$$

The vertex is (3, –16).

313. **(3, –15)**

First, add –3 to both sides of the equation.

$$y - 3 = 2x^2 - 12x$$

Next, factor the 2 from the terms on the right.

$$y - 3 = 2(x^2 - 6x)$$

Now *complete the square* in the parentheses by adding 9 in the parentheses. You add 18 to the other side, because that 9 is being multiplied by 2; in effect, you're adding 18 to both sides.

$$y + 15 = 2(x^2 - 6x + 9)$$

The right side can be written as a perfect square.

$$y + 15 = 2(x - 3)^2$$

The vertex is (3, –15).

314. **(–4, 0)**

The original equation is already a perfect square trinomial.

The right side can be written as a perfect square.

$$y = (x + 4)^2$$

The vertex is (–4, 0). Think of the standard form as being $y - 0 = (x + 4)^2$ in this case.

315. **(5, –79)**

First, add 4 to both sides of the equation.

$$y + 4 = 3x^2 - 30x$$

Next, factor the 3 from the terms on the right.

$$y + 4 = 3(x^2 - 10x)$$

Now *complete the square* in the parentheses by adding 25 in the parentheses. You add 75 to the other side, because that 25 is being multiplied by 3; in effect, you're adding 75 to both sides.

$$y + 79 = 3(x^2 - 10x + 25)$$

The right side in the parentheses can be written as a perfect square.

$$y + 79 = 3(x - 5)^2$$

The vertex is (5, –79).

316. $\left(\frac{1}{2}, -\frac{11}{4}\right)$

First, add 3 to both sides of the equation.

$$y + 3 = -x^2 + x$$

Next, factor the –1 from the terms on the right.

$$y + 3 = -1(x^2 - x)$$

Now *complete the square* in the parentheses by adding $\frac{1}{4}$ in the parentheses. You add $-\frac{1}{4}$ to the other side, because that $\frac{1}{4}$ is being multiplied by –1; in effect, you're adding $-\frac{1}{4}$ to both sides.

$$y + \frac{11}{4} = -\left(x^2 - x + \frac{1}{4}\right)$$

The right side in the parentheses can be written as a perfect square.

$$y + \frac{11}{4} = -\left(x - \frac{1}{2}\right)^2$$

The vertex is $\left(\frac{1}{2}, -\frac{11}{4}\right)$.

317. **(6, –36)**

The equation is ready for you to *complete the square* on the right by adding 36 to both sides.

$$y + 36 = x^2 - 12x + 36$$

The right side can be written as a perfect square.

$$y + 36 = (x - 6)^2$$

The vertex is (6, –36).

318. $\left(\frac{3}{4}, \frac{25}{8}\right)$

First, add –2 to both sides of the equation.

$$y - 2 = -2x^2 + 3x$$

Next, factor the –2 from the terms on the right.

$$y - 2 = -2\left(x^2 - \frac{3}{2}x\right)$$

Now *complete the square* in the parentheses by adding $\frac{9}{16}$ in the parentheses. You add $-\frac{9}{8}$ to the other side, because that $\frac{9}{16}$ is being multiplied by –2; in effect, you're adding $-\frac{9}{8}$ to both sides.

$$y - \frac{25}{8} = -2\left(x^2 - \frac{3}{2}x + \frac{9}{16}\right)$$

The right side in the parentheses can be written as a perfect square.

$$y - \frac{25}{8} = -2\left(x - \frac{3}{4}\right)^2$$

The vertex is $\left(\frac{3}{4}, \frac{25}{8}\right)$.

319. **(3, 10)**

First, add 80 to both sides of the equation.

$$y + 80 = -10x^2 + 60x$$

Next, factor the –10 from the terms on the right.

$$y + 80 = -10(x^2 - 6x)$$

Now *complete the square* in the parentheses by adding 9 in the parentheses. You add –90 to the other side, because that 9 is being multiplied by –10; in effect, you're adding –90 to both sides.

$$y - 10 = -10(x^2 - 6x + 9)$$

The right side in the parentheses can be written as a perfect square.

$$y - 10 = -10(x - 3)^2$$

The vertex is (3, 10).

320. $\left(-\frac{1}{3}, \frac{7}{3}\right)$

First, add –3 to both sides of the equation.

$$y - 3 = 6x^2 + 4x$$

Next, factor the 6 from the terms on the right.

$$y - 3 = 6\left(x^2 + \frac{2}{3}x\right)$$

Now *complete the square* in the parentheses by adding $\frac{1}{9}$ in the parentheses. You add $\frac{2}{3}$ to the other side, because that $\frac{1}{9}$ is being multiplied by 6; in effect, you're adding $\frac{2}{3}$ to both sides.

$$y - \frac{7}{3} = 6\left(x^2 + \frac{2}{3}x + \frac{1}{9}\right)$$

The right side in the parentheses can be written as a perfect square.

$$y - \frac{7}{3} = 6\left(x + \frac{1}{3}\right)^2$$

The vertex is $\left(-\frac{1}{3}, \frac{7}{3}\right)$.

321. **Opens upward; through (3, –7)**

The *y*-intercept is (0, –16), and the *x*-intercepts are (4, 0) and (–4, 0). The vertex is at (0, –16). The parabola opens upward. To help with the graphing, choose another point in the parabola — for example, (3, –7).

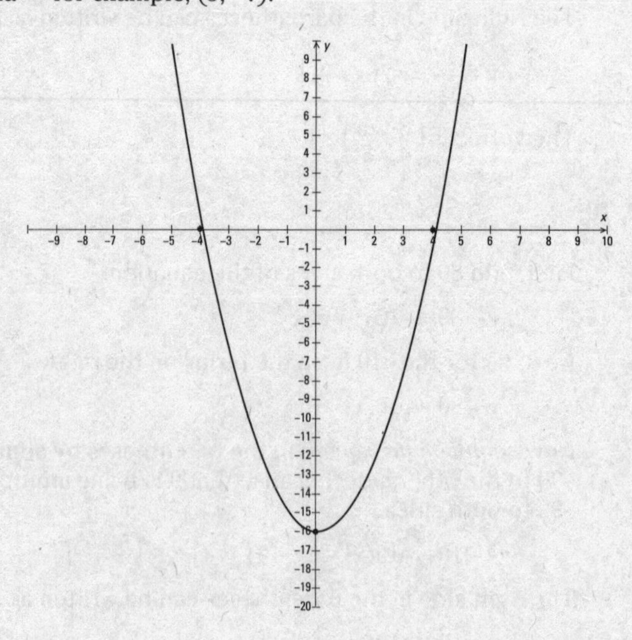

322. Opens downward; through (–3, 16)

The *y*-intercept is (0, 25), and the *x*-intercepts are (5, 0) and (–5, 0). The vertex is at (0, 25). The parabola opens downward. To help with the graphing, choose another point in the parabola — for example, (–3, –16).

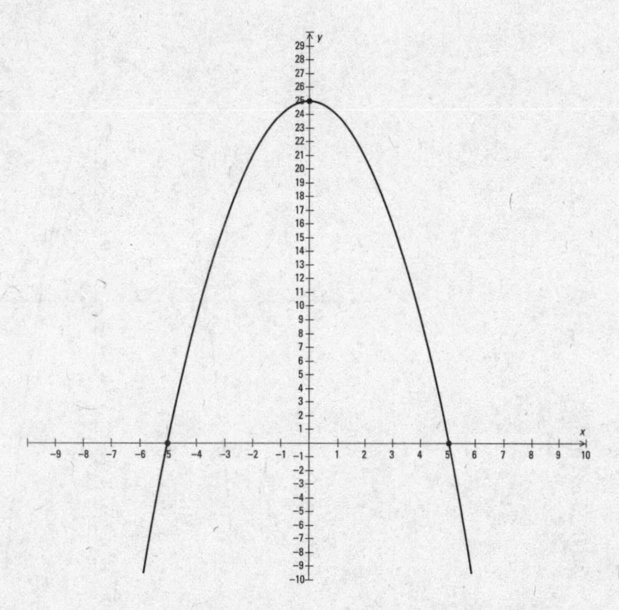

323. Opens upward; through (2, –7)

The y-intercept is (0, –15), and the x-intercepts are (–5, 0) and (3, 0). The vertex is at (–1, –16). The parabola opens upward. To help with the graphing, choose another point in the parabola — for example, (2, –7).

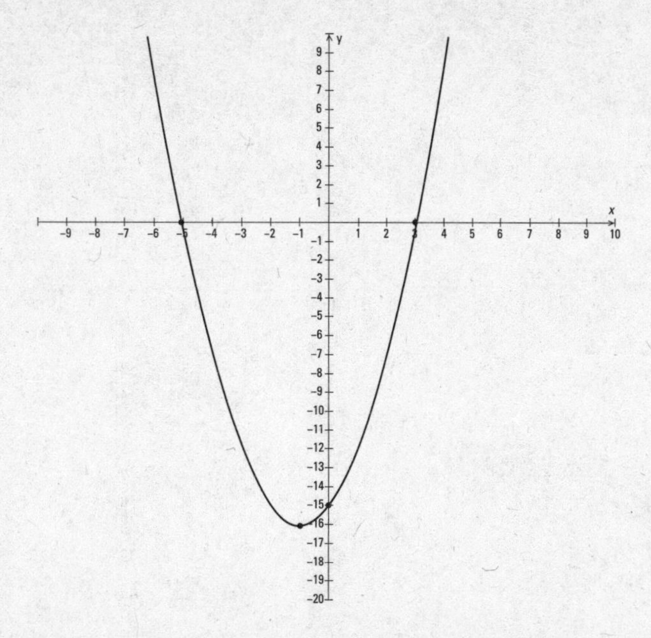

324. Opens downward; through (3, –7)

The y-intercept is (0, 8), and the x-intercepts are (–4, 0) and (2, 0). The vertex is at (–1, 9). The parabola opens downward. To help with the graphing, choose another point in the parabola — for example, (3, –11).

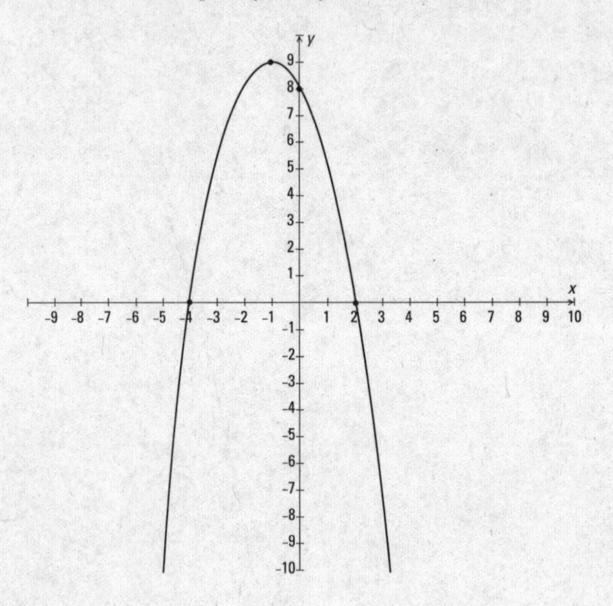

325. Opens upward; through (1, 8)

The y-intercept is (0, 3), and the x-intercepts are (–1, 0) and (–3, 0). The vertex is at (–2, –1). The parabola opens upward. To help with the graphing, choose another point in the parabola — for example, (1, 8).

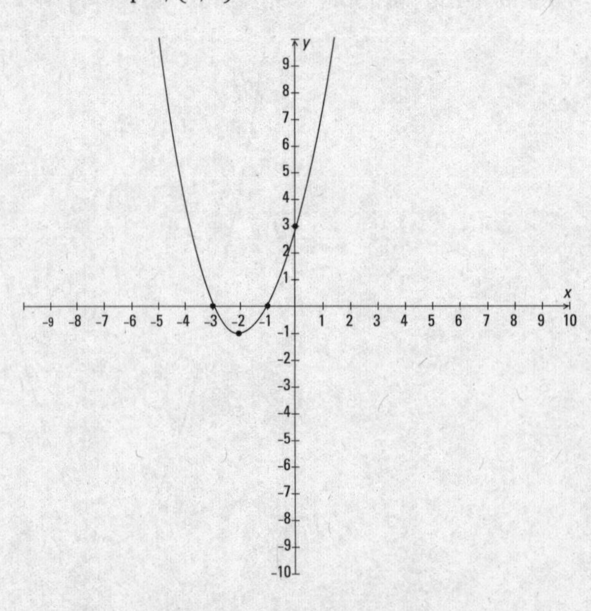

326. Opens upward; through (1, –4)

The y-intercept is (0, –3), and the x-intercepts are $\left(-\frac{1}{3}, 0\right)$ and $\left(\frac{3}{2}, 0\right)$. The vertex is at $\left(\frac{7}{12}, -\frac{121}{24}\right)$. The parabola opens upward. To help with the graphing, choose another point in the parabola — for example, (1, –4).

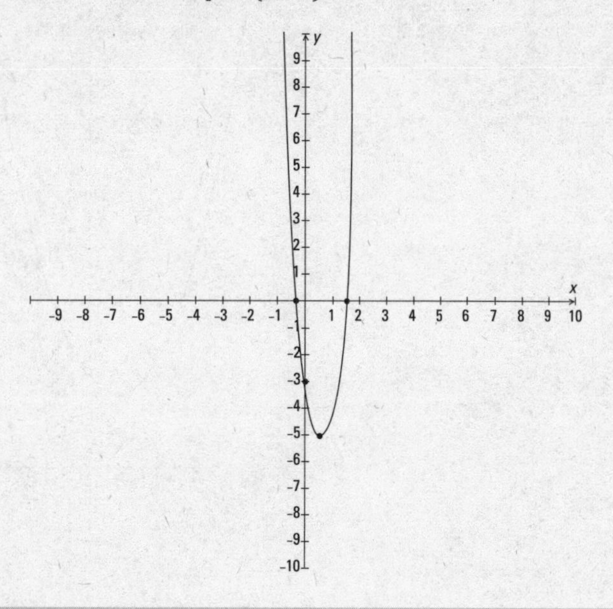

327. Opens downward; through (–1, 12)

The y-intercept is (0, 20), and the x-intercepts are $\left(-\frac{5}{3}, 0\right)$ and (2, 0). The vertex is at $\left(\frac{1}{6}, \frac{121}{6}\right)$. The parabola opens downward. To help with the graphing, choose another point in the parabola — for example, (–1, 12).

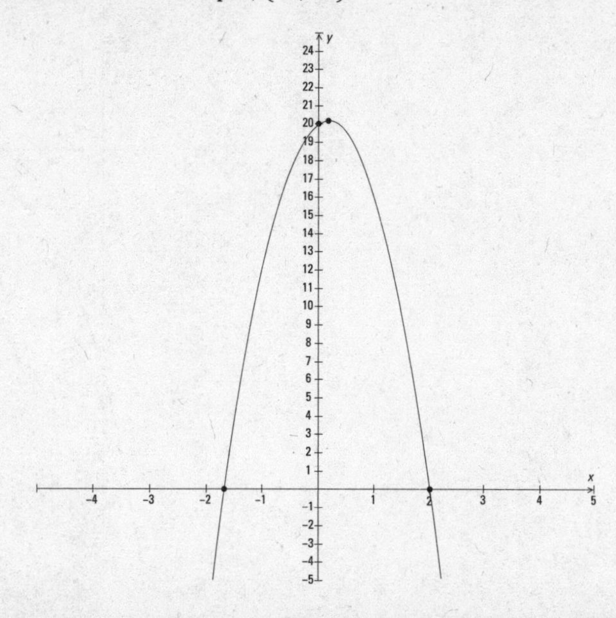

328. Opens upward; through (–2, 5)

The y-intercept is (0, 5), and there are no x-intercepts. The vertex is at (–1, 4). The parabola opens upward.

Two other points that help in the graphing are (3, 20) and (–3, 8).

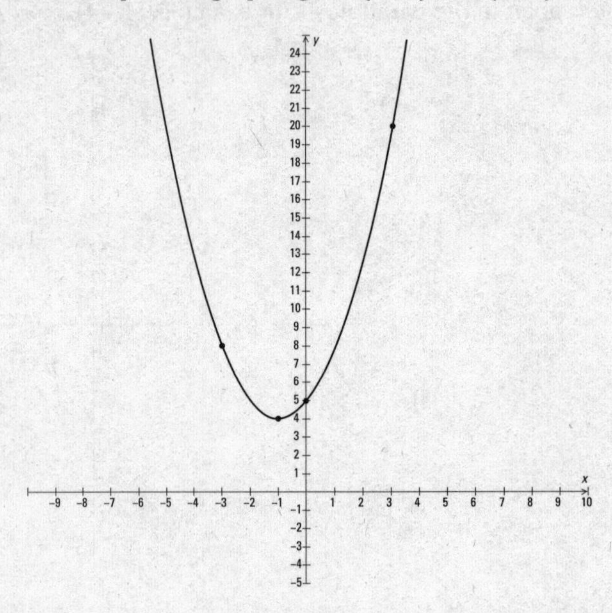

329. Opens downward; through (3, –5)

The y-intercept is (0, –8), and there are no x-intercepts. The vertex is at (2, –4). The parabola opens downward.

Two other points that help in the graphing are (–1, –13) and (3, –8).

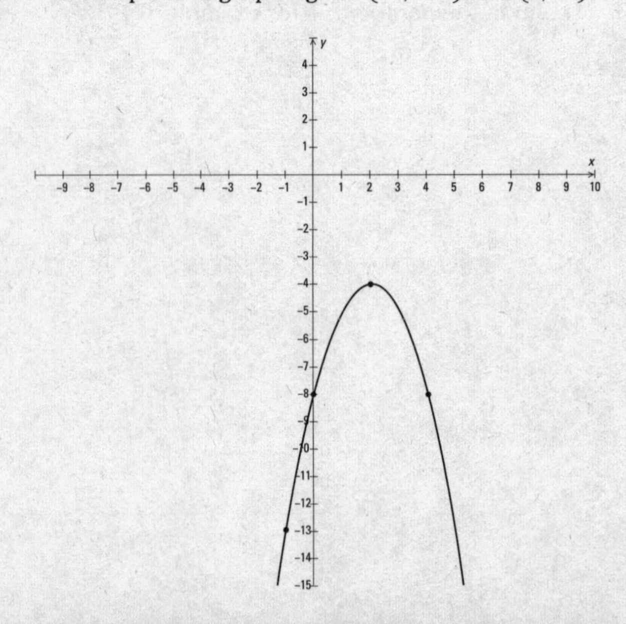

330. Opens upward; through (–2, 3)

The y-intercept is (0, 3), and the one x-intercept is (–1, 0). The vertex is at (–1, 0). The parabola opens upward.

Other points that help in the graphing are (–2, 3) and (–3, 12) .

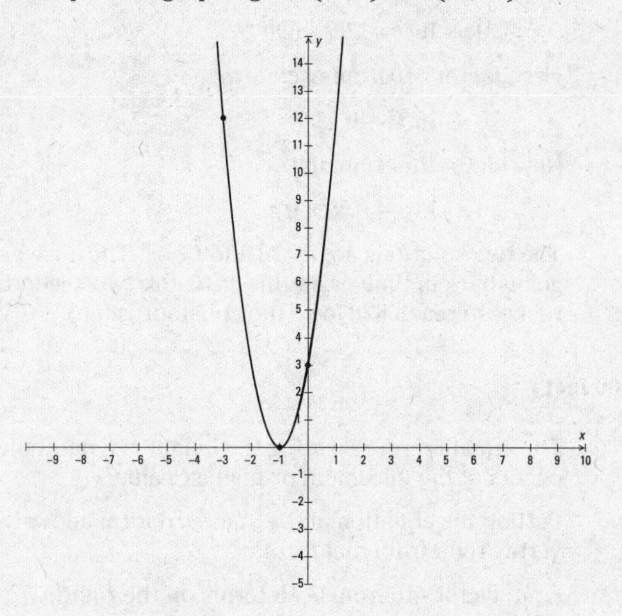

331. 576 feet

The equation representing the height is a parabola opening downward. The vertex occurs at the maximum or highest value.

Putting the equation in the standard form allows you to read the coordinates of the vertex right from that form.

First, add –320 to both sides of the equation.

$$h(t) - 320 = -16t^2 + 128t$$

Now factor –16 from both terms on the right.

$$h(t) - 320 = -16(t^2 - 8t)$$

Complete the square in the parentheses by adding 16 inside the parentheses. The 16 is being multiplied by the –16, also, so you have to add 16(–16) = –256 to the left side of the equation.

$$h(t) - 576 = -16(t^2 - 8t + 16)$$

Factoring on the right, $h(t) - 576 = -16(t - 4)^2$.

The vertex occurs at (4, 576). You read that as: When $t = 4$ (after four seconds), the height is 576 feet.

332. **10 seconds**

The equation representing the height is a parabola opening downward. The *x*-intercepts represent positions where the height is 0 (when the rocket hits the ground).

To solve for the *x*-intercepts, set the height, $h(t) = 0$ and solve for *t*.

$$0 = -16t^2 + 128t + 320$$

First, factor –16 from each term.

$$0 = -16(t^2 - 8t - 20)$$

Now factor the trinomial.

$$0 = -16(t - 10)(t + 2)$$

The two solutions are $t = 10$ and $t = -2$. The answer is 10 seconds. The $t = -2$ represents going back in time or, in this case, the two seconds that it would have taken for the rocket to reach 320 feet (the constant value).

333. **100 feet**

The equation representing the height is a parabola opening downward. The vertex occurs at the maximum or highest value.

Putting the equation in the standard form allows you to read the coordinates of the vertex right from that form.

First, factor –16 from both terms on the right.

$$h(t) = -16(t^2 - 5t)$$

Complete the square in the parentheses by adding $\frac{25}{4}$ inside the parentheses. The $\frac{25}{4}$ is being multiplied by the –16, also, so you have to add –100 to the left side of the equation.

$$h(t) - 100 = -16\left(t^2 - 5t + \frac{25}{4}\right)$$

Factoring on the right, $h(t) - 100 = -16\left(t - \frac{5}{2}\right)^2$.

The vertex occurs at $\left(\frac{5}{2}, 100\right)$. You read that as: When $t = 2.5$ (after 2.5 seconds), the height is 100 feet.

334. **5 seconds**

The equation representing the height is a parabola opening downward. The *x*-intercepts represent positions where the height is 0 (when the ball hits the ground).

To solve for the *x*-intercepts, set the height, $h(t) = 0$, and solve for *t*.

$$0 = -16t^2 + 80t$$

First, factor $-16t$ from each term.

$$0 = -16t(t - 5)$$

The two solutions are $t = 0$ and $t = 5$. The answer is 5 seconds. The $t = 0$ represents the ground height, also, just as the ball is shot.

335. 11 items

The equation representing the profit is a parabola opening downward. The x-intercepts represent positions where the profit changes from a negative number to a positive number and then back to a negative number.

To solve for the x-intercepts, set the profit, $p(x) = 0$, and solve for x.

$$0 = -10x^2 + 1{,}000x - 9{,}000$$

First, factor –10 from each term.

$$0 = -10(x^2 - 100x + 900)$$

Now factor the trinomial.

$$0 = -10(x - 10)(x - 90)$$

The solutions are $x = 10$ and $x = 90$. It takes 10 sold items for the profit values seen in the graph (the y values) to become positive, and at 90 sold items, the profit values become negative again. So, if the number of items sold can't be fractional, it would be the 11th item that gives a positive product.

336. $16,000

The equation representing the profit is a parabola opening downward. The vertex occurs at the maximum or highest value.

Putting the equation in the standard form allows you to read the coordinates of the vertex right from that form.

First, add 9,000 to both sides of the equation.

$$p(x) + 9{,}000 = -10x^2 + 1{,}000x$$

Now factor –10 from both terms on the right.

$$p(x) + 9{,}000 = -10(x^2 - 100x)$$

Complete the square in the parentheses by adding 2,500 inside the parentheses. The 2,500 is being multiplied by the –10, also, so you have to add 2,500(–10) = –25,000 to the left side of the equation.

$$p(x) - 16{,}000 = -10(x^2 - 100x + 2{,}500)$$

Factoring on the right, $p(x) - 16{,}000 = -10(x - 50)^2$.

The vertex occurs at (50, 16,000). You read that as: When $x = 50$ (50 items are sold), the profit is $16,000.

337. January 12th

The equation representing the number of skis sold is a parabola opening downward. The vertex occurs at the maximum or highest value.

Putting the equation in the standard form allows you to read the coordinates of the vertex right from that form.

First, add –100 to both sides of the equation.

$$k(n)-100=-\frac{2}{15}n^2+\frac{10}{3}n$$

Now factor $-\frac{2}{15}$ from both terms on the right.

$$k(n)-100=-\frac{2}{15}\left(n^2-25n\right)$$

Complete the square in the parentheses by adding $\frac{625}{4}$ inside the parentheses. The $\frac{625}{4}$ is being multiplied by the $-\frac{2}{15}$, also, so you have to add $-\frac{2}{15}\left(\frac{625}{4}\right)=-\frac{125}{6}$ to the left side of the equation.

$$k(n)-\frac{725}{6}=-\frac{2}{15}\left(n^2-25n+\frac{625}{4}\right)$$

Factoring on the right, $k(n)-\frac{725}{6}=-\frac{2}{15}\left(n-\frac{25}{2}\right)^2$.

The vertex occurs at $\left(\frac{25}{2},\frac{725}{6}\right)$. You read that as: When $n=\frac{25}{2}=12\frac{1}{2}$ (halfway through the twelfth day, January 12th), the number of skis expected to be sold is $\frac{725}{6}=120\frac{5}{6}$, or between 120 and 121 pairs.

Answers 301–400

338. January 31st

Because you can't sell a negative number of skis, the least number on any particular day would be 0. Find the intercepts when $k(n)$ is 0.

$$0=-\frac{2}{15}n^2+\frac{10}{3}n+100$$

As it turns out, this quadratic doesn't factor and doesn't have any real roots. The number of skis is never 0. So, instead, you look at the two endpoints of the month — January 1st and January 31st. Compare them to determine which is the smaller number.

$$k(1)=-\frac{2}{15}(1)^2+\frac{10}{3}(1)+100=103.2$$

$$k(31)=-\frac{2}{15}(31)^2+\frac{10}{3}(31)+100=75.2$$

So it's the 31st when the least number of skis is sold.

339. 30 years

In this case, the *fastest* means the person with the least time. Using this equation, which is a parabola opening upward, you want the lowest point, which is the vertex.

You only need the age, or the *x*-coordinate of the vertex; so, rather than changing the equation to the standard form, it's easier to just use the formula for the *x*-coordinate (in this model the *g*-coordinate) of the vertex.

Using $g=-\frac{b}{2a}$, you have $g=-\frac{(-2,400)}{2(40)}=30$.

So a person at age 30 should be able to complete the obstacle course in the least amount of time.

340. **About 28 minutes**

You need to know how much time the equation predicts each person will take. Substitute 10 and 20 into the equation:

$$t(10)=\frac{1}{7}\left(40(10)^2-2{,}400(10)+36{,}070\right)=2{,}295\frac{5}{7}$$

$$t(20)=\frac{1}{7}\left(40(20)^2-2{,}400(20)+36{,}070\right)=581\frac{3}{7}$$

The difference between the two is $1{,}714\frac{2}{7}$ seconds. So a 20-year-old can complete the course in about 28 minutes less than the 10-year old.

341. **(0, 0), (1, 0), (–1, 0)**

To find the *y*-intercept, let all the *x*'s in the equation be equal to 0 and solve for *y*.

$$y = 0(0-1)(0+1) = 0$$

To find the *x*-intercepts, let *y* be equal to 0 and solve for *x*.

$$0 = x(x-1)(x+1)$$

Set the factors equal to 0 to get $x = 0$, $x = 1$, $x = -1$.

The point (0, 0) is both an *x* and a *y* intercept.

342. **(0, 0), (–2, 0), (–3, 0)**

To find the *y*-intercept, let all the *x*'s in the equation be equal to 0 and solve for *y*.

$$y = 0^2(0+2)(0+3)^2 = 0$$

To find the *x*-intercepts, let *y* be equal to 0 and solve for *x*.

$$0 = x^2(x+2)(x+3)^2$$

Set the factors equal to 0 to get $x = 0$, $x = -2$, $x = -3$.

The solutions $x = 0$ and $x = -3$ are double roots.

The point (0, 0) is both an *x* and a *y* intercept.

343. **(0, 24), (–4, 0), (3, 0), (2, 0)**

To find the *y*-intercept, let all the *x*'s in the equation be equal to 0 and solve for *y*.

$$y = (0+4)(0-3)(0-2) = 24$$

To find the *x*-intercepts, let *y* be equal to 0 and solve for *x*.

$$0 = (x+4)(x-3)(x-2)$$

Set the factors equal to 0 to get $x = -4$, $x = 3$, $x = 2$.

344. **(0, 36), (–3, 0), (2, 0)**

To find the *y*-intercept, let all the *x*'s in the equation be equal to 0 and solve for *y*.

$$y = (0+3)^2(0-2)^2 = 36$$

To find the *x*-intercepts, let *y* be equal to 0 and solve for *x*.

$$0 = (x+3)^2(x-2)^2$$

Set the factors equal to 0 to get $x = -3$, $x = 2$.

The solutions are both double roots.

345. **(0, 0), (–6, 0)**

To find the *y*-intercept, let all the *x*'s in the equation be equal to 0 and solve for *y*.

$$y = 0^2 + 6(0) = 0$$

To find the *x*-intercepts, let *y* be equal to 0 and solve for *x*.

$$0 = x^2 + 6x = x(x+6)$$

Set the factors equal to 0 to get $x = 0$, $x = -6$.

The point (0, 0) is both an *x* and a *y* intercept.

346. **(0, 0), (–2, 0)**

To find the *y*-intercept, let all the *x*'s in the equation be equal to 0 and solve for *y*.

$$y = 0^3 + 2(0)^2 = 0$$

To find the *x*-intercepts, let *y* be equal to 0 and solve for *x*.

$$0 = x^3 + 2x^2 = x^2(x+2)$$

Set the factors equal to 0 to get $x = 0$, $x = -2$.

The point (0, 0) is both an *x* and a *y* intercept. And the solution $x = 0$ is a double root.

347. **(0, 1)**

To find the *y*-intercept, let all the *x*'s in the equation be equal to 0 and solve for *y*.

$$y = 0^2 + 0 + 1 = 1$$

To find the *x*-intercepts, let *y* be equal to 0 and solve for *x*.

$$0 = x^2 + x + 1$$

The quadratic doesn't factor, so use the quadratic formula.

$$x = \frac{-1 \pm \sqrt{1^2 - 4(1)(1)}}{2(1)} = \frac{-1 \pm \sqrt{-3}}{2}$$

There are no real solutions to the equation, so there are no *x*-intercepts.

348. **(0, –4)**

To find the *y*-intercept, let all the *x*'s in the equation be equal to 0 and solve for *y*.

$$y = -0^2 - 2(0) - 4 = -4$$

To find the *x*-intercepts, let *y* be equal to 0 and solve for *x*.

$$0 = -x^2 - 2x - 4$$

The quadratic doesn't factor, so use the quadratic formula.

$$x = \frac{-(-2) \pm \sqrt{(-2)^2 - 4(-1)(-4)}}{2(-1)} = \frac{2 \pm \sqrt{-12}}{-2}$$

There are no real solutions to the equation, so there are no *x*-intercepts. But, simplifying the expression further, you get $\frac{2 \pm \sqrt{4}\sqrt{-1}\sqrt{3}}{-2} = \frac{2 \pm 2i\sqrt{3}}{-2} = -1 \pm i\sqrt{3}$.

349. **(0, 0), (–3, 0), (–1, 0)**

To find the *y*-intercept, let all the *x*'s in the equation be equal to 0 and solve for *y*.

$$y = 0^3 + 4(0)^2 + 3(0) = 0$$

To find the *x*-intercepts, let *y* be equal to 0 and solve for *x*.

$$0 = x^3 + 4x^2 + 3x = x(x^2 + 4x + 3) = x(x + 3)(x + 1)$$

Set the factors equal to 0 to get $x = 0$, $x = -3$, $x = -1$.

The point (0, 0) is both an *x* and a *y* intercept.

350. **(0, –1), (1, 0)**

To find the *y*-intercept, let all the *x*'s in the equation be equal to 0 and solve for *y*.

$$y = 0^3 - 1 = -1$$

To find the *x*-intercepts, let *y* be equal to 0 and solve for *x*.

$$0 = x^3 - 1$$

Factoring the binomial gives you $0 = (x - 1)(x^2 + x + 1)$; the trinomial factor does not have real solutions. So the best route is to use the original equation, add 1 to each side, and find the cube root of each side.

$$0 = x^3 - 1$$

$$1 = x^3$$

$$1 = x$$

351. **(0, –3), (–3, 0), (1, 0), (–1, 0)**

To find the *y*-intercept, let all the *x*'s in the equation be equal to 0 and solve for *y*.

$$y = 0^3 + 3(0)^2 - 0 - 3 = -3$$

To find the *x*-intercepts, let $y = 0$ and solve for *x* by factoring by grouping.

Factor x^2 from the first two terms and –1 from the last two terms.

$$0 = x^2(x + 3) - 1(x + 3)$$

Now factor out the common binomial, $(x + 3)$.

$$0 = (x + 3)(x^2 - 1)$$

Factor the difference of two squares.

$$0 = (x + 3)(x - 1)(x + 1)$$

Setting each factor equal to 0, you get $x = -3$, $x = 1$, $x = -1$.

352. **(0, 8), (2, 0), (–2, 0)**

To find the *y*-intercept, let all the *x*'s in the equation be equal to 0 and solve for *y*.

$$y = 0^3 - 2(0)^2 - 4(0) + 8 = 8$$

To find the *x*-intercepts, let $y = 0$ and solve for x by factoring by grouping.

Factor x^2 from the first two terms and –4 from the last two terms.

$$0 = x^2(x - 2) - 4(x - 2)$$

Now factor out the common binomial, $(x - 2)$.

$$0 = (x - 2)(x^2 - 4)$$

Factor the difference of two squares.

$$0 = (x - 2)(x - 2)(x + 2)$$

Setting each factor equal to 0, you get $x = 2$, $x = 2$, $x = -2$.

The solution $x = 2$ is a double root.

353. **(0, –1,000), (–10, 0), (10, 0)**

To find the *y*-intercept, let all the *x*'s in the equation be equal to 0 and solve for *y*.

$$y = 0^3 + 10(0)^2 - 100(0) - 1{,}000 = -1{,}000$$

To find the *x*-intercepts, let $y = 0$ and solve for x by factoring by grouping.

Factor x^2 from the first two terms and –100 from the last two terms.

$$0 = x^2(x + 10) - 100(x + 10)$$

Now factor out the common binomial, $(x + 10)$.

$$0 = (x + 10)(x^2 - 100)$$

Factor the difference of two squares.

$$0 = (x + 10)(x - 10)(x + 10)$$

Setting each factor equal to 0, you get $x = -10$, $x = 10$, $x = -10$.

The solution $x = -10$ is a double root.

354. **(0, 9), (–1, 0), (3, 0), (–3, 0)**

To find the *y*-intercept, let all the *x*'s in the equation be equal to 0 and solve for *y*.

$$y = -0^3 - (0)^2 + 9(0) + 9 = 9$$

To find the *x*-intercepts, let $y = 0$ and solve for *x* by factoring by grouping.

Factor $-x^2$ from the first two terms and 9 from the last two terms.

$$0 = -x^2(x + 1) + 9(x + 1)$$

Now factor out the common binomial, $(x + 1)$.

$$0 = (x + 1)(-x^2 + 9)$$

Factor the difference of two squares.

$$0 = (x + 1)(3 - x)(3 + x)$$

Setting each factor equal to 0, you get $x = -1$, $x = 3$, $x = -3$.

355. $\mathbf{(0, 48), \left(-\frac{3}{2}, 0\right), (4, 0), (-4, 0)}$

To find the *y*-intercept, let all the *x*'s in the equation be equal to 0 and solve for *y*.

$$y = -2(0)^3 - 3(0)^2 + 32(0) + 48 = 48$$

To find the *x*-intercepts, let $y = 0$ and solve for *x* by factoring by grouping.

Factor $-x^2$ from the first two terms and 16 from the last two terms.

$$0 = -x^2(2x + 3) + 16(2x + 3)$$

Now factor out the common binomial, $(2x + 3)$.

$$0 = (2x + 3)(-x^2 + 16)$$

Factor the difference of two squares.

$$0 = (2x + 3)(4 - x)(4 + x)$$

Setting each factor equal to 0, you get $x = -\frac{3}{2}$, $x = 4$, $x = -4$.

356. $\mathbf{(0, -7), (-7, 0), \left(\frac{1}{2}, 0\right), \left(-\frac{1}{2}, 0\right)}$

To find the *y*-intercept, let all the *x*'s in the equation be equal to 0 and solve for *y*.

$$y = 4(0)^3 + 28(0)^2 - 0 - 7 = -7$$

To find the *x*-intercepts, let $y = 0$ and solve for *x* by factoring by grouping.

Factor $4x^2$ from the first two terms and –1 from the last two terms.

$$0 = 4x^2(x + 7) - 1(x + 7)$$

Now factor out the common binomial, $(x + 7)$.

$$0 = (x + 7)(4x^2 - 1)$$

Factor the difference of two squares.

$$0 = (x + 7)(2x - 1)(2x + 1)$$

Setting each factor equal to 0, you get $x = -7$, $x = \frac{1}{2}$, $x = -\frac{1}{2}$.

357. **(0, –36), $\left(\frac{1}{3}, 0\right)$, (6, 0), (–6, 0)**

To find the y-intercept, let all the x's in the equation be equal to 0 and solve for y.

$$y = -3(0)^3 + 0^2 + 108(0) - 36 = -36$$

To find the x-intercepts, let $y = 0$ and solve for x by factoring by grouping.

Factor $-x^2$ from the first two terms and 36 from the last two terms.

$$0 = -x^2(3x - 1) + 36(3x - 1)$$

Now factor out the common binomial, $(3x - 1)$.

$$0 = (3x - 1)(-x^2 + 36)$$

Factor the difference of two squares.

$$0 = (3x - 1)(6 - x)(6 + x)$$

Setting each factor equal to 0, you get $x = \frac{1}{3}$, $x = 6$, $x = -6$.

358. **(0, 343), (7, 0), (–7, 0)**

To find the y-intercept, let all the x's in the equation be equal to 0 and solve for y.

$$y = 0^3 - 7(0)^2 - 49(0) + 343 = 343$$

To find the x-intercepts, let $y = 0$ and solve for x by factoring by grouping.

Factor x^2 from the first two terms and –49 from the last two terms.

$$0 = x^2(x - 7) - 49(x - 7)$$

Now factor out the common binomial, $(x - 7)$.

$$0 = (x - 7)(x^2 - 49)$$

Factor the difference of two squares.

$$0 = (x - 7)(x - 7)(x + 7)$$

Setting each factor equal to 0, you get $x = 7$, $x = 7$, $x = -7$.

The solution $x = 7$ is a double root.

359. **(0, –24), (–3, 0), (2, 0)**

To find the y-intercept, let all the x's in the equation be equal to 0 and solve for y.

$$y = 0^4 + 3(0)^3 - 8(0) - 24 = -24$$

To find the x-intercepts, let $y = 0$ and solve for x by factoring by grouping.

Factor x^3 from the first two terms and –8 from the last two terms.

$$0 = x^3(x + 3) - 8(x + 3)$$

Now factor out the common binomial, $(x + 3)$.

$$0 = (x + 3)(x^3 - 8)$$

Factor the difference of two cubes.

$$0 = (x + 3)(x - 2)(x^2 + 2x + 4)$$

Setting each factor equal to 0, you get $x = -3$, $x = 2$. The trinomial has no real roots when set equal to 0.

360. **(0, –4), (4, 0), (–1, 0)**

To find the *y*-intercept, let all the *x*'s in the equation be equal to 0 and solve for *y*.

$$y = 0^4 - 4(0)^3 + 0 - 4 = -4$$

To find the *x*-intercepts, let $y = 0$ and solve for *x* by factoring by grouping.

Factor x^3 from the first two terms and 1 from the last two terms.

$$0 = x^3(x - 4) + 1(x - 4)$$

Now factor out the common binomial, $(x - 4)$.

$$0 = (x - 4)(x^3 + 1)$$

Factor the sum of two cubes.

$$0 = (x - 4)(x + 1)(x^2 - x + 1)$$

Setting each factor equal to 0, you get $x = 4$, $x = -1$. The trinomial has no real roots when set equal to 0.

361. **–1, –2, –3, –6**

Because the leading coefficient is 1, the possible rational roots are the same as the factors of 6: ±1, ±2, ±3, ±6.

There are no sign changes in the polynomial, so there are no positive real roots. So none of the positive roots listed are possible roots of the polynomial.

Replacing all *x*'s with $-x$, you have $0 = x^4 - 3x^3 - 2x + 6$. Because there are two sign changes, there are either two or zero negative real roots. So any of the negative roots listed are possible roots of the polynomial.

362. **±1, ±2, ±4, ±8**

Because the leading coefficient is 1, the possible rational roots are the same as the factors of 8: ±1, ±2, ±4, ±8.

There are three sign changes in the polynomial, so there are three or one positive real roots.

Replacing all *x*'s with $-x$, you have $0 = -x^5 - 2x^4 + 3x^2 - 2x - 8$. Because there are two sign changes, there are either two or zero negative real roots. So any of the positive or negative roots listed are possible roots of the polynomial.

363. **±1, ±2, ±3, ±4, ±6, ±9, ±12, ±18, ±36**

Because the leading coefficient is 1, the possible rational roots are the same as the factors of 36: ±1, ±2, ±3, ±4, ±6, ±9, ±12, ±18, ±36.

There are two sign changes in the polynomial, so there are two or zero positive real roots.

Replacing all x's with $-x$, you have $0 = -x^5 - 2x^4 + 3x^2 - 2x + 36$. Because there are three sign changes, there are either three or one negative real roots. So any of the positive or negative roots listed are possible roots of the polynomial.

364. **1, 2, 5, 10, 25, 50**

Because the leading coefficient is 1, the possible rational roots are the same as the factors of 50: ±1, ±2, ±5, ±10, ±25, ±50.

There are three sign changes in the polynomial, so there are three or one positive real roots.

Replacing all x's with $-x$, you have $0 = -x^5 - 2x^4 - 3x^2 - 2x - 50$. Because there are no sign changes, there are no negative real roots. So none of the negative roots listed are possible roots of the polynomial, leaving only the positive possible roots as the correct answer.

365. $\mathbf{1, 5, \frac{1}{2}, \frac{5}{2}}$

The possible rational roots are the factors of 5 plus each of those factors divided by the factors of 2: $\pm 1, \pm 5, \pm\frac{1}{2}, \pm\frac{5}{2}$.

There are two sign changes in the polynomial, so there are two or zero positive real roots.

Replacing all x's with $-x$, you have $0 = 2x^4 + 3x^2 + 2x + 5$. Because there are no sign changes, there are no negative real roots. So none of the negative roots listed are possible roots of the polynomial, leaving only the positive possible roots as the correct answer.

366. $\mathbf{\pm 1, \pm 2, \pm 3, \pm 6, \pm\frac{1}{3}, \pm\frac{2}{3}}$

The possible rational roots are the factors of 6 plus each of those factors divided by the factors of 3: $\pm 1, \pm 2, \pm 3, \pm 6, \pm\frac{1}{3}, \pm\frac{2}{3}$.

There is one sign change in the polynomial, so there is one positive real root.

Replacing all x's with $-x$, you have $0 = 3x^4 + 3x^2 + 2x - 6$. Because there is one sign change, there is one negative real root. So any of the positive or negative roots listed are possible roots of the polynomial.

367. $\mathbf{\pm 1, \pm 2, \pm 3, \pm 6, \pm\frac{1}{2}, \pm\frac{3}{2}, \pm\frac{1}{4}, \pm\frac{3}{4}}$

The possible rational roots are the factors of 6 plus each of those factors divided by the factors of 4: $\pm 1, \pm 2, \pm 3, \pm 6, \pm\frac{1}{2}, \pm\frac{3}{2}, \pm\frac{1}{4}, \pm\frac{3}{4}$.

There is one sign change in the polynomial, so there is one positive real root.

Replacing all x's with $-x$, you have $0 = 4x^4 + 3x^2 + 2x - 6$. Because there is one sign change, there is one negative real root. So any of the positive or negative roots listed are possible roots of the polynomial.

368. $\pm 1, \pm 2, \pm 5, \pm 10, \pm \frac{1}{2}, \pm \frac{5}{2}$

The possible rational roots are the factors of 10 plus each of those factors divided by the factors of 2: $\pm 1, \pm 2, \pm 5, \pm 10, \pm \frac{1}{2}, \pm \frac{5}{2}$.

There are two sign changes in the polynomial, so there are two or zero positive real roots.

Replacing all x's with $-x$, you have $0 = -2x^5 - 2x^4 + 3x^2 - 2x + 10$. Because there are three sign changes, there are three or one negative real roots. So any of the positive or negative roots listed are possible roots of the polynomial.

369. $\pm 1, \pm 2, \pm \frac{1}{2}, \pm \frac{1}{3}, \pm \frac{2}{3}, \pm \frac{1}{6}$

The possible rational roots are the factors of 2 plus each of those factors divided by the factors of 6: $\pm 1, \pm 2, \pm \frac{1}{2}, \pm \frac{1}{3}, \pm \frac{2}{3}, \pm \frac{1}{6}$.

There are two sign changes in the polynomial, so there are two or zero positive real roots.

Replacing all x's with $-x$, you have $0 = -6x^5 - 2x^4 + 3x^2 - 2x + 2$. Because there are three sign changes, there are three or one negative real roots. So any of the positive or negative roots listed are possible roots of the polynomial.

370. $1, 2, 3, 4, 6, 12, \frac{1}{2}, \frac{3}{2}, \frac{1}{4}, \frac{3}{4}, \frac{1}{8}, \frac{3}{8}$

The possible rational roots are the factors of 12 plus each of those factors divided by the factors of 8: $\pm 1, \pm 2, \pm 3, \pm 4, \pm 6, \pm 12, \pm \frac{1}{2}, \pm \frac{3}{2}, \pm \frac{1}{4}, \pm \frac{3}{4}, \pm \frac{1}{8}, \pm \frac{3}{8}$.

There are two sign changes in the polynomial, so there are two or zero positive real roots.

Replacing all x's with $-x$, you have $0 = 8x^4 + 3x^2 + 2x + 12$. Because there are no sign changes, there are no negative real roots. So none of the negative roots listed are possible roots of the polynomial, leaving only the positive possible roots as the correct answer.

371. $(x-2)(x+6)(x-5)$

There are two possible positive real roots and just one negative. Try one of the positive possibilities:

$$\begin{array}{r|rrrr} 2 & 1 & -1 & -32 & 60 \\ & & 2 & 2 & -60 \\ \hline & 1 & 1 & -30 & 0 \end{array}$$

Because 2 is a root, $(x-2)$ is a factor. Writing the coefficients in the bottom row in the corresponding trinomial, you have

$$x^3 - x^2 - 32x + 60 = (x-2)(x^2 + x - 30)$$

Now factor the quadratic trinomial.

$$= (x-2)(x+6)(x-5)$$

Answers 301–400

372. $(x + 1)(x + 8)(x - 7)$

There is only one possible positive real root, so try for a negative root first.

$$\begin{array}{r|rrrr} -1 & 1 & 2 & -55 & -56 \\ & & -1 & -1 & 56 \\ \hline & 1 & 1 & -56 & 0 \end{array}$$

Because –1 is a root, $(x + 1)$ is a factor. Writing the coefficients in the bottom row in the corresponding trinomial, you have

$$x^3 + 2x^2 - 55x - 56 = (x + 1)(x^2 + x - 56)$$

Now factor the quadratic trinomial.

$$= (x + 1)(x + 8)(x - 7)$$

373. $(x + 3)(x + 9)(x - 1)$

There is only one possible positive real root, so try for a negative root first.

$$\begin{array}{r|rrrr} -3 & 1 & 11 & 15 & -27 \\ & & -3 & -24 & 27 \\ \hline & 1 & 8 & -9 & 0 \end{array}$$

Because –3 is a root, $(x + 3)$ is a factor. Writing the coefficients in the bottom row in the corresponding trinomial, you have

$$x^3 + 11x^2 + 15x - 27 = (x + 3)(x^2 + 8x - 9)$$

Now factor the quadratic trinomial.

$$= (x + 3)(x + 9)(x - 1)$$

374. $(x + 2)(x + 6)(x + 1)$

There are no possible positive real roots, so negative roots are your only option.

$$\begin{array}{r|rrrr} -2 & 1 & 9 & 20 & 12 \\ & & -2 & -14 & -12 \\ \hline & 1 & 7 & 6 & 0 \end{array}$$

Because –2 is a root, $(x + 2)$ is a factor. Writing the coefficients in the bottom row in the corresponding trinomial, you have

$$x^3 + 9x^2 + 20x + 12 = (x + 2)(x^2 + 7x + 6)$$

Now factor the quadratic trinomial.

$$= (x + 2)(x + 6)(x + 1)$$

375. $(x-3)^2(x+7)$

There are two or no possible positive real roots, so try a positive root first.

3	1	1	−33	63
		3	12	−63
	1	4	−21	0

Because 3 is a root, $(x-3)$ is a factor. Writing the coefficients in the bottom row in the corresponding trinomial, you have

$$x^3+x^2-33x+63=(x-3)(x^2+4x-21)$$

Now factor the quadratic trinomial.

$$=(x-3)(x+7)(x-3)$$

The solution $x=3$ is a double root, so you can write the factorization as $(x-3)^2(x+7)$.

376. $(x+4)^2(x-6)$

There is only one possible positive real root, so try for a negative root first.

−4	1	2	−32	−96
		−4	8	96
	1	−2	−24	0

Because −4 is a root, $(x+4)$ is a factor. Writing the coefficients in the bottom row in the corresponding trinomial, you have

$$x^3+2x^2-32x-96=(x+4)(x^2-2x-24)$$

Now factor the quadratic trinomial.

$$=(x+4)(x+4)(x-6)$$

The solution $x=-4$ is a double root, so you can write the factorization as $(x+4)^2(x-6)$.

377. $(x-1)(x+2)(x+5)(x-3)$

There are two or no possible positive real roots, so there's no advantage to trying one type of root or the other. But, trying a positive number first, you have

1	1	3	−15	−19	30
		1	4	−11	−30
	1	4	−11	−30	0

Now, using the coefficients in the bottom row, try a negative number (because you now have two negative and one positive remaining):

−2	1	4	−11	−30
		−2	−4	30
	1	2	−15	0

Answers 301–400

Because 1 and –2 are roots, $(x - 1)$ and $(x + 2)$ are factors. Writing the coefficients in the bottom row in the corresponding trinomial, you have

$$x^4 + 3x^3 - 15x^2 - 19x + 30$$

$$= (x - 1)(x + 2)(x^2 + 2x - 15)$$

Now factor the quadratic trinomial.

$$= (x - 1)(x + 2)(x + 5)(x - 3)$$

378. $(x + 2)(x + 3)(x - 1)(x + 4)$

There is only one possible positive real root, so try for a negative root first.

$$\begin{array}{r|rrrrr} -2 & 1 & 8 & 17 & -2 & -24 \\ & & -2 & -12 & -10 & 24 \\ \hline & 1 & 6 & 5 & -12 & 0 \end{array}$$

Now, using the coefficients in the bottom row, try another possible negative root:

$$\begin{array}{r|rrrr} -3 & 1 & 6 & 5 & -12 \\ & & -3 & -9 & 12 \\ \hline & 1 & 3 & -4 & 0 \end{array}$$

Because –2 and –3 are roots, $(x + 2)$ and $(x + 3)$ are factors. Writing the coefficients in the bottom row in the corresponding trinomial, you have

$$x^4 + 8x^3 + 17x^2 - 2x - 24$$

$$= (x + 2)(x + 3)(x^2 + 3x - 4)$$

Now factor the quadratic trinomial.

$$= (x + 2)(x + 3)(x - 1)(x + 4)$$

379. $(x + 5)(x + 2)(2x + 1)(x - 2)$

There is only one possible positive real root, so try for a negative root first.

$$\begin{array}{r|rrrrr} -5 & 2 & 11 & -3 & -44 & -20 \\ & & -10 & -5 & 40 & 20 \\ \hline & 2 & 1 & -8 & -4 & 0 \end{array}$$

Now, using the coefficients in the bottom row, try another possible negative root:

$$\begin{array}{r|rrrr} -2 & 2 & 1 & -8 & -4 \\ & & -4 & 6 & 4 \\ \hline & 2 & -3 & -2 & 0 \end{array}$$

Because –5 and –2 are roots, $(x + 5)$ and $(x + 2)$ are factors. Writing the coefficients in the bottom row in the corresponding trinomial, you have

$$2x^4 + 11x^3 - 3x^2 - 44x - 20$$

$$= (x + 5)(x + 2)(2x^2 - 3x - 2)$$

Now factor the quadratic trinomial.

$$= (x + 5)(x + 2)(2x + 1)(x - 2)$$

380. $-(x-3)(x-1)(3x-2)(x+3)$

There are three or one possible positive real roots, so try for a positive root first.

$$\begin{array}{r|rrrrr} 3 & -3 & 5 & 25 & -45 & 18 \\ & & -9 & -12 & 39 & -18 \\ \hline & -3 & -4 & 13 & -6 & 0 \end{array}$$

Now, using the coefficients in the bottom row, try another possible positive root:

$$\begin{array}{r|rrrr} 1 & -3 & -4 & 13 & -6 \\ & & -3 & -7 & 6 \\ \hline & -3 & -7 & 6 & 0 \end{array}$$

Because 3 and 1 are roots, $(x-3)$ and $(x-1)$ are factors. Writing the coefficients in the bottom row in the corresponding trinomial, you have

$$-3x^4 + 5x^3 + 25x^2 - 45x + 18$$

$$= (x-3)(x-1)(-3x^2 - 7x + 6)$$

Now factor the quadratic trinomial.

$$= (x-3)(x-1)(-3x+2)(x+3) \text{ or}$$

$$= -(x-3)(x-1)(3x-2)(x+3)$$

381. 28

Use synthetic division to find the remainder.

$$\begin{array}{r|rrrr} 1 & 1 & -1 & -32 & 60 \\ & & 1 & 0 & -32 \\ \hline & 1 & 0 & -32 & \boxed{28} \end{array}$$

So $f(1) = 28$.

382. −108

Use synthetic division to find the remainder.

$$\begin{array}{r|rrrr} 1 & 1 & 2 & -55 & -56 \\ & & 1 & 3 & -52 \\ \hline & 1 & 3 & -52 & \boxed{-108} \end{array}$$

So $f(1) = -108$.

383. 0

Use synthetic division to find the remainder.

$$\begin{array}{r|rrrr} 1 & 1 & 11 & 15 & -27 \\ & & 1 & 12 & 27 \\ \hline & 1 & 12 & 27 & \boxed{0} \end{array}$$

So $f(1) = 0$.

384. **0**

Use synthetic division to find the remainder.

$$\begin{array}{r|rrrr} -1 & 1 & 9 & 20 & 12 \\ & & -1 & -8 & -12 \\ \hline & 1 & 8 & 12 & \boxed{0} \end{array}$$

So $f(-1) = 0$.

385. **9**

Use synthetic division to find the remainder.

$$\begin{array}{r|rrrr} 2 & 1 & 1 & -33 & 63 \\ & & 2 & 6 & -54 \\ \hline & 1 & 3 & -27 & \boxed{9} \end{array}$$

So $f(2) = 9$.

386. **–32**

Use synthetic division to find the remainder.

$$\begin{array}{r|rrrr} -2 & 1 & 2 & -32 & -96 \\ & & -2 & 0 & 64 \\ \hline & 1 & 0 & -32 & \boxed{-32} \end{array}$$

So $f(-2) = -32$.

387. **32**

Use synthetic division to find the remainder.

$$\begin{array}{r|rrrrr} -1 & 1 & 3 & -15 & -19 & 30 \\ & & -1 & -2 & 17 & 2 \\ \hline & 1 & 2 & -17 & -2 & \boxed{32} \end{array}$$

So $f(-1) = 32$.

388. **0**

Use synthetic division to find the remainder.

$$\begin{array}{r|rrrrr} -2 & 1 & 8 & 17 & -2 & -24 \\ & & -2 & -12 & -10 & 24 \\ \hline & 1 & 6 & 5 & -12 & \boxed{0} \end{array}$$

So $f(-2) = 0$.

389. **–55**

Use synthetic division to find the remainder.

1⌋	2	11	–3	–44	–20
		1	12	9	–35
	1	12	9	–35	**–55**

So $f(1) = -55$.

390. **128**

Use synthetic division to find the remainder.

2⌋	–3	5	25	9	18
		–6	–2	46	110
	–3	–1	23	55	**128**

So $f(2) = 128$.

391. **Falls infinitely left, rises infinitely right**

As $x \to +\infty, y \to +\infty$.

As $x \to -\infty, y \to -\infty$.

This is what you expect with a cubic (odd-powered) polynomial whose lead coefficient is positive: rising to the right and falling to the left.

Evaluating the function for smaller and smaller values of x, and then for larger and larger values of x, you see the conclusions demonstrated.

x	y	x	y
–20	–4,388	20	12,412
–100	–901,988	100	1,102,012
–200	–7,603,988	200	8,404,012

392. **Rises infinitely left, falls infinitely right**

As $x \to +\infty, y \to -\infty$.

As $x \to -\infty, y \to +\infty$.

This is what you expect with a cubic (odd-powered) polynomial whose lead coefficient is negative: rising to the left and falling to the right.

Evaluating the function for smaller and smaller values of x, and then for larger and larger values of x, you see the conclusions demonstrated.

x	y	x	y
–20	8,997	20	–8,323
–100	1,013,237	100	–993,363
–200	8,046,537	200	–7,966,663

393. Falls infinitely left, rises infinitely right

As $x \to +\infty$, $y \to +\infty$.

As $x \to -\infty$, $y \to -\infty$.

This is what you expect with a cubic (odd-powered) polynomial whose lead coefficient is positive: rising to the right and falling to the left.

Evaluating the function for smaller and smaller values of x, and then for larger and larger values of x, you see the conclusions demonstrated.

x	y	x	y
−20	−88,340	20	85,900
−100	−11,025,620	100	10,965,580
−200	−88,111,220	200	87,871,180

394. Falls infinitely left, falls infinitely right

As $x \to +\infty$, $y \to -\infty$.

As $x \to -\infty$, $y \to -\infty$.

This is what you expect with a quartic (even-powered) polynomial whose lead coefficient is negative: falling to the right and falling to the left.

Evaluating the function for smaller and smaller values of x, and then for larger and larger values of x, you see the conclusions demonstrated.

x	y	x	y
−20	−407,020	20	−233,020
−100	−210,995,020	100	−189,005,020
−200	−3,287,990,020	200	−3,112,010,020

395. Falls infinitely left, falls infinitely right

As $x \to +\infty$, $y \to -\infty$.

As $x \to -\infty$, $y \to -\infty$.

This is what you expect with a quartic (even-powered) polynomial whose lead coefficient is negative: falling to the right and falling to the left.

Evaluating the function for smaller and smaller values of x, and then for larger and larger values of x, you see the conclusions demonstrated.

x	y	x	y
−20	−519,982	20	−439,982
−100	−304,999,982	100	−294,999,982
−200	−4,839,999,982	200	−4,759,999,982

396. Rises infinitely left, rises infinitely right

As $x \to +\infty$, $y \to +\infty$.

As $x \to -\infty$, $y \to +\infty$.

This is what you expect with a quartic (even-powered) polynomial whose lead coefficient is positive: rising to the right and rising to the left.

Evaluating the function for smaller and smaller values of *x*, and then for larger and larger values of *x*, you see the conclusions demonstrated.

x	y	x	y
–20	641,234	20	641,154
–100	400,030,194	100	400,029,794
–200	6,400,120,394	200	6,400,119,594

397. Falls infinitely left, rises infinitely right

As $x \to +\infty$, $y \to +\infty$.

As $x \to -\infty$, $y \to -\infty$.

This is what you expect with a fifth-degree (odd-powered) polynomial whose lead coefficient is positive: rising to the right and falling to the left.

Evaluating the function for smaller and smaller values of *x*, and then for larger and larger values of *x*, you see the conclusions demonstrated.

x	y	x	y
–20	–6,718,830	20	6,081,250
–100	-2.020×10^{10}	100	1.980×10^{10}
–200	-6.432×10^{11}	200	6.368×10^{11}

Notice that some of the numbers have gotten so small (or large) that I've reverted to scientific notation.

398. Rises infinitely left, falls infinitely right

As $x \to +\infty$, $y \to -\infty$.

As $x \to -\infty$, $y \to +\infty$.

This is what you expect with a fifth-degree polynomial whose lead coefficient is negative: rising to the left and falling to the right.

Evaluating the function for smaller and smaller values of *x*, and then for larger and larger values of *x*, you see the conclusions demonstrated.

x	y	x	y
–20	2,879,970	20	–3,519,950
–100	9,799,999,810	100	-1.020×10^{10}
–200	$3,168 \times 10^{11}$	200	-3.232×10^{11}

Notice that some of the numbers have gotten so small (or large) that I've reverted to scientific notation.

399. Rises infinitely left, rises infinitely right

As $x \to +\infty, y \to +\infty$.

As $x \to -\infty, y \to +\infty$.

This is what you expect with a sixth-degree polynomial whose lead coefficient is positive: rising to the right and rising to the left.

Evaluating the function for smaller and smaller values of x, and then for larger and larger values of x, you see the conclusions demonstrated.

x	y	x	y
–20	127,999,990	20	127,999,990
–100	2.000×10^{12}	100	2.000×10^{12}
–200	1.28×10^{14}	200	1.28×10^{14}

Notice that the y-values for each x and its opposite are the same. The function is *even,* so it's symmetric about the y-axis.

400. Falls infinitely left, falls infinitely right

As $x \to +\infty, y \to -\infty$.

As $x \to -\infty, y \to -\infty$.

This is what you expect with a sixth-degree polynomial whose lead coefficient is negative: falling to the right and falling to the left.

Evaluating the function for smaller and smaller values of x, and then for larger and larger values of x, you see the conclusions demonstrated.

x	y	x	y
–20	–63,998,750	20	–63,998,750
–100	-2.000×10^{12}	100	-2.000×10^{12}
–200	-6.4×10^{13}	200	-6.4×10^{13}

Notice that the y-values for each x and its opposite are the same. The function is *even,* so it's symmetric about the y-axis.

401. Falls infinitely left, rises infinitely right; intercepts when $x = -3, 0, 5$

The x-intercepts are found by setting y equal to 0 and solving for x by factoring.

$$0 = x(x-5)(x+3).$$

The y-intercept is found by letting the x's equal 0 and solving for y.

$$y = 0^3 - 2(0)^2 - 15(0) = 0$$

Because the polynomial is cubic with a positive lead coefficient, it falls to the left and rises to the right.

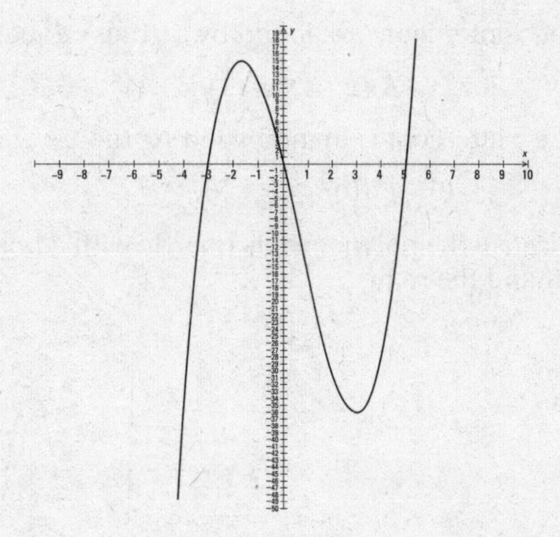

402. Falls infinitely left, rises infinitely right; intercepts when $x = 0, 2, 4$

The x-intercepts are found by setting y equal to 0 and solving for x by factoring.

$$0 = x(x-2)(x-4)$$

The y-intercept is found by letting the x's equal 0 and solving for y.

$$y = 0^3 - 6(0)^2 + 8(0) = 0$$

Because the polynomial is cubic with a positive lead coefficient, it falls to the left and rises to the right.

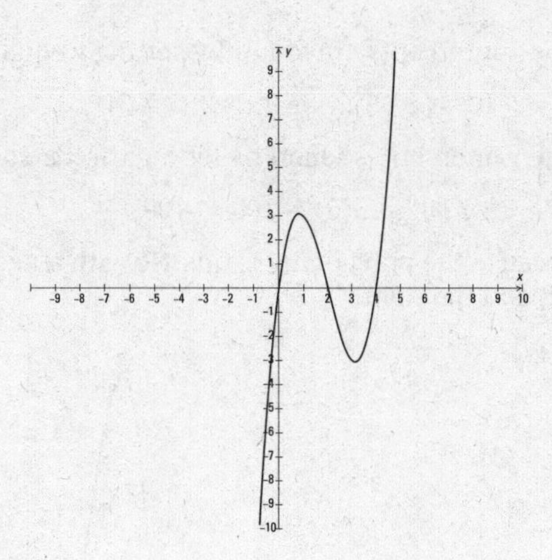

403. Rises infinitely both left and right; x-intercepts when $x = \pm 1, \pm 4$; y-intercept when $y = 16$

The x-intercepts are found by setting y equal to 0 and solving for x by factoring.

$$0 = (x - 4)(x + 4)(x - 1)(x + 1)$$

The y-intercept is found by letting the x's equal 0 and solving for y.

$$y = 0^4 - 17(0)^2 + 16 = 16$$

Because the polynomial is quartic with a positive lead coefficient, it rises both to the left and the right.

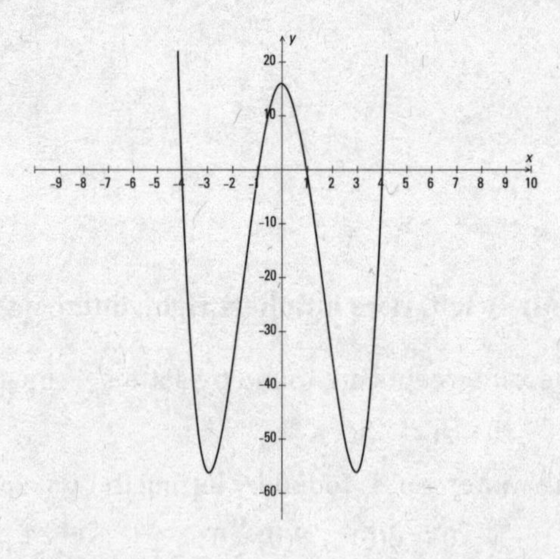

404. Falls infinitely both left and right; x-intercepts when $x = \pm 2, \pm 5$; y-intercept when $y = -100$

The x-intercepts are found by setting y equal to 0 and solving for x by factoring.

$$0 = -(x - 5)(x + 5)(x - 2)(x + 2)$$

The y-intercept is found by letting the x's equal 0 and solving for y.

$$y = -0^4 + 29(0)^2 - 100 = -100$$

Because the polynomial is quartic with a negative lead coefficient, it falls both to the left and the right.

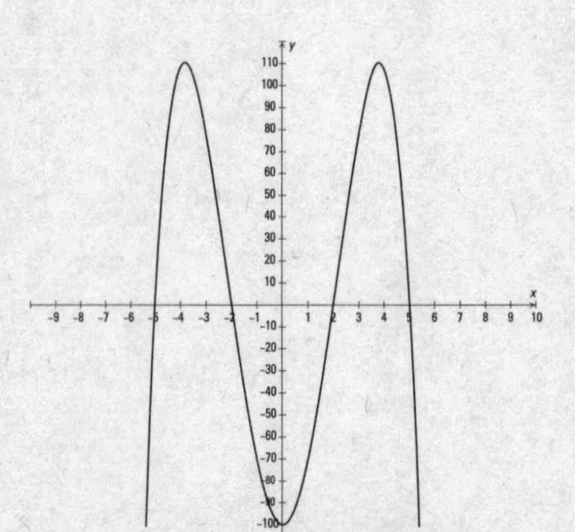

405. Rises infinitely right, falls infinitely left; x-intercepts when $x = 3, 5, 6$; y-intercept when $y = -90$

The x-intercepts are found by setting y equal to 0 and solving for x using the factor theorem and synthetic division.

$$\begin{array}{r|rrrr} 3 & 1 & -14 & 63 & -90 \\ & & 3 & -33 & 90 \\ \hline & 1 & -11 & 30 & 0 \end{array}$$

Now, factoring the quadratic formed by the quotient, start with the factor obtained from $x = 3$ and write:

$$0 = (x - 3)(x - 5)(x - 6).$$

The y-intercept is found by letting the x's equal 0 and solving for y.

$$y = -0^3 - 14(0)^2 + 63(0) - 90 = -90$$

Because the polynomial is cubic with a positive lead coefficient, it falls to the left and rises to the right.

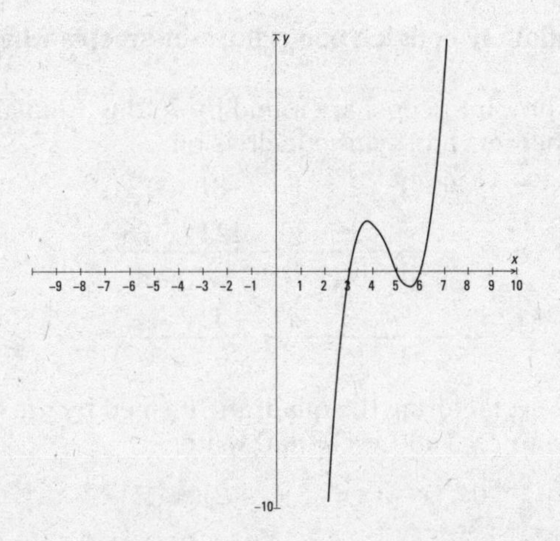

406. Rises infinitely both left and right; x-intercepts when $x = -3, -2, 2, 5$; y-intercept when $y = 60$

The x-intercepts are found by setting y equal to 0 and solving for x using the factor theorem and synthetic division.

$$\begin{array}{r|rrrrr} 2 & 1 & -2 & -19 & 8 & 60 \\ & & 2 & 0 & -38 & -60 \\ \hline -2 & 1 & 0 & -19 & -30 & 0 \\ & & -2 & 4 & 30 & \\ \hline & 1 & -2 & -15 & 0 & \end{array}$$

Now, factoring the quadratic formed by the quotient, start with the two factors formed from $x = 2$ and $x = -2$ and write:

$$0 = (x - 2)(x + 2)(x - 5)(x + 3).$$

The y-intercept is found by letting the x's equal 0 and solving for y.

$$y = 0^4 - 2(0)^3 - 19(0)^2 + 8(0) + 60 = 60$$

Because the polynomial is quartic with a positive lead coefficient, it rises both to the left and to the right.

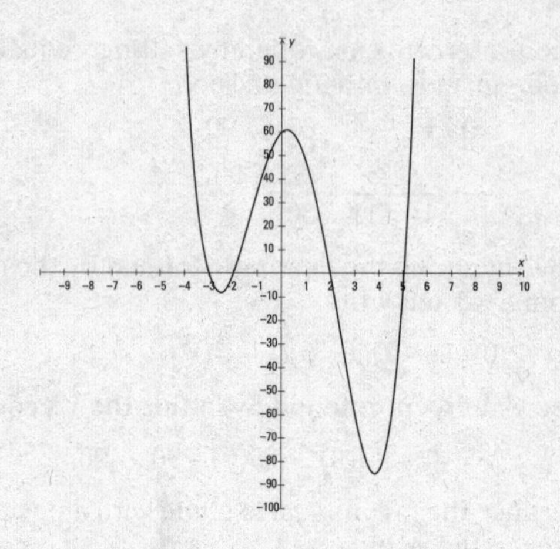

407. Rises infinitely both left and right; x-intercepts when $x = -7, -3, -2, 3$; y-intercept when $y = -126$

The x-intercepts are found by setting y equal to 0 and solving for x using the factor theorem and synthetic division.

3⌋	1	9	5	–81	–126
		3	36	123	126
–3⌋	1	12	41	42	0
		–3	–27	–42	
	1	9	14	0	

Now, factoring the quadratic formed by the quotient, start with the two factors formed from $x = 3$ and $x = -3$ and write:

$$0 = (x - 3)(x + 3)(x + 2)(x + 7).$$

The y-intercept is found by letting the x's equal 0 and solving for y.

$$y = 0^4 + 9(0)^3 + 5(0)^2 - 81(0) - 126 = -126$$

Because the polynomial is quartic with a positive lead coefficient, it rises both to the left and to the right.

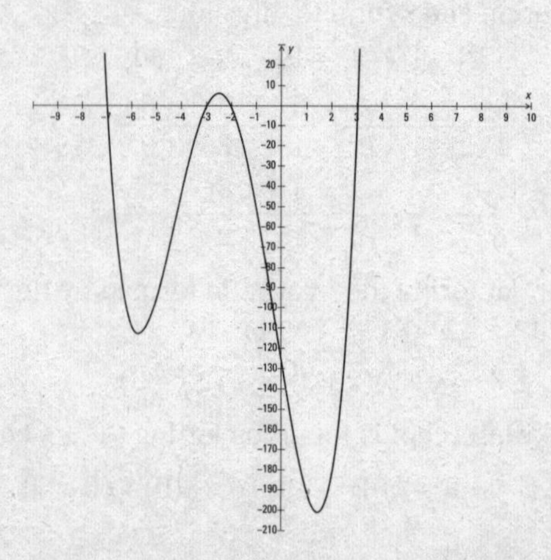

408. Rises infinitely both left and right; x-intercepts when $x = -4, -\frac{1}{2}, 2, 4$; y-intercept when $y = 32$

The x-intercepts are found by setting y equal to 0 and solving for x using the factor theorem and synthetic division.

2⌋	2	−3	−34	48	32
		4	2	−64	−32
4⌋	2	1	−32	−16	0
		8	36	16	
	2	9	4	0	

Now, factoring the quadratic formed by the quotient, start with the two factors formed from $x = 2$ and $x = 4$ and write:

$$0 = (x-2)(x-4)(2x+1)(x+4).$$

The y-intercept is found by letting the x's equal 0 and solving for y.

$$y = 2(0)^4 - 3(0)^3 - 34(0)^2 + 4(0) + 32 = 32$$

Because the polynomial is quartic with a positive lead coefficient, it rises both to the left and to the right.

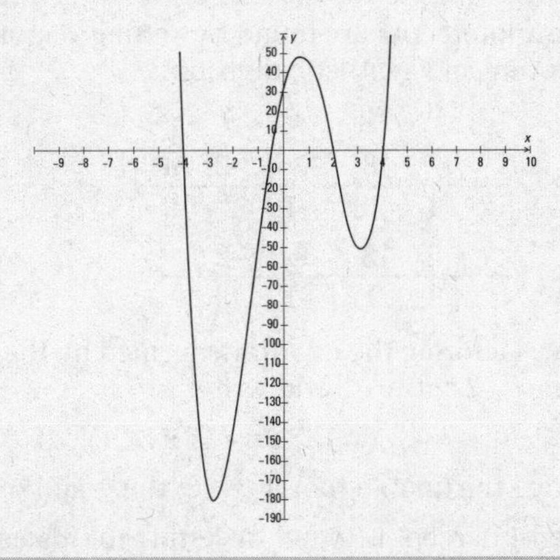

409. Rises infinitely both left and right; x-intercepts when $x = -\frac{3}{2}, \frac{3}{2}, 5$; y-intercept when $y = -225$

The x-intercepts are found by setting y equal to 0 and solving for x using the factor theorem and synthetic division.

5⌋	4	−40	91	90	−225
		20	−100	−45	225
5⌋	4	−20	−9	45	0
		20	0	−45	
	4	0	−9	0	

Now, factoring the quadratic formed by the quotient, start with the two factors formed from $x = 5$ and $x = 5$ and write:

$$0 = (x-5)(x-5)(2x-3)(2x+3).$$

Notice the double root at $x = 5$; the graph of the curve will just *touch* at that intercept.

The y-intercept is found by letting the x's equal 0 and solving for y.

$$y = 4(0)^4 - 40(0)^3 + 91(0)^2 + 90(0) - 225 = -225$$

Because the polynomial is quartic with a positive lead coefficient, it rises both to the left and to the right.

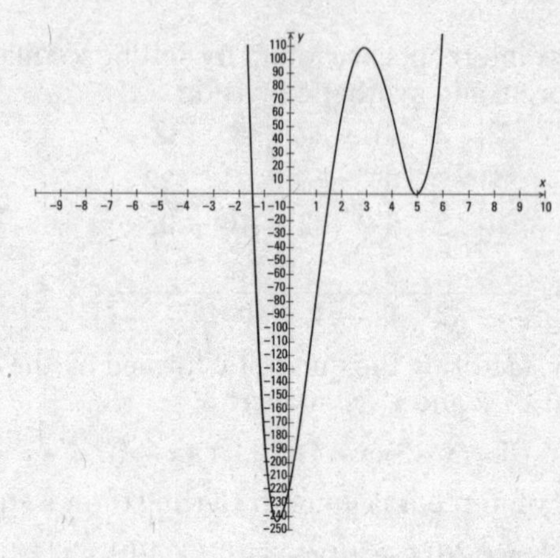

410. **Rises infinitely both left and right; x-intercepts when $x = -\frac{1}{3}, \frac{1}{3}, 2$; y-intercept when $y = -4$**

The x-intercepts are found by setting y equal to 0 and solving for x using the factor theorem and synthetic division.

2⌋	9	–36	35	4	–4
		18	–36	–2	4
2⌋	9	–18	–1	2	0
		18	0	–2	
	9	0	–1	0	

Now, factoring the quadratic formed by the quotient, start with the two factors formed from $x = 2$ and $x = 2$ and write:

$$0 = (x-2)(x-2)(9x-1)(9x+1).$$

Notice the double root at $x = 2$; the graph of the curve will just *touch* at that intercept.

The y-intercept is found by letting the x's equal 0 and solving for y.

$$y = 9(0)^4 - 36(0)^3 + 35(0)^2 + 4(0) - 4 = -4$$

Because the polynomial is quartic with a positive lead coefficient, it rises both to the left and to the right.

411. **All real numbers except that $x \neq -1$**

For a number to be in the domain of this function, the number cannot create a value of 0 in the denominator. The denominator $x + 1 = 0$ when $x = -1$. So $x = -1$ is excluded.

In interval notation, the domain is written:

$$(-\infty, -1) \cup (-1, \infty).$$

412. **All real numbers except that $x \neq 5$**

For a number to be in the domain of this function, the number cannot create a value of 0 in the denominator. The denominator $x - 5 = 0$ when $x = 5$. So $x = 5$ is excluded.

In interval notation, the domain is written:

$$(-\infty, 5) \cup (5, \infty).$$

413. **All real numbers except that $x \neq -4$**

For a number to be in the domain of this function, the number cannot create a value of 0 in the denominator. The denominator $x + 4 = 0$ when $x = -4$. So $x = -4$ is excluded.

In interval notation, the domain is written:

$$(-\infty, -4) \cup (-4, \infty).$$

414. **All real numbers except that $x \neq 2$**

For a number to be in the domain of this function, the number cannot create a value of 0 in the denominator. The denominator $x - 2 = 0$ when $x = 2$. So $x = 2$ is excluded.

In interval notation, the domain is written:

$$(-\infty, 2) \cup (2, \infty).$$

415. **All real numbers except that $x \neq 1$ or 2**

For a number to be in the domain of this function, the number cannot create a value of 0 in the denominator. The denominator $x^2 - 3x + 2 = (x - 1)(x - 2) = 0$ when $x = 1$ or $x = 2$. So both are excluded.

In interval notation, the domain is written:

$$(-\infty, 1) \cup (1, 2) \cup (2, \infty).$$

416. **All real numbers except that $x \neq 5$ or -5**

For a number to be in the domain of this function, the number cannot create a value of 0 in the denominator. The denominator $x^2 - 25 = (x - 5)(x + 5) = 0$ when $x = 5$ or $x = -5$. So both are excluded.

In interval notation, the domain is written:

$$(-\infty, -5) \cup (-5, 5) \cup (5, \infty).$$

417. **All real numbers except that $x \neq 0$ or 5**

For a number to be in the domain of this function, the number cannot create a value of 0 in the denominator. The denominator $x^2 - 5x = x(x - 5) = 0$ when $x = 0$ or $x = 5$. So both are excluded.

In interval notation, the domain is written:

$$(-\infty, 0) \cup (0, 5) \cup (5, \infty).$$

418. **All real numbers except that $x \neq -1$**

For a number to be in the domain of this function, the number cannot create a value of 0 in the denominator. The denominator $x^2 + 2x + 1 = (x + 1)^2 = 0$ when $x = -1$. So $x = -1$ is excluded.

In interval notation, the domain is written:

$$(-\infty, -1) \cup (-1, \infty).$$

419. **All real numbers except that $x \neq 3$ or -3**

For a number to be in the domain of this function, the number cannot create a value of 0 in the denominator. The denominator $x^2 - 9 = (x - 3)(x + 3) = 0$ when $x = 3$ or $x = -3$. So both are excluded.

In interval notation, the domain is written:

$$(-\infty, -3) \cup (-3, 3) \cup (3, \infty).$$

420. **All real numbers**

For a number to be in the domain of this function, the number cannot create a value of 0 in the denominator. The denominator $4x^2 + 1$ cannot be factored and $4x^2 + 1 = 0$ has no real solutions. So the domain contains all real numbers.

In interval notation, the domain is written:

$$(-\infty, \infty).$$

421. **When $x = 1$**

Factor both numerator and denominator.

$$\frac{x^2 - 1}{x^2 - 3x + 2} = \frac{(x-1)(x+1)}{(x-1)(x-2)}$$

Divide out the common factor, $(x - 1)$.

$$\frac{\cancel{(x-1)}(x+1)}{\cancel{(x-1)}(x-2)} = \frac{x+1}{x-2}$$

The domain still cannot contain $x = 1$, from the factor $x - 1 = 0$. The discontinuity is considered to be *removable* and shown in a graph with a hole at the point $(1, -2)$.

422. When $x = -4$

Factor the denominator.

$$\frac{x+4}{x^2-16} = \frac{x+4}{(x-4)(x+4)}$$

Divide out the common factor, $(x + 4)$.

$$\frac{\cancel{x+4}}{(x-4)\cancel{(x+4)}} = \frac{1}{x-4}$$

The domain still cannot contain $x = -4$, from the factor $x + 4 = 0$. The discontinuity is considered to be *removable* and shown in a graph with a hole at the point $\left(-4, -\frac{1}{8}\right)$.

423. When $x = -2$

Factor both numerator and denominator.

$$\frac{x^2+5x+6}{x^2-4} = \frac{(x+2)(x+3)}{(x-2)(x+2)}$$

Divide out the common factor, $(x + 2)$.

$$\frac{\cancel{(x+2)}(x+3)}{(x-2)\cancel{(x+2)}} = \frac{x+3}{x-2}$$

The domain still cannot contain $x = -2$, from the factor $x + 2 = 0$. The discontinuity is considered to be *removable* and shown in a graph with a hole at the point $\left(-2, -\frac{1}{4}\right)$.

424. When $x = -\frac{3}{2}$

Factor both numerator and denominator.

$$\frac{2x^2+9x+9}{2x^2-5x-12} = \frac{(2x+3)(x+3)}{(2x+3)(x-4)}$$

Divide out the common factor, $(2x + 3)$.

$$\frac{\cancel{(2x+3)}(x+3)}{\cancel{(2x+3)}(x-4)} = \frac{x+3}{x-4}$$

The domain still cannot contain $x = -\frac{3}{2}$, from the factor $2x + 3 = 0$. The discontinuity is considered to be *removable* and shown in a graph with a hole at the point $\left(-\frac{3}{2}, -\frac{3}{11}\right)$.

425. When $x = 3$

Factor both numerator and denominator.

$$\frac{2x^2-5x-3}{x^2+x-12} = \frac{(2x+1)(x-3)}{(x-3)(x+4)}$$

Divide out the common factor, $(x-3)$.

$$\frac{(2x+1)\cancel{(x-3)}}{\cancel{(x-3)}(x+4)}=\frac{2x+1}{x+4}$$

The domain still cannot contain $x = 3$, from the factor $x - 3 = 0$. The discontinuity is considered to be *removable* and shown in a graph with a hole at the point (3, 1).

426. **When $x = 5$**

Factor both numerator and denominator.

$$\frac{x^2-5x}{x^2-25}=\frac{x(x-5)}{(x-5)(x+5)}$$

Divide out the common factor, $(x-5)$.

$$\frac{x\cancel{(x-5)}}{\cancel{(x-5)}(x+5)}=\frac{x}{x+5}$$

The domain still cannot contain $x = 5$, from the factor $x - 5 = 0$. The discontinuity is considered to be *removable* and shown in a graph with a hole at the point $\left(5, \frac{1}{2}\right)$.

427. **When $x = 4$**

Factor both numerator and denominator.

$$\frac{x^2-3x-4}{5x^2-22x+8}=\frac{(x+1)(x-4)}{(5x-2)(x-4)}$$

Divide out the common factor, $(x-4)$.

$$\frac{(x+1)\cancel{(x-4)}}{(5x-2)\cancel{(x-4)}}=\frac{x+1}{5x-2}$$

The domain still cannot contain $x = 4$, from the factor $x - 4 = 0$. The discontinuity is considered to be *removable* and shown in a graph with a hole at the point $\left(4, \frac{5}{18}\right)$.

428. **When $x = 1$**

Factor both numerator and denominator.

$$\frac{x^3-x^2+3x-3}{x^3+2x^2-x-2}=\frac{(x-1)(x^2+3)}{(x+2)(x-1)(x+1)}$$

Divide out the common factor, $(x-1)$.

$$\frac{\cancel{(x-1)}(x^2+3)}{(x+2)\cancel{(x-1)}(x+1)}=\frac{x^2+3}{(x+2)(x+1)}$$

The domain still cannot contain $x = 1$, from the factor $x - 1 = 0$. The discontinuity is considered to be *removable* and shown in a graph with a hole at the point $\left(1, \frac{2}{3}\right)$.

429. **When $x = -1$**

Factor both numerator and denominator.

$$\frac{x^4 - 4x^3 + x - 4}{x^4 - 10x^2 + 9} = \frac{(x-4)(x+1)(x^2 - x + 1)}{(x-1)(x+1)(x-3)(x+3)}$$

Divide out the common factor, $(x + 1)$.

$$\frac{(x-4)\cancel{(x+1)}(x^2 - x + 1)}{(x-1)\cancel{(x+1)}(x-3)(x+3)} = \frac{(x-4)(x^2 - x + 1)}{(x-1)(x-3)(x+3)}$$

The domain still cannot contain $x = -1$, from the factor $x + 1 = 0$. The discontinuity is considered to be *removable* and shown in a graph with a hole at the point $\left(-1, -\frac{15}{16}\right)$.

430. **When $x = -2$**

Factor both numerator and denominator.

$$\frac{x^4 - 9x^2 + 20}{x^4 + 7x^3 + 8x + 56} = \frac{(x-2)(x+2)(x^2 - 5)}{(x+7)(x+2)(x^2 - 2x + 4)}$$

Divide out the common factor, $(x + 2)$.

$$\frac{(x-2)\cancel{(x+2)}(x^2 - 5)}{(x+7)\cancel{(x+2)}(x^2 - 2x + 4)} = \frac{(x-2)(x^2 - 5)}{(x+7)(x^2 - 2x + 4)}$$

The domain still cannot contain $x = -2$, from the factor $x + 2 = 0$. The discontinuity is considered to be *removable* and shown in a graph with a hole at the point $\left(-2, \frac{1}{15}\right)$.

431. **1**

Evaluating $\lim\limits_{x \to \infty} \frac{x+3}{x-2}$, divide each term in the fraction by x.

$$\lim_{x \to \infty} \frac{\frac{x}{x} + \frac{3}{x}}{\frac{x}{x} - \frac{2}{x}} = \lim_{x \to \infty} \frac{1 + \frac{3}{x}}{1 - \frac{2}{x}}$$

Because $\lim\limits_{x \to \infty} \frac{c}{x^n} = 0$,

$$\lim_{x \to \infty} \frac{1 + \frac{3}{x}}{1 - \frac{2}{x}} = \frac{1+0}{1-0} = 1.$$

432. **2**

Evaluating $\lim\limits_{x \to \infty} \frac{2x-4}{x+3}$, divide each term in the fraction by x.

$$\lim_{x \to \infty} \frac{\frac{2x}{x} - \frac{4}{x}}{\frac{x}{x} + \frac{3}{x}} = \lim_{x \to \infty} \frac{2 - \frac{4}{x}}{1 + \frac{3}{x}}$$

Because $\lim_{x\to\infty} \frac{c}{x^n} = 0$,

$$\lim_{x\to\infty} \frac{2-\frac{4}{x}}{1+\frac{3}{x}} = \frac{2-0}{1+0} = 2.$$

433. **3**

Evaluating $\lim_{x\to\infty} \frac{3x^2-4x+1}{x^2-2x+3}$, divide each term in the fraction by x^2.

$$\lim_{x\to\infty} \frac{\frac{3x^2}{x^2}-\frac{4x}{x^2}+\frac{1}{x^2}}{\frac{x^2}{x^2}-\frac{2x}{x^2}+\frac{3}{x^2}} = \lim_{x\to\infty} \frac{3-\frac{4}{x}+\frac{1}{x^2}}{1-\frac{2}{x}+\frac{3}{x^2}}$$

Because $\lim_{x\to\infty} \frac{c}{x^n} = 0$,

$$\lim_{x\to\infty} \frac{3-\frac{4}{x}+\frac{1}{x^2}}{1-\frac{2}{x}+\frac{3}{x^2}} = \frac{3-0+0}{1-0+0} = 3.$$

434. $\frac{1}{3}$

Evaluating $\lim_{x\to\infty} \frac{x^2-1}{3x^2+7}$, divide each term in the fraction by x^2.

$$\lim_{x\to\infty} \frac{\frac{x^2}{x^2}-\frac{1}{x^2}}{\frac{3x^2}{x^2}+\frac{7}{x^2}} = \lim_{x\to\infty} \frac{1-\frac{1}{x^2}}{3+\frac{7}{x^2}}$$

Because $\lim_{x\to\infty} \frac{c}{x^n} = 0$,

$$\lim_{x\to\infty} \frac{1-\frac{1}{x^2}}{3+\frac{7}{x^2}} = \frac{1-0}{3+0} = \frac{1}{3}.$$

435. **2**

Evaluating $\lim_{x\to\infty} \frac{4x^2+3x-2}{2x^2-1}$, divide each term in the fraction by x^2.

$$\lim_{x\to\infty} \frac{\frac{4x^2}{x^2}+\frac{3x}{x^2}-\frac{2}{x^2}}{\frac{2x^2}{x^2}-\frac{1}{x^2}} = \lim_{x\to\infty} \frac{4+\frac{3}{x}-\frac{2}{x^2}}{2-\frac{1}{x^2}}$$

Because $\lim_{x\to\infty} \frac{c}{x^n} = 0$,

$$\lim_{x\to\infty} \frac{4+\frac{3}{x}-\frac{2}{x^2}}{2-\frac{1}{x^2}} = \frac{4+0-0}{2-0} = 2.$$

436. $\frac{1}{2}$

Evaluating $\lim_{x\to\infty}\frac{x^2+x-3}{2x^2+5x+3}$, divide each term in the fraction by x^2.

$$\lim_{x\to\infty}\frac{\frac{x^2}{x^2}+\frac{x}{x^2}-\frac{3}{x^2}}{\frac{2x^2}{x^2}+\frac{5x}{x^2}+\frac{3}{x^2}}=\lim_{x\to\infty}\frac{1+\frac{1}{x}-\frac{3}{x^2}}{2+\frac{5}{x}+\frac{3}{x^2}}$$

Because $\lim_{x\to\infty}\frac{c}{x^n}=0$,

$$\lim_{x\to\infty}\frac{1+\frac{1}{x}-\frac{3}{x^2}}{2+\frac{5}{x}+\frac{3}{x^2}}=\frac{1+0-0}{2+0+0}=\frac{1}{2}.$$

437. 0

Evaluating $\lim_{x\to\infty}\frac{x+1}{x^2-2x+4}$, divide each term in the fraction by x^2.

$$\lim_{x\to\infty}\frac{\frac{x}{x^2}+\frac{1}{x^2}}{\frac{x^2}{x^2}-\frac{2x}{x^2}+\frac{4}{x^2}}=\lim_{x\to\infty}\frac{\frac{1}{x}+\frac{1}{x^2}}{1-\frac{2}{x}+\frac{4}{x^2}}$$

Because $\lim_{x\to\infty}\frac{c}{x^n}=0$,

$$\lim_{x\to\infty}\frac{\frac{1}{x}+\frac{1}{x^2}}{1-\frac{2}{x}+\frac{4}{x^2}}=\frac{0+0}{1-0+0}=\frac{0}{1}=0.$$

438. 0

Evaluating $\lim_{x\to\infty}\frac{4x+3}{x^3-4}$, divide each term in the fraction by x^3.

$$\lim_{x\to\infty}\frac{\frac{4x}{x^3}+\frac{3}{x^3}}{\frac{x^3}{x^3}-\frac{4}{x^3}}=\lim_{x\to\infty}\frac{\frac{4}{x^2}+\frac{3}{x^3}}{1-\frac{4}{x^3}}$$

Because $\lim_{x\to\infty}\frac{c}{x^n}=0$,

$$\lim_{x\to\infty}\frac{\frac{4}{x^2}+\frac{3}{x^3}}{1-\frac{4}{x^3}}=\frac{0+0}{1-0}=\frac{0}{1}=0.$$

439. **No limit**

Evaluating $\lim_{x\to\infty}\frac{x^2+4}{7x+3}$, divide each term in the fraction by x^2.

$$\lim_{x\to\infty}\frac{\frac{x^2}{x^2}+\frac{4}{x^2}}{\frac{7x}{x^2}+\frac{3}{x^2}}=\lim_{x\to\infty}\frac{1+\frac{4}{x^2}}{\frac{7}{x}+\frac{3}{x^2}}$$

Because $\lim_{x\to\infty} \frac{c}{x^n} = 0$,

$$\lim_{x\to\infty} \frac{1+\frac{4}{x^2}}{\frac{7}{x}+\frac{3}{x^2}} = \frac{1+0}{0+0} = \frac{1}{0},$$ which is *undefined.* There is no limit.

440. **No limit**

Evaluating $\lim_{x\to\infty} \frac{x^3-4x^2+2}{x^2-4x+1}$, divide each term in the fraction by x^3.

$$\lim_{x\to\infty} \frac{\frac{x^3}{x^3}-\frac{4x^2}{x^3}+\frac{2}{x^3}}{\frac{x^2}{x^3}-\frac{4x}{x^3}+\frac{1}{x^3}} = \lim_{x\to\infty} \frac{1-\frac{4}{x}+\frac{2}{x^3}}{\frac{1}{x}-\frac{4}{x^2}+\frac{1}{x^3}}$$

Because $\lim_{x\to\infty} \frac{c}{x^n} = 0$,

$$\lim_{x\to\infty} \frac{1-\frac{4}{x}+\frac{2}{x^3}}{\frac{1}{x}-\frac{4}{x^2}+\frac{1}{x^3}} = \frac{1-0+0}{0-0+0} = \frac{1}{0},$$ which is *undefined.* There is no limit.

441. $+\infty$

Approaching 4 from the right, consider the following arbitrary values.

x	$\frac{x+1}{x-4}$
4.1	$\frac{5.1}{0.1} = 51$
4.01	$\frac{5.01}{0.01} = 501$
4.001	$\frac{5.001}{0.001} = 5{,}001$

The closer x gets to 4, the larger the function value becomes. The limit approaches $+\infty$.

442. $+\infty$

Approaching –6 from the right, consider the following arbitrary values.

x	$\frac{5-x}{x+6}$
–5.9	$\frac{10.9}{0.1} = 109$
–5.99	$\frac{10.99}{0.01} = 1{,}099$
–5.999	$\frac{10.999}{0.001} = 10{,}999$

The closer x gets to –6, the larger the function value becomes. The limit approaches $+\infty$.

443. $-\infty$

Approaching –3 from the left, consider the following arbitrary values.

x	$\frac{2-x}{x+3}$
–3.1	$\frac{5.1}{-0.1}=-51$
–3.01	$\frac{5.01}{-0.01}=-501$
–3.001	$\frac{5.001}{-0.001}=-5{,}001$

The closer x gets to –3, the smaller the function value becomes. The limit approaches $-\infty$.

444. $-\infty$

Approaching 2 from the left, consider the following arbitrary values.

x	$\frac{x+1}{x-2}$
1.9	$\frac{2.9}{-0.1}=-29$
1.99	$\frac{2.99}{-0.01}=-299$
1.999	$\frac{2.999}{-0.001}=-2{,}999$

The closer x gets to 2, the smaller the function value becomes. The limit approaches $-\infty$.

445. $-\infty$

Approaching 3 from the left, consider the following arbitrary values.

x	$\frac{x}{(x-3)(x+2)}$
2.9	$\frac{2.9}{(-0.1)(4.9)}\approx -5.918$
2.99	$\frac{2.99}{(-0.01)(4.99)}\approx -59.92$
2.999	$\frac{2.999}{(-0.001)(4.999)}\approx -599.9$

The closer x gets to 3, the smaller the function value becomes. The limit approaches $-\infty$.

446. $+\infty$

Approaching 3 from the right, consider the following arbitrary values.

x	$\frac{x^2}{(x-3)(x+3)}$
3.1	$\frac{9.61}{0.61} \approx 15.8$
3.01	$\frac{9.0601}{0.0601} \approx 150.8$
3.001	$\frac{9.006001}{0.006001} \approx 1{,}500.8$

The closer x gets to 3, the larger the function value becomes. The limit approaches $+\infty$.

447. $-\infty$

Approaching –4 from the right, consider the following arbitrary values.

x	$\frac{x+3}{x(x-1)(x+4)}$
–3.9	$\frac{-0.9}{1.911} \approx -0.47$
–3.99	$\frac{-0.99}{0.99101} \approx -4.97$
–3.999	$\frac{-0.999}{0.019991001} \approx -49.97$

The closer x gets to –4, the smaller the function value becomes. The limit approaches $-\infty$.

448. $+\infty$

Approaching 5 from the right, consider the following arbitrary values.

x	$\frac{x^2-3x+2}{x^2-25}$
5.1	$\frac{12.71}{1.01} \approx 12.58$
5.01	$\frac{12.0701}{0.1001} \approx 120.58$
5.001	$\frac{12.007001}{0.010001} \approx 1{,}200.58$

The closer x gets to 5, the larger the function value becomes. The limit approaches $+\infty$.

449. $-\infty$

Approaching –4 from the left, consider the following arbitrary values.

x	$\frac{x+3}{4x+x^2}$
–4.1	$\frac{-1.1}{0.41} \approx -2.68$
–4.01	$\frac{-1.01}{0.0401} \approx -25.19$
–4.001	$\frac{-1.001}{0.004001} \approx -250.19$

The closer x gets to –4, the smaller the function value becomes. The limit approaches $-\infty$.

450. $-\infty$

Approaching 0 from the right, consider the following arbitrary values.

x	$\frac{x+3}{x^2-6x}$
0.1	$\frac{3.1}{-0.59} \approx -5.25$
0.01	$\frac{3.01}{-0.0599} \approx -50.25$
0.001	$\frac{3.001}{-0.005999} \approx -500.25$

The closer x gets to 0, the smaller the function value becomes. The limit approaches $-\infty$.

451. $x = 2, y = 1$

There are no common factors in the numerator and denominator.

To solve for vertical asymptotes, set the denominator equal to 0.

$x - 2 = 0$ gives you $x = 2$, the equation of the vertical asymptote.

When the highest powers in the numerator and denominator are the same, you find the horizontal asymptote by making a fraction of the coefficients of those highest powers. Because both coefficients are 1, you have $\frac{1}{1} = 1$, so $y = 1$ is the equation of the horizontal asymptote.

452. $x = -7, y = 2$

There are no common factors in the numerator and denominator.

To solve for vertical asymptotes, set the denominator equal to 0.

$x + 7 = 0$ gives you $x = -7$, the equation of the vertical asymptote.

When the highest powers in the numerator and denominator are the same, you find the horizontal asymptote by making a fraction of the coefficients of those highest powers. Because the coefficient in the numerator is 2 and the coefficient in the denominator is 1, you have $\frac{2}{1} = 2$, so $y = 2$ is the equation of the horizontal asymptote.

453. $x = -2, y = 1$

There are no common factors in the numerator and denominator.

To solve for vertical asymptotes, set the denominator equal to 0.

$x + 2 = 0$ gives you $x = -2$, the equation of the vertical asymptote.

When the highest powers in the numerator and denominator are the same, you find the horizontal asymptote by making a fraction of the coefficients of those highest powers. Because the coefficients in the numerator and denominator are both 1, you have $\frac{1}{1} = 1$, so $y = 1$ is the equation of the horizontal asymptote.

454. $x = 2, x = -2, y = 1$

First factor the numerator and denominator.

$$y = \frac{x^2 - 3x}{x^2 - 4} = \frac{x(x-3)}{(x-2)(x+2)}$$

There are no common factors in the numerator and denominator.

To solve for vertical asymptotes, set the denominator equal to 0.

$(x - 2)(x + 2) = 0$ gives you $x = 2$ and $x = -2$, the equations of the vertical asymptotes.

When the highest powers in the numerator and denominator are the same, you find the horizontal asymptote by making a fraction of the coefficients of those highest powers. Because the coefficients in the numerator and denominator are both 1, you have $\frac{1}{1} = 1$, so $y = 1$ is the equation of the horizontal asymptote.

455. $x = 1, x = -5, y = 3$

First factor the numerator and denominator.

$$y = \frac{3x^2 + 2x - 1}{x^2 + 4x - 5} = \frac{(3x-1)(x+1)}{(x-1)(x+5)}$$

There are no common factors in the numerator and denominator.

To solve for vertical asymptotes, set the denominator equal to 0.

$(x - 1)(x + 5) = 0$ gives you $x = 1$ and $x = -5$, the equations of the vertical asymptotes.

When the highest powers in the numerator and denominator are the same, you find the horizontal asymptote by making a fraction of the coefficients of those highest powers. Because the coefficient in the numerator is 3 and the coefficient in the denominator is 1, you have $\frac{3}{1} = 3$, so $y = 3$ is the equation of the horizontal asymptote.

456. $x = -3, y = 1$

First factor the numerator and denominator.

$$y = \frac{x^2 + 3x - 4}{x^2 + 7x + 12} = \frac{(x+4)(x-1)}{(x+4)(x+3)}$$

Divide by the common factor, $(x + 4)$. This means that there's a removable discontinuity when $x = -4$.

$$y = \frac{\cancel{(x+4)}(x-1)}{\cancel{(x+4)}(x+3)} = \frac{x-1}{x+3}$$

Now use the reduced version to finish.

To solve for vertical asymptotes, set the denominator equal to 0.

$x + 3 = 0$ gives you $x = -3$, the equation of the vertical asymptote.

When the highest powers in the numerator and denominator are the same, you find the horizontal asymptote by making a fraction of the coefficients of those highest powers. Because the coefficients in the numerator and denominator are both 1, you have $\frac{1}{1} = 1$, so $y = 1$ is the equation of the horizontal asymptote.

457. $x = -1, y = 1$

First factor the numerator and denominator.

$$y = \frac{x^2 - 5x + 4}{x^2 - 1} = \frac{(x-4)(x-1)}{(x-1)(x+1)}$$

Divide by the common factor, $(x - 1)$. This means that there's a removable discontinuity when $x = 1$.

$$y = \frac{(x-4)\cancel{(x-1)}}{\cancel{(x-1)}(x+1)} = \frac{x-4}{x+1}$$

Now use the reduced version to finish.

To solve for vertical asymptotes, set the denominator equal to 0.

$x + 1 = 0$ gives you $x = -1$, the equation of the vertical asymptote.

When the highest powers in the numerator and denominator are the same, you find the horizontal asymptote by making a fraction of the coefficients of those highest powers. Because the coefficients in the numerator and denominator are both 1, you have $\frac{1}{1} = 1$, so $y = 1$ is the equation of the horizontal asymptote.

458. $y = 2$

There are no common factors in the numerator and denominator.

To solve for vertical asymptotes, set the denominator equal to 0.

$x^2 + 3 = 0$ has no real solutions, so there is no vertical asymptote.

When the highest powers in the numerator and denominator are the same, you find the horizontal asymptote by making a fraction of the coefficients of those highest powers. Because the coefficient in the numerator is 2 and the coefficient in the denominator is 1, you have $\frac{2}{1} = 2$, so $y = 2$ is the equation of the horizontal asymptote.

459. $x = 0, y = \frac{1}{2}$

There are no common factors in the numerator and denominator.

To solve for vertical asymptotes, set the denominator equal to 0.

$2x^2 = 0$ gives you $x = 0$, the equation of the vertical asymptote.

When the highest powers in the numerator and denominator are the same, you find the horizontal asymptote by making a fraction of the coefficients of those highest powers. Because the coefficient in the numerator is 1 and the coefficient in the denominator is 2, you have $\frac{1}{2}$, so $y = \frac{1}{2}$ is the equation of the horizontal asymptote.

460. $x = 3, x = -3, y = 1$

First factor the numerator and denominator.

$$y = \frac{x^3 - 16x}{x^3 - 9x} = \frac{x(x-4)(x+4)}{x(x-3)(x+3)}$$

Divide by the common factor, x. This means that there's a removable discontinuity when $x = 0$.

$$y = \frac{\cancel{x}(x-4)(x+4)}{\cancel{x}(x-3)(x+3)} = \frac{(x-4)(x+4)}{(x-3)(x+3)}$$

Now use the reduced version to finish.

To solve for vertical asymptotes, set the denominator equal to 0.

$(x - 3)(x + 3) = 0$ gives you $x = 3$ and $x = -3$, the equations of the vertical asymptotes.

When the highest powers in the numerator and denominator are the same, you find the horizontal asymptote by making a fraction of the coefficients of those highest powers. Because the coefficients in both the numerator and denominator are 1, you have $\frac{1}{1} = 1$, so $y = 1$ is the equation of the horizontal asymptote.

461. $x = 1, y = x + 2$

There are no common factors in the numerator and denominator.

To solve for vertical asymptotes, set the denominator equal to 0.

$x - 1 = 0$ gives you $x = 1$, the equation of the vertical asymptote.

To solve for the oblique asymptote, divide the numerator by the denominator.

$$\begin{array}{r} x + 2 \\ x-1 \overline{\smash{)}\, x^2 + x - 6} \\ \underline{-(x^2 - x)} \\ 2x - 6 \\ \underline{-(2x - 2)} \\ -4 \end{array}$$

The quotient (without the remainder) gives you the equation of the oblique asymptote: $y = x + 2$.

462. $x = -4, y = x - 6$

There are no common factors in the numerator and denominator.

To solve for vertical asymptotes, set the denominator equal to 0.

$x + 4 = 0$ gives you $x = -4$, the equation of the vertical asymptote.

To solve for the oblique asymptote, divide the numerator by the denominator.

$$\begin{array}{r}
x \quad -6 \\
x+4\overline{)x^2-2x-3} \\
\underline{-(x^2+4x)} \\
-6x-3 \\
\underline{-(-6x-24)} \\
+21
\end{array}$$

The quotient (without the remainder) gives you the equation of the oblique asymptote: $y = x - 6$.

463. $x = 5, y = x + 9$

There are no common factors in the numerator and denominator.

To solve for vertical asymptotes, set the denominator equal to 0.

$x - 5 = 0$ gives you $x = 5$, the equation of the vertical asymptote.

To solve for the oblique asymptote, divide the numerator by the denominator.

$$\begin{array}{r}
x \;+\; 9 \\
x-5\overline{)\;x^2+4x} \\
\underline{-(x^2-5x)} \\
+9x \\
\underline{-(9x-45)} \\
+45
\end{array}$$

The quotient (without the remainder) gives you the equation of the oblique asymptote: $y = x + 9$.

464. $x = 2, x = -2, y = x - 3$

There are no common factors in the numerator and denominator.

To solve for vertical asymptotes, set the denominator equal to 0.

$x^2 - 4 = (x - 2)(x + 2) = 0$ gives you $x = 2$ and $x = -2$, the equations of the vertical asymptotes.

To solve for the oblique asymptote, divide the numerator by the denominator.

$$\begin{array}{r} x \;-\; 3 \\ x^2-4\overline{\big)\, x^3 - 3x^2 \qquad\quad +1} \\ \underline{-\left(x^3 \qquad -4x\right)} \\ -3x^2 \;+4x \;\;+1 \\ \underline{-\left(-3x^2 \qquad +12\right)} \\ +4x \;\; -11 \end{array}$$

The quotient (without the remainder) gives you the equation of the oblique asymptote: $y = x - 3$.

465. $y = x$

There are no common factors in the numerator and denominator.

To solve for vertical asymptotes, set the denominator equal to 0.

$x^2 + 1 = 0$ has no real solution, so there are no vertical asymptotes.

To solve for the oblique asymptote, divide the numerator by the denominator.

$$\begin{array}{r} x \\ x^2+1\overline{\big)\,x^3 } \\ \underline{-\left(x^3+x\right)} \\ -x \end{array}$$

The quotient (without the remainder) gives you the equation of the oblique asymptote: $y = x$.

466. Asymptotes: $x = 3$, $y = 1$

The y-intercept is (0, 0). This is the x-intercept, also.

The vertical asymptote is $x = 3$.

The horizontal asymptote is $y = 1$.

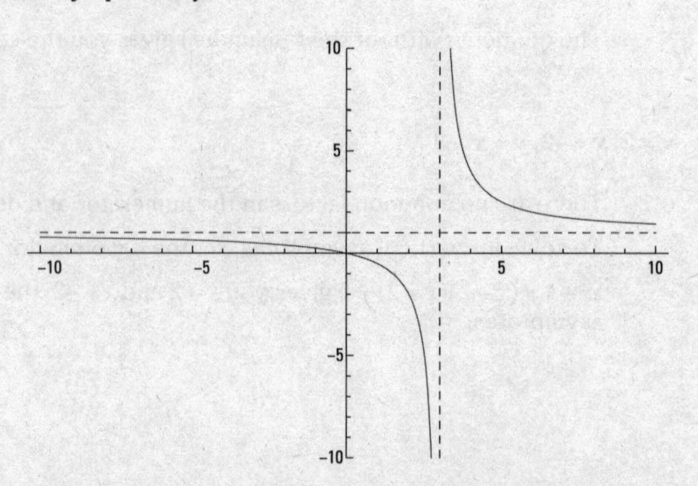

Answers 401–500

467. **Asymptotes: $x = -1$, $y = 2$**

The y-intercept is (0, 0). This is the x-intercept, also.

The vertical asymptote is $x = -1$.

The horizontal asymptote is $y = 2$.

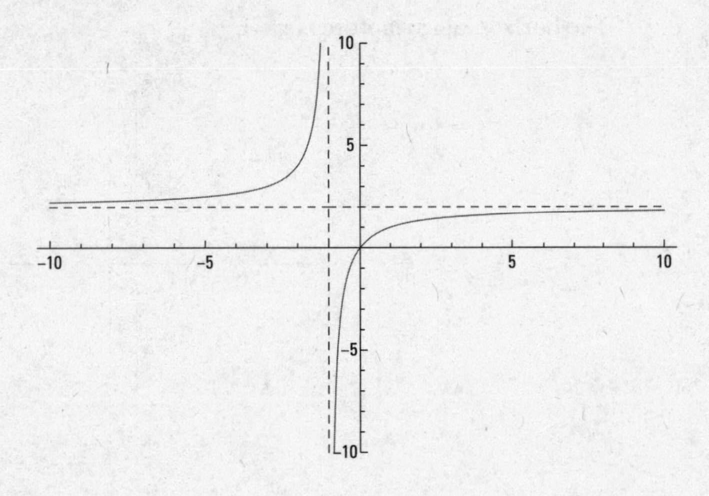

468. **Asymptotes: $x = 2$, $y = 1$**

The y-intercept is (0, 2).

The x-intercept is (4, 0).

The vertical asymptote is $x = 2$.

The horizontal asymptote is $y = 1$.

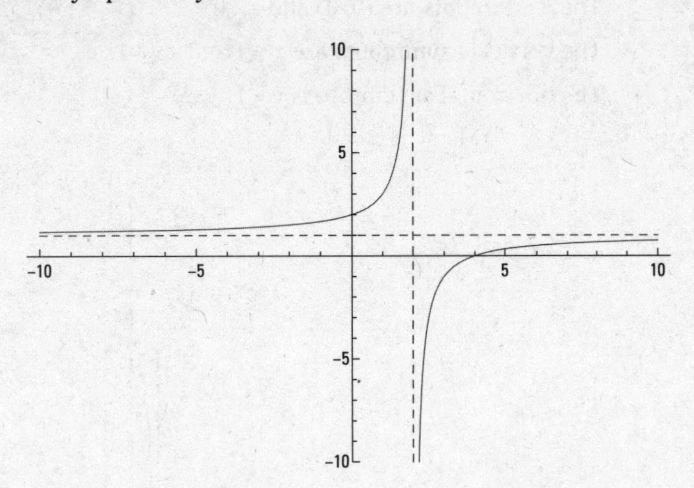

469. Asymptotes: $x = 1, y = 1$

The y-intercept is $(0, -7)$.

The x-intercept is $(-7, 0)$.

The vertical asymptote is $x = 1$.

The horizontal asymptote is $y = 1$.

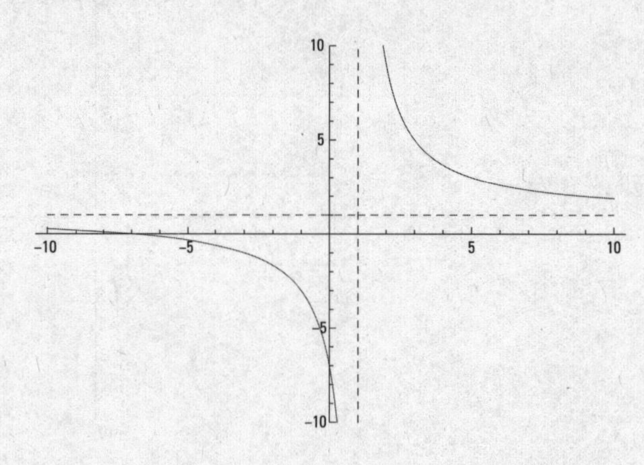

470. Asymptotes: $x = 1, x = -1, y = 1$

First, factor the numerator and denominator:

$$y = \frac{x^2 - 5x}{x^2 - 1} = \frac{x(x-5)}{(x-1)(x+1)}$$

The y-intercept is $(0, 0)$.

The x-intercepts are $(0, 0)$ and $(5, 0)$.

The vertical asymptotes are $x = 1$ and $x = -1$.

The horizontal asymptote is $y = 1$.

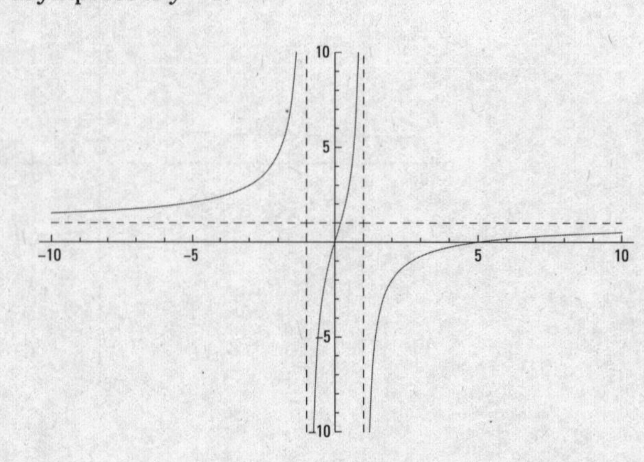

471. Asymptotes: $x = 2, x = -2, y = 1$

First, factor the numerator and denominator:

$$y = \frac{x^2 + 3x}{x^2 - 4} = \frac{x(x+3)}{(x-2)(x+2)}$$

The y-intercept is (0, 0).

The x-intercepts are (0, 0) and (–3, 0).

The vertical asymptotes are $x = 2$ and $x = -2$.

The horizontal asymptote is $y = 1$.

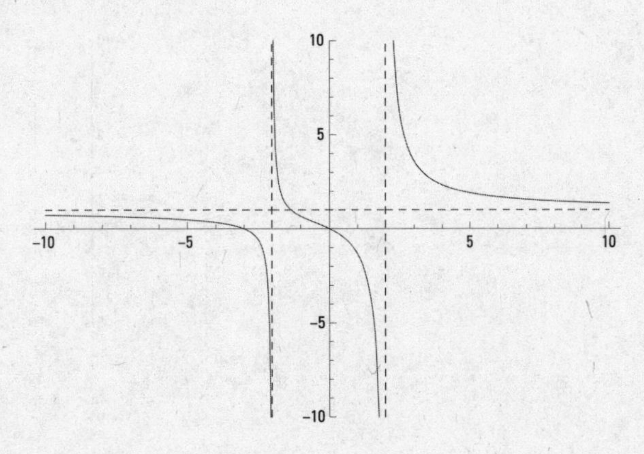

472. Asymptotes: $x = 6, x = -1, y = 2$

First, factor the numerator and denominator:

$$y = \frac{2x^2 - 5x - 12}{x^2 - 5x - 6} = \frac{(2x+3)(x-4)}{(x-6)(x+1)}$$

The y-intercept is (0, 2).

The x-intercepts are $\left(-\frac{3}{2}, 0\right)$ and (4, 0).

The vertical asymptotes are $x = 6$ and $x = -1$.

The horizontal asymptote is $y = 2$.

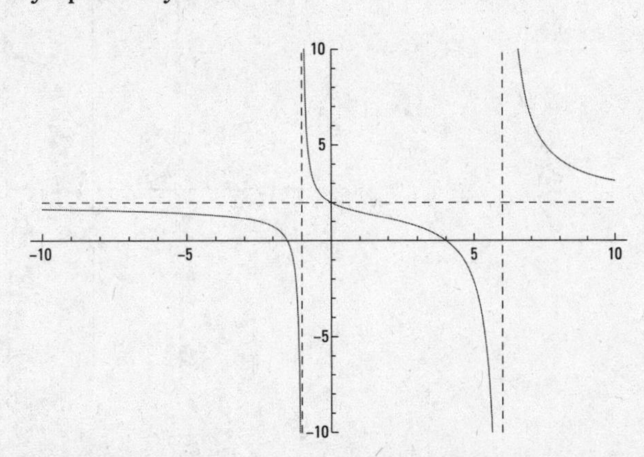

473. Asymptotes: $x = 2, x = -4, y = 3$

First, factor the numerator and denominator:

$$y = \frac{3x^2 - 3}{x^2 + 2x - 8} = \frac{3(x-1)(x+1)}{(x+4)(x-2)}$$

The *y*-intercept is $\left(0, \frac{3}{8}\right)$.

The *x*-intercepts are (1, 0) and (–1, 0).

The vertical asymptotes are $x = -4$ and $x = 2$.

The horizontal asymptote is $y = 3$.

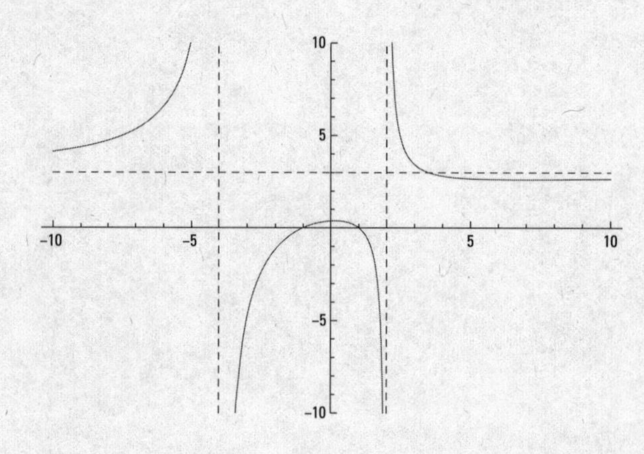

474. Asymptotes: $x = 3, x = -3, y = 0$

First, factor the denominator:

$$y = \frac{x}{x^2 - 9} = \frac{x}{(x-3)(x+3)}$$

The *y*-intercept is (0, 0); this is the *x*-intercept, also.

The vertical asymptotes are $x = 3$ and $x = -3$.

The horizontal asymptote is $y = 0$.

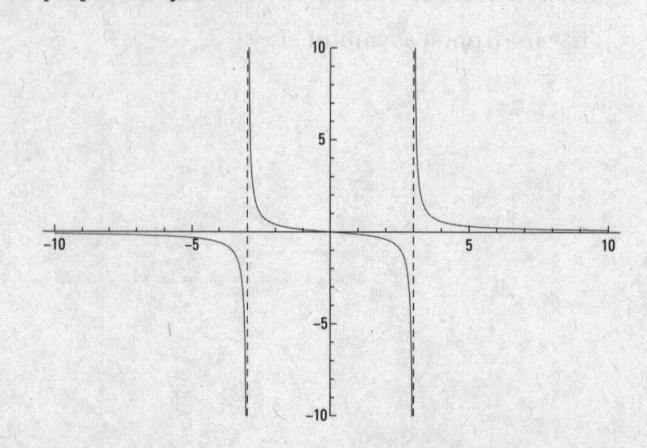

475. Asymptotes: $x = 1, x = -1, y = 0$

First, factor the denominator:

$$y = \frac{4x}{1-x^2} = \frac{4x}{(1-x)(1+x)}$$

The y-intercept is (0, 0); this is the x-intercept, also.

The vertical asymptotes are $x = 1$ and $x = -1$.

The horizontal asymptote is $y = 0$.

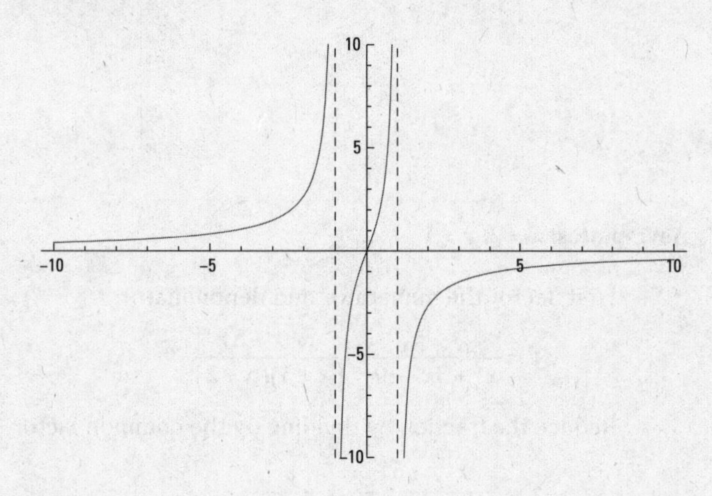

476. Asymptotes: $x = -3, y = 1$

First, factor the numerator and denominator:

$$y = \frac{x^2 - 3x - 4}{x^2 + 4x + 3} = \frac{(x-4)(x+1)}{(x+3)(x+1)}$$

Reduce the fraction by dividing by the common factor, $(x + 1)$.

$$y = \frac{(x-4)\cancel{(x+1)}}{(x+3)\cancel{(x+1)}} = \frac{x-4}{x+3}$$

This indicates a removable discontinuity when $x = -1$. Show this on the graph with a *hole* at the point $\left(-1, -\frac{5}{2}\right)$.

The y-intercept is $\left(0, -\frac{4}{3}\right)$.

The x-intercept is (4, 0).

The vertical asymptote is $x = -3$.

The horizontal asymptote is $y = 1$.

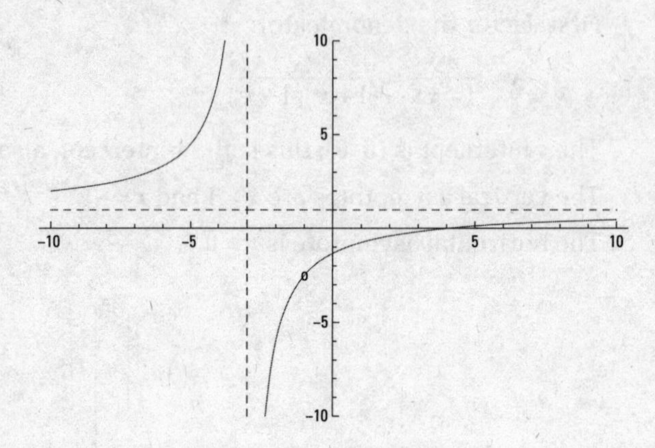

477. **Asymptotes: $x = 2$, $y = 1$**

First, factor the numerator and denominator:

$$y = \frac{x^2 + 5x}{x^2 + 3x - 10} = \frac{x(x+5)}{(x+5)(x-2)}$$

Reduce the fraction by dividing by the common factor, $(x + 5)$.

$$y = \frac{x\cancel{(x+5)}}{\cancel{(x+5)}(x-2)} = \frac{x}{x-2}$$

This indicates a removable discontinuity when $x = -5$. Show this on the graph with a *hole* at the point $\left(-5, \frac{5}{7}\right)$.

The y-intercept is $(0, 0)$; this is also the x-intercept.

The vertical asymptote is $x = 2$.

The horizontal asymptote is $y = 1$.

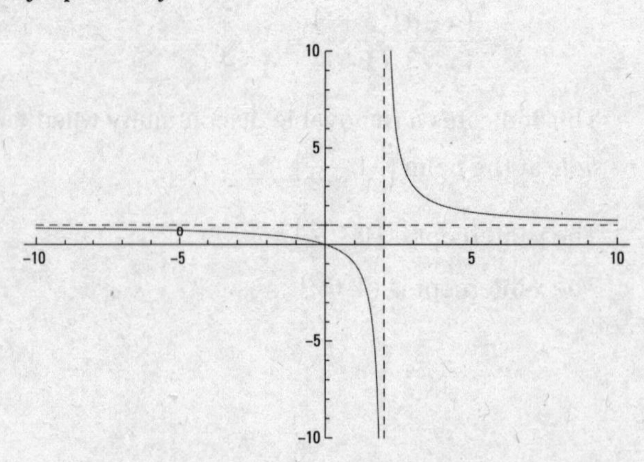

478. Asymptotes: $x = 1, y = x + 2$

First, factor the numerator:

$$y = \frac{x^2 + x}{x - 1} = \frac{x(x+1)}{x-1}$$

There are no common factors in the numerator and denominator.

The y-intercept is (0, 0); this is one of the x-intercepts, too. The other x-intercept is (–1, 0).

The vertical asymptote is $x = 1$.

The oblique asymptote is $y = x + 2$.

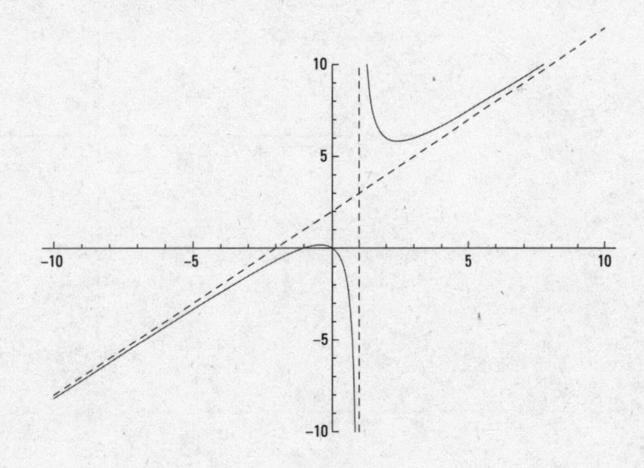

Answers 401–500

479. Asymptotes: $x = -3, y = x$

There are no common factors in the numerator and denominator.

The y-intercept is $\left(0, -\frac{5}{3}\right)$.

The x-intercepts are at about (1.2, 0) and (–4.2, 0).

The vertical asymptote is $x = -3$.

The oblique asymptote is $y = x$.

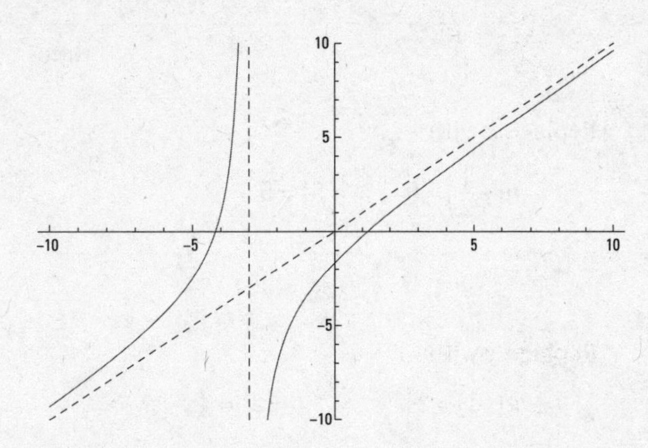

480. **Asymptote: $y = 0$**

There are no common factors in the numerator and denominator.

The y-intercept is (0, 0); this is also the x-intercept.

There is no vertical asymptote.

The horizontal asymptote is $y = 0$.

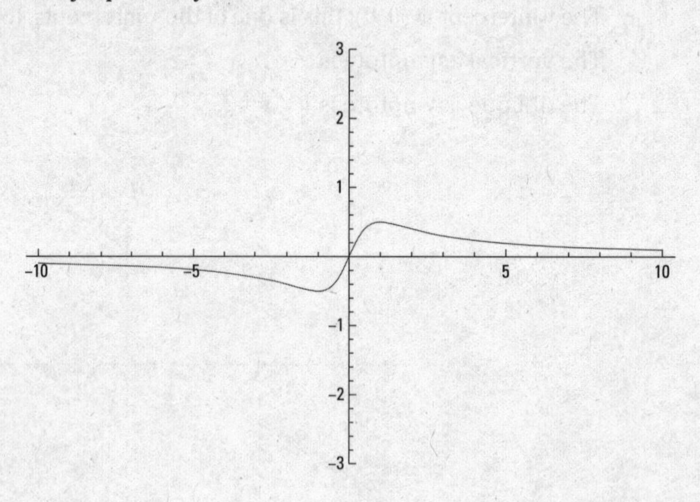

481. **16**

Replace x with 2.

$$f(2) = 4^2 = 16$$

482. $\mathbf{\frac{1}{27}}$

Replace x with –3.

$$g(-3) = 3^{-3} = \frac{1}{3^3} = \frac{1}{27}$$

483. **5**

Replace x with $-\frac{1}{2}$.

$$h\left(-\frac{1}{2}\right) = 5^{-2\left(-\frac{1}{2}\right)} = 5^1 = 5$$

484. **4**

Replace x with –1.

$$k(-1) = 2^{(-1)^2+1} = 2^{1+1} = 2^2 = 4$$

485. **4**

Replace x with 0.

$$f(0)=\left(\frac{1}{4}\right)^{0-1}=\left(\frac{1}{4}\right)^{-1}=\left(\frac{4}{1}\right)^{1}=4^1=4$$

486. **16**

Replace x with 4.

$$g(4)=\left(\frac{1}{2}\right)^{-4}=\left(\frac{2}{1}\right)^{4}=2^4=16$$

487. **80**

Replace x with 3.

$$k(3)=5(2)^{3+1}=5(2)^4=5\cdot 16=80$$

488. **156,250**

Replace x with –3.

$$\begin{aligned}h(-3)&=10\left(\frac{1}{5}\right)^{2(-3)}=10\left(\frac{1}{5}\right)^{-6}\\&=10\left(\frac{5}{1}\right)^{6}=10\cdot 5^6\\&=10\cdot 15{,}625=156{,}250\end{aligned}$$

489. **329**

Replace x with 2.

$$\begin{aligned}m(2)&=4(3)^{2(2)}+5\\&=4(3)^4+5=4(81)+5\\&=324+5=329\end{aligned}$$

490. **9,999,999,997**

Replace x with 2.

$$\begin{aligned}n(2)&=100(0.01)^{-2(2)}-3\\&=100\left(\frac{1}{100}\right)^{-4}-3\\&=100\left(\frac{100}{1}\right)^{4}-3\\&=100(100{,}000{,}000)-3\\&=10{,}000{,}000{,}000-3\\&=9{,}999{,}999{,}997\end{aligned}$$

491. **1**

Replace x with 1.

$$f(1) = e^{1-1} = e^0 = 1$$

492. $\mathbf{\frac{1}{e^2}}$

Replace x with –1.

$$g(-1) = e^{2(-1)} = e^{-2} = \frac{1}{e^2}$$

493. $\mathbf{e^2}$

Replace x with –2.

$$h(-2) = e^{(-2)^2-2} = e^{4-2} = e^2$$

494. $\mathbf{e^2}$

Replace x with 1.

$$k(1) = e^{4(1)-2} = e^{4-2} = e^2$$

495. $\mathbf{\frac{2}{e^4}}$

Replace x with –4.

$$m(-4) = 2e^{-4} = \frac{2}{e^4}$$

496. $\mathbf{-3e^3}$

Replace x with –1.

$$n(-1) = -3e^{-3(-1)} = -3e^3$$

497. $\mathbf{e^3}$

Replace x with –2.

$$p(-2) = \left(\frac{1}{e}\right)^{-2-1} = \left(\frac{1}{e}\right)^{-3}$$
$$= \left(\frac{e}{1}\right)^3 = e^3$$

498. $\mathbf{\frac{1}{4e^2}}$

Replace x with 2.

$$q(2) = \left(\frac{1}{2e}\right)^2 = \frac{1}{4e^2}$$

499. $\frac{e}{3}$

Replace x with 2.

$$r(2) = \left(\frac{3}{e}\right)^{1-2} = \left(\frac{3}{e}\right)^{-1}$$
$$= \left(\frac{e}{3}\right)^{1} = \frac{e}{3}$$

500. $\frac{4}{e^2}$

Replace x with –1.

$$t(-1) = \left(2e^{-1}\right)^2 = 4e^{-2} = \frac{4}{e^2}$$

501. **Rises; through (2, 4)**

The curve rises from left to right, with the x-axis acting as an asymptote toward negative infinity. The y-intercept is (0, 1).

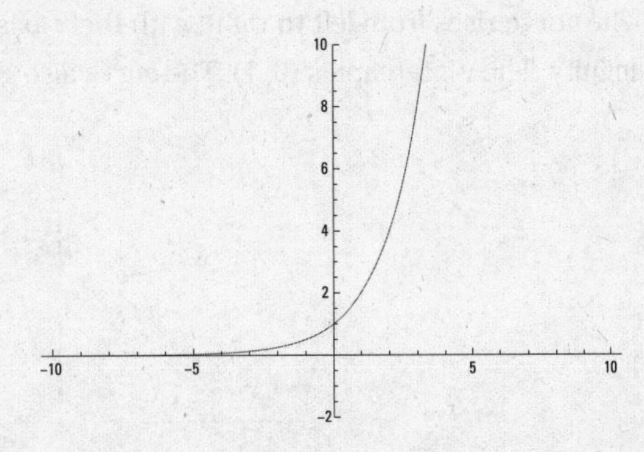

502. **Rises; through (1, 1)**

The curve rises from left to right, with the x-axis acting as an asymptote toward negative infinity. The graph is translated one unit to the right from the basic graph $y = 3^x$, resulting in a y-intercept of $\left(0, \frac{1}{3}\right)$ and points (1, 1), (2, 3), and so on.

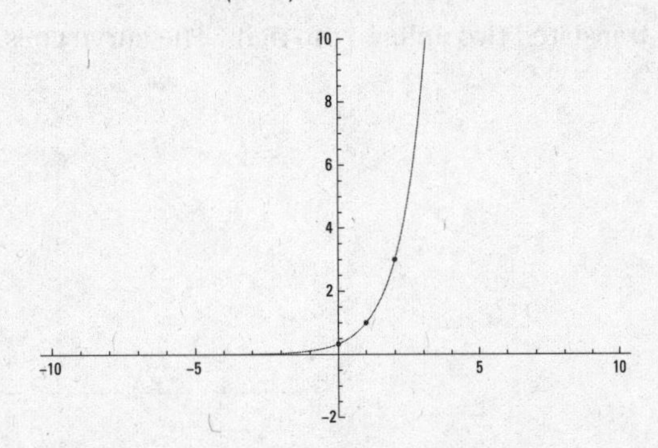

503. Falls; through (–3, 2)

The curve falls from left to right, with the x-axis acting as an asymptote toward positive infinity. The graph is a variation on the graph of $y = 2^x$; it's reflected over the y-axis and translated two units to the left. The curve goes through the points (–2, 1), (–3, 2) and $\left(0, \frac{1}{4}\right)$.

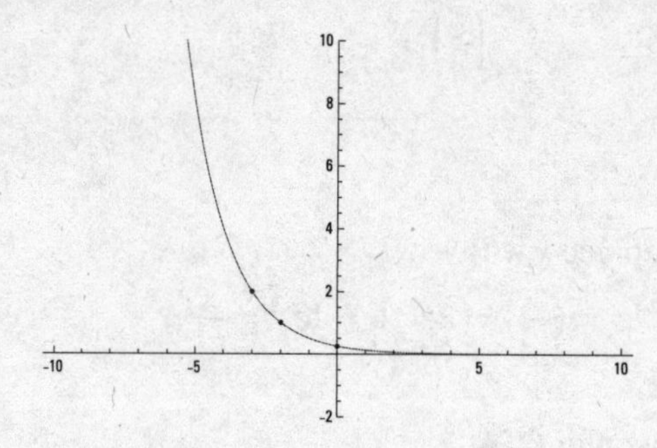

504. Rises; through (1, e)

The curve rises from left to right, with the x-axis acting as an asymptote toward negative infinity. The y-intercept is (0, 1). The curve also goes through the points (1, e) and $\left(-1, \frac{1}{e}\right)$.

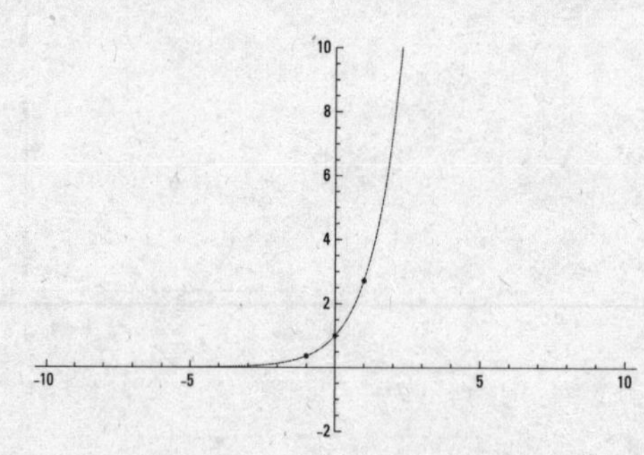

505. Falls; through (2, 1)

The curve falls from left to right, with the x-axis acting as an asymptote toward positive infinity. The graph is a variation on the graph of $y = e^x$; it's reflected over the y-axis and translated two units to the right. The curve goes through (1, e), (2, 1) and $\left(3, \frac{1}{e}\right)$.

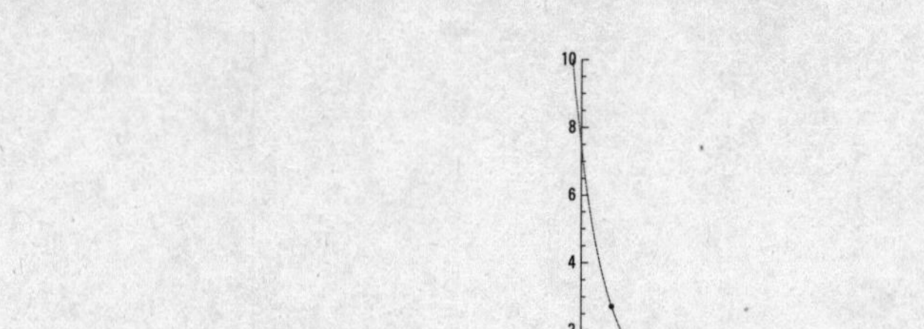

506. **3**

Rewriting the expression as $\log_2(8) = x$, apply the log-exponential equivalence to get $2^x = 8$.

Change the 8 to a power of 2.

$$2^x = 2^3$$

The exponents are equal.

$$x = 3$$

507. **–2**

Rewriting the expression as $\log_3\left(\frac{1}{9}\right) = x$, apply the log-exponential equivalence to get $3^x = \frac{1}{9}$.

Change the $\frac{1}{9}$ to a power of 3.

$$3^x = \frac{1}{3^2} = 3^{-2}$$

The exponents are equal.

$$x = -2$$

508. $-\frac{1}{2}$

Rewriting the expression as $\log_4\left(\frac{1}{2}\right) = x$, apply the log-exponential equivalence to get $4^x = \frac{1}{2}$.

Change both the 4 and the $\frac{1}{2}$ to powers of 2.

$$4^x = \frac{1}{2}$$

$$\left(2^2\right)^x = 2^{-1}$$

$$2^{2x} = 2^{-1}$$

The exponents are equal.

$$2x = -1,\ x = -\frac{1}{2}$$

509. **–3**

Rewriting the expression as $\log_5\left(\frac{1}{125}\right) = x$, apply the log-exponential equivalence to get $5^x = \frac{1}{125}$.

Change the $\frac{1}{125}$ to a power of 5.

$$5^x = \frac{1}{125}$$
$$5^x = \frac{1}{5^3}$$
$$5^x = 5^{-3}$$

The exponents are equal.

$$x = -3$$

510. $\frac{1}{6}$

Rewriting the expression as $\log_8(\sqrt{2}) = x$, apply the log-exponential equivalence to get $8^x = \sqrt{2}$.

Change both the 8 and the $\sqrt{2}$ to powers of 2.

$$8^x = \sqrt{2}$$
$$(2^3)^x = 2^{1/2}$$
$$2^{3x} = 2^{1/2}$$

The exponents are equal.

$$3x = \frac{1}{2},\ x = \frac{1}{6}$$

511. −3

When no base is shown on the log, you assume that it is base 10. Rewriting the expression as $\log_{10}(0.001) = x$, apply the log-exponential equivalence to get $10^x = 0.001$.

Change the 0.001 to a power of 10.

$$10^x = 0.001$$
$$10^x = 10^{-3}$$

The exponents are equal.

$$x = -3$$

512. $\frac{3}{2}$

Rewriting the expression as $\log_9(27) = x$, apply the log-exponential equivalence to get $9^x = 27$.

Change both the 9 and the 27 to powers of 3.

$$9^x = 27$$
$$(3^2)^x = 3^3$$
$$3^{2x} = 3^3$$

The exponents are equal.

$$2x = 3,\ x = \frac{3}{2}$$

513. $\frac{3}{4}$

Rewriting the expression as $\log_{16}(8) = x$, apply the log-exponential equivalence to get $16^x = 8$.

Change both the 16 and the 8 to powers of 2.

$$16^x = 8$$
$$(2^4)^x = 2^3$$
$$2^{4x} = 2^3$$

The exponents are equal.

$$4x = 3,\ x = \frac{3}{4}$$

514. $\frac{5}{2}$

Rewriting the expression as $\log_4(32) = x$, apply the log-exponential equivalence to get $4^x = 32$.

Change both the 4 and the 32 to powers of 2.

$$4^x = 32$$
$$(2^2)^x = 2^5$$
$$2^{2x} = 2^5$$

The exponents are equal.

$$2x = 5,\ x = \frac{5}{2}$$

515. $\frac{3}{2}$

When no base is shown on the log, you assume that it is base 10. Rewriting the expression as $\log_{10}\left(\sqrt{1{,}000}\right) = x$, apply the log-exponential equivalence to get $10^x = \sqrt{1{,}000}$.

Change the $\sqrt{1{,}000}$ to a power of 10.

$$10^x = \sqrt{1{,}000}$$
$$10^x = 1{,}000^{1/2}$$
$$10^x = \left(10^3\right)^{1/2}$$
$$10^x = 10^{3/2}$$

The exponents are equal.

$$x = \frac{3}{2}$$

516. 2

Rewriting the expression as $\ln e^2 = x$, apply the log-exponential equivalence, using the base e, to get $e^x = e^2$.

The exponents are equal.

$$x = 2$$

517. $-\frac{1}{2}$

Rewriting the expression as $\ln e^{-1/2} = x$, apply the log-exponential equivalence, using the base e, to get $e^x = e^{-1/2}$.

The exponents are equal.

$$x = -\frac{1}{2}$$

518. –1

Rewriting the expression as $\ln \frac{1}{e} = x$, apply the log-exponential equivalence, using the base e, to get $e^x = \frac{1}{e}$.

Now write the term on the right as a power of e.

$$e^x = \frac{1}{e}$$

$$e^x = e^{-1}$$

The exponents are equal.

$$x = -1$$

519. –5

Rewriting the expression as $\ln \frac{1}{e^5} = x$, apply the log-exponential equivalence, using the base e, to get $e^x = \frac{1}{e^5}$.

Now write the term on the right as a power of e.

$$e^x = \frac{1}{e^5}$$

$$e^x = e^{-5}$$

The exponents are equal.

$$x = -5$$

520. 0

Rewriting the expression as $\ln 1 = x$, apply the log-exponential equivalence, using the base e, to get $e^x = 1$.

Now write the term on the right as a power of e.

$$e^x = 1$$

$$e^x = e^0$$

The exponents are equal.

$$x = 0$$

521. 1

Rewriting the expression as $\ln e = x$, apply the log-exponential equivalence, using the base e, to get $e^x = e$.

Now write the term on the right as a power of e.

$$e^x = e$$

$$e^x = e^1$$

The exponents are equal.

$$x = 1$$

522. $\frac{1}{2}$

Rewriting the expression as $\ln\sqrt{e} = x$, apply the log-exponential equivalence, using the base e, to get $e^x = \sqrt{e}$.

Now write the term on the right as a power of e.

$$e^x = \sqrt{e}$$

$$e^x = e^{1/2}$$

The exponents are equal.

$$x = \frac{1}{2}$$

523. $\frac{5}{2}$

Rewriting the expression as $\ln\sqrt{e^5} = x$, apply the log-exponential equivalence, using the base e, to get $e^x = \sqrt{e^5}$.

Now write the term on the right as a power of e.

$$e^x = \sqrt{e^5}$$

$$e^x = e^{5/2}$$

The exponents are equal.

$$x = \frac{5}{2}$$

524. $\frac{3}{2}$

Rewriting the expression as $\ln\frac{e^2}{\sqrt{e}} = x$, apply the log-exponential equivalence, using the base e, to get $e^x = \frac{e^2}{\sqrt{e}}$.

Now write the term on the right as powers of e.

$$e^x = \frac{e^2}{\sqrt{e}}$$

$$e^x = \frac{e^2}{e^{1/2}}$$

Simplify the fraction on the right by subtracting the exponents.

$$e^x = e^{2-1/2}$$

$$e^x = e^{3/2}$$

The exponents are equal.

$$x = \frac{3}{2}$$

525. $-\frac{14}{3}$

Rewriting the expression as $\ln\frac{\sqrt[3]{e}}{e^5} = x$, apply the log-exponential equivalence, using the base e, to get $e^x = \frac{\sqrt[3]{e}}{e^5}$.

Now write the term on the right as powers of e.

$$e^x = \frac{\sqrt[3]{e}}{e^5}$$

$$e^x = \frac{e^{1/3}}{e^5}$$

Simplify the fraction on the right by subtracting the exponents.

$$e^x = e^{1/3-5}$$

$$e^x = e^{-14/3}$$

The exponents are equal.

$$x = -\frac{14}{3}$$

526. **Rises; through (1, 0)**

The curve rises next to its vertical asymptote, the y-axis. It crosses the x-axis at (1, 0) and goes through the points (2, 1) and (4, 2).

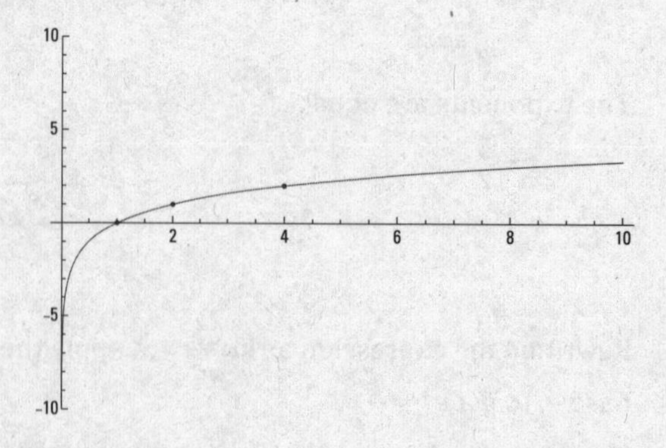

527. **Rises; through (2, 0)**

The curve rises next to its vertical asymptote, the line $x = 1$. It crosses the x-axis at (2, 0) and goes through the points (3, 1) and (5, 2). Note that this graph is a translation of the basic graph of $y = \log_2 x$ one unit to the right.

528. Rises; through (–1, 0)

The curve rises next to its vertical asymptote, the line $x = -2$. It crosses the x-axis at (–1, 0) and also goes through the point (8, 1). Note that this graph is a translation of the basic graph of $y = \log_{10} x$ two units to the left.

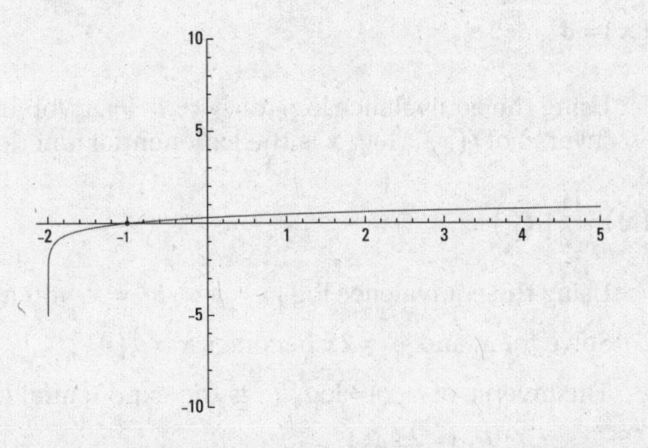

529. Rises; through (1, 0)

The curve rises next to its vertical asymptote, the y-axis. It crosses the x-axis at (1, 0) and goes through the points (e, 1) and (e^2, 2).

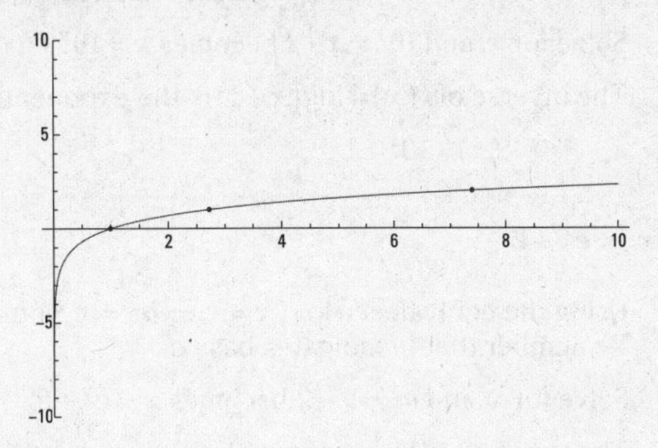

530. **Rises; through (*e*, 2)**

The curve rises next to its vertical asymptote, the *y*-axis. It crosses the *x*-axis at (1, 0) and goes through the points $(e, 2)$ and $(e^2, 4)$.

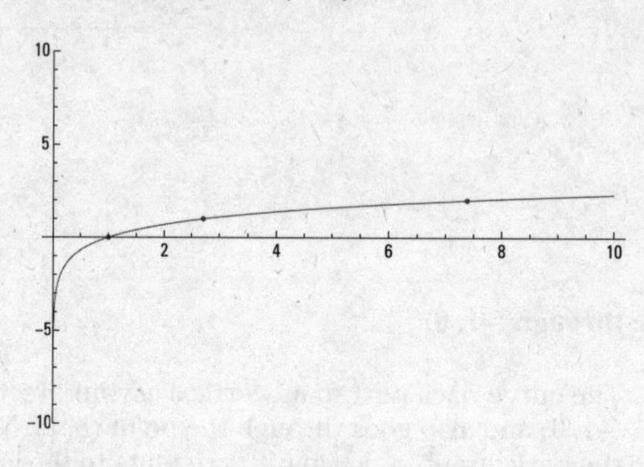

531. $f^{-1}(x)=3^x$

Using the equivalence $\log_b x = y \Leftrightarrow b^y = x$, you have $\log_3 x = y \Leftrightarrow 3^y = x$. From this, the inverse of $f(x)=\log_3 x$ is the exponential function with base 3, $f^{-1}(x)=3^x$.

532. $f^{-1}(x)=\frac{1}{2}(4^x)$

Using the equivalence $\log_b x = y \Leftrightarrow b^y = x$, you have $\log_4 2x = y \Leftrightarrow 4^y = 2x$.

Solve for *x*, and $4^y = 2x$ becomes $x=\frac{1}{2}(4^y)$.

The inverse of $f(x)=\log_4 2x$ is the exponential function with base 4,

$$f^{-1}(x)=\frac{1}{2}(4^x).$$

533. $f^{-1}(x)=10^x-1$

Using the equivalence $\log_b x = y \Leftrightarrow b^y = x$, you have $\log(x+1)=y \Leftrightarrow 10^y = x+1$. Remember that when no log base is indicated, it's base 10.

Solve for *x*, and $10^y = x+1$ becomes $x=10^y-1$.

The inverse of $f(x)=\log(x+1)$ is the exponential function with base 10,

$$f^{-1}(x)=10^x-1.$$

534. $f^{-1}(x)=e^x+2$

Using the equivalence $\log_b x = y \Leftrightarrow b^y = x$, you have $\ln(x-2)=y \Leftrightarrow e^y = x-2$. Remember that ln indicates base *e*.

Solve for *x*, and $e^y = x-2$ becomes $x=e^y+2$.

The inverse of $f(x)=\ln(x-2)$ is the exponential function with base *e*,

$$f^{-1}(x)=e^x+2.$$

535. $f^{-1}(x) = e^{x/3}$

Before using the equivalence $\log_b x = y \Leftrightarrow b^y = x$, first write the function with y instead of $f(x)$, and then divide each side of the equation by 3.

$$3\ln x = y$$
$$\ln x = \frac{y}{3}$$
$$\ln x = \frac{y}{3} \Leftrightarrow e^{y/3} = x$$

Remember that ln indicates base e.

The inverse of $f(x) = 3\ln x$ is the exponential function with base e,

$$f^{-1}(x) = e^{x/3}.$$

536. $f^{-1}(x) = \log_2 x$

Applying the equivalence $b^x = y \Leftrightarrow \log_b y = x$, you have $2^x = y \Leftrightarrow \log_2 y = x$.

The inverse of the exponential function $f(x) = 2^x$ is the logarithmic function $f^{-1}(x) = \log_2 x$.

537. $f^{-1}(x) = -1 + \log_3 x$

Applying the equivalence $b^x = y \Leftrightarrow \log_b y = x$, you have $3^{x+1} = y \Leftrightarrow \log_3 y = x + 1$.

Solving for x, you find that $x = -1 + \log_3 y$. Notice that the –1 is written in front to prevent any misinterpretation — thinking that 1 is subtracted from y.

The inverse of the exponential function $f(x) = 3^{x+1}$ is the logarithmic function $f^{-1}(x) = -1 + \log_3 x$.

538. $f^{-1}(x) = 1 + \ln x$

Applying the equivalence $b^x = y \Leftrightarrow \log_b y = x$, you have $e^{x-1} = y \Leftrightarrow \log_e y = x - 1$.

Solving for x and changing $\log_e$ to ln, you first get $1 + \ln y = x$.

Notice that the 1 is written in front to prevent any misinterpretation — thinking that 1 is added to y.

The inverse of the exponential function $f(x) = e^{x-1}$ is the logarithmic function $f^{-1}(x) = 1 + \ln x$.

539. $f^{-1}(x) = \frac{1}{2}\ln\frac{x}{3}$

Before using the equivalence $b^x = y \Leftrightarrow \log_b y = x$, first write the function with y instead of $f(x)$, and then divide each side of the equation by 3.

$$y = 3e^{2x}$$
$$\frac{y}{3} = e^{2x}$$

Now, apply the equivalence.

$$e^{2x} = \frac{y}{3} \Leftrightarrow \log_e \frac{y}{3} = 2x$$

Solving for x and changing $\log_e$ to ln, you first get

$$\ln\frac{y}{3} = 2x$$
$$\frac{1}{2}\ln\frac{y}{3} = x.$$

The inverse of the exponential function $f(x) = 3e^{2x}$ is the logarithmic function $f^{-1}(x) = \frac{1}{2}\ln\frac{x}{3}$.

540. $f^{-1}(x) = -\ln x$

Before using the equivalence $b^x = y \Leftrightarrow \log_b y = x$, first write the function rule with a negative exponent. You have

$$f(x) = e^{-x}$$

Now, apply the equivalence.

$$e^{-x} = y \Leftrightarrow \log_e y = -x$$

Solving for x and changing $\log_e$ to ln, you first get $-\ln y = x$.

The inverse of the exponential function $f(x) = \frac{1}{e^x}$ is the logarithmic function $f^{-1}(x) = -\ln x$.

541. $x = \frac{7}{5}$

Rewrite the equation, changing both of the bases to powers of 2.

$$4^{x+1} = 8^{3-x}$$
$$\left(2^2\right)^{x+1} = \left(2^3\right)^{3-x}$$

You raise a power to a power by multiplying the exponents.

$$2^{2(x+1)} = 2^{3(3-x)}$$
$$2^{2x+2} = 2^{9-3x}$$

The bases are the same, so the exponents are equal.

$$2x + 2 = 9 - 3x$$
$$5x = 7$$
$$x = \frac{7}{5}$$

542. 9

Rewrite the equation changing both of the bases to powers of 5.

$$25^{2x-3} = 125^{x+1}$$
$$\left(5^2\right)^{2x-3} = \left(5^3\right)^{x+1}$$

You raise a power to a power by multiplying the exponents.

$$5^{2(x-3)} = 5^{3x(+1)}$$
$$5^{4x-6} = 5^{3x+3}$$

The bases are the same, so the exponents are equal.

$$4x - 6 = 3x + 3$$
$$x = 9$$

543. $x = -\frac{1}{4}$

Rewrite the equation, changing both of the bases to powers of 3.

$$9^{x-5} = 27^{2x-3}$$
$$\left(3^2\right)^{x-5} = \left(3^3\right)^{2x-3}$$

You raise a power to a power by multiplying the exponents.

$$3^{2(x-5)} = 3^{3(2x-3)}$$
$$3^{2x-10} = 3^{6x-9}$$

The bases are the same, so the exponents are equal.

$$2x - 10 = 6x - 9$$
$$-4x = 1$$
$$x = -\frac{1}{4}$$

544. $x = 2$

Rewrite the equation, changing both of the bases to powers of 2.

$$16^{x-2} = 32^{2-x}$$
$$\left(2^4\right)^{x-2} = \left(2^5\right)^{2-x}$$

You raise a power to a power by multiplying the exponents.

$$2^{4(x-2)} = 2^{5(2-x)}$$
$$2^{4x-8} = 2^{10-5x}$$

The bases are the same, so the exponents are equal.

$$4x - 8 = 10 - 5x$$
$$9x = 18$$
$$x = 2$$

545. $x = \frac{9}{5}$

Rewrite the equation, changing both of the bases to powers of 3.

$$\left(\frac{1}{9}\right)^x = 27^{x-3}$$
$$\left(3^{-2}\right)^x = \left(3^3\right)^{x-3}$$

You raise a power to a power by multiplying the exponents.

$$3^{-2(x)} = 3^{3(x-3)}$$
$$3^{-2x} = 3^{3x-9}$$

The bases are the same, so the exponents are equal.

$$-2x = 3x - 9$$
$$-5x = -9$$
$$x = \frac{9}{5}$$

546. $x = 0$

Rewrite the equation, changing both of the bases to powers of 2.

$$\left(\frac{1}{8}\right)^{2-x} = \left(\frac{1}{4}\right)^{x+3}$$
$$\left(2^{-3}\right)^{2-x} = \left(2^{-2}\right)^{x+3}$$

You raise a power to a power by multiplying the exponents.

$$2^{-3(2-x)} = 2^{-2(x+3)}$$
$$2^{-6+3x} = 2^{-2x-6}$$

The bases are the same, so the exponents are equal.

$$-6 + 3x = -2x - 6$$
$$5x = 0$$
$$x = 0$$

547. $x = -\frac{4}{9}$

Rewrite the equation, changing both of the bases to powers of 2.

$$\sqrt{8^x} = 4^{3x+1}$$
$$\left(\left(2^3\right)^x\right)^{1/2} = \left(2^2\right)^{3x+1}$$

You raise a power to a power by multiplying the exponents.

$$2^{3(x)(1/2)} = 2^{2(3x+1)}$$
$$2^{3x/2} = 2^{6x+2}$$

The bases are the same, so the exponents are equal.

$$\frac{3x}{2} = 6x + 2$$
$$3x = 12x + 4$$
$$-9x = 4$$
$$x = -\frac{4}{9}$$

548. $x = -\frac{1}{6}$

Rewrite the equation, changing the right term to a power of *e*.

$$e^{5x+2} = \left(\frac{1}{e}\right)^{x-1}$$
$$e^{5x+2} = \left(e^{-1}\right)^{x-1}$$

You raise a power to a power by multiplying the exponents.

$$e^{5x+2} = e^{-1(x-1)}$$
$$e^{5x+2} = e^{-x+1}$$

The bases are the same, so the exponents are equal.

$$5x + 2 = -x + 1$$
$$6x = -1$$
$$x = -\frac{1}{6}$$

549. $x = \frac{14}{3}$

Rewrite the equation, changing the left term to a power of *e*.

$$\sqrt{e^{x-4}} = e^{5-x}$$
$$\left(e^{x-4}\right)^{1/2} = e^{5-x}$$

You raise a power to a power by multiplying the exponents.

$$e^{(x-4)(1/2)} = e^{5-x}$$
$$e^{x/2-2} = e^{5-x}$$

The bases are the same, so the exponents are equal.

$$\frac{x}{2} - 2 = 5 - x$$
$$\frac{x}{2} = 7 - x$$
$$\frac{3}{2}x = 7$$
$$x = \frac{14}{3}$$

550. $x = -1$ or $x = -2$

Rewrite the equation, changing the right term to a power of *e*.

$$e^{x^2} = \frac{1}{e^{3x+2}}$$
$$e^{x^2} = e^{-(3x+2)}$$

The bases are the same, so the exponents are equal.

$$x^2 = -(3x + 2)$$

$$x^2 = -3x - 2$$

$$x^2 + 3x + 2 = 0$$

$$(x + 1)(x + 2) = 0$$

$x = -1$ or $x = -2$

Check $x = -1$ in the original equation.

$$e^{(-1)^2} \stackrel{?}{=} \frac{1}{e^{3(-1)+2}}$$

$$e^1 \stackrel{?}{=} \frac{1}{e^{-1}}$$

$$e^1 = e^1$$

The solution $x = -1$ works.

Now, try $x = -2$.

$$e^{(-2)^2} \stackrel{?}{=} \frac{1}{e^{3(-2)+2}}$$

$$e^4 \stackrel{?}{=} \frac{1}{e^{-4}}$$

$$e^4 = e^4$$

This solution works, also.

551. $x = 8$

Applying the equivalence $\log_b x = y \Leftrightarrow b^y = x$, you have $2^3 = x$.

So $x = 8$.

552. $x = 9$

Applying the equivalence $\log_b x = y \Leftrightarrow b^y = x$, you have $3^2 = x$.

So $x = 9$.

553. $x = 3$

Applying the equivalence $\log_b x = y \Leftrightarrow b^y = x$, you have $9^{1/2} = x$.

Rewrite the fractional exponent as a radical.

$$\sqrt{9} = x$$

So $x = 3$.

554. $x = \frac{1}{2}$

Applying the equivalence $\log_b x = y \Leftrightarrow b^y = x$, you have $16^{-1/4} = x$.

Rewrite the fractional exponent as a radical in a fraction.

$$\frac{1}{\sqrt[4]{16}} = x$$

$$\frac{1}{2} = x$$

555. $x = \frac{1}{7}$

Applying the equivalence $\log_b x = y \Leftrightarrow b^y = x$, you have $7^{-1} = x$.

Rewrite the negative exponent as a fraction.

$$\frac{1}{7} = x$$

556. $x = \frac{1}{216}$

Applying the equivalence $\log_b x = y \Leftrightarrow b^y = x$, you have $6^{-3} = x$.

Rewrite the negative exponent as a fraction.

$$\frac{1}{6^3} = \frac{1}{216} = x$$

557. $x = \frac{1}{e^5}$

Applying the equivalence $\ln x = y \Leftrightarrow e^y = x$, you have $e^{-5} = x$.

Rewrite the negative exponent as a fraction.

$$\frac{1}{e^5} = x$$

558. $x = e^2 - 1$

Applying the equivalence $\ln x = y \Leftrightarrow e^y = x$, you have $e^2 = x + 1$.

Solve for x.

$$x = e^2 - 1$$

559. $x = \frac{e^{-4} + 2}{2}$

Applying the equivalence $\ln x = y \Leftrightarrow e^y = x$, you have $e^{-4} = 2x - 2$.

Solve for x.

$$2x = e^{-4} + 2$$

$$x = \frac{e^{-4} + 2}{2}$$

If you prefer not having a negative exponent in the numerator, then multiply both numerator and denominator by e^4, giving you

$$x = \frac{e^{-4} + 2}{2} \cdot \frac{e^4}{e^4} = \frac{e^0 + 2e^4}{2e^4} = \frac{1 + 2e^4}{2e^4}.$$

560. $x = 4$

Applying the equivalence $\ln x = y \Leftrightarrow e^y = x$, you have $e^5 = e^{x+1}$.

The bases are the same, so the exponents are equal.

$$5 = x + 1$$

$$4 = x$$

561. $x = 2$

Rewrite the expression on the left, using the law of logarithms involving the log of a product.

$$\log_4 x(x+2) = \frac{3}{2}$$

Now apply the equivalence $\log_b x = y \Leftrightarrow b^y = x$ to get

$$4^{3/2} = x(x+2).$$

Compute the power of 4 on the left, and distribute on the right.

$$8 = x^2 + 2x$$

Solve for x.

$$x^2 + 2x - 8 = 0$$

$$(x + 4)(x - 2) = 0$$

$x = -4$ or $x = 2$

Check $x = -4$ in the original equation.

$$\log_4(-4) + \log_4(-4+2) = \frac{3}{2}$$

Throw this one out immediately. You can't have the log of a negative number.

Now try $x = 2$.

$$\log_4(2) + \log_4(2+2) = \frac{3}{2}$$

$$\frac{1}{2} + 1 = \frac{3}{2}$$

This one works.

562. $x = 3$

Rewrite the expression on the left, using the law of logarithms involving the log of a product.

$$\log_3(x-2)(x+6) = 2$$

Now apply the equivalence $\log_b x = y \Leftrightarrow b^y = x$ to get

$$3^2 = (x-2)(x+6).$$

Compute the power of 3 on the left, and multiply on the right.

$$9 = x^2 + 4x - 12$$

Solve for x.

$$x^2 + 4x - 21 = 0$$

$$(x + 7)(x - 3) = 0$$

$x = -7$ or $x = 3$

Check $x = -7$ in the original equation.

$$\log_3(-7-2)+\log_3(-7+6)=2$$

Throw this one out immediately. You can't have the log of a negative number.

Now try $x = 3$.

$$\log_3(3-2)+\log_3(3+6)=2$$

$$0+2=2$$

The 3 works.

563. $x = 2$

Rewrite the expression on the left, using the law of logarithms involving the log of a quotient.

$$\log_2 \frac{x+14}{x} = 3$$

Now apply the equivalence $\log_b x = y \Leftrightarrow b^y = x$ to get

$$2^3 = \frac{x+14}{x}.$$

Compute the power of 2 on the left, and then multiply each side by x.

$$8 = \frac{x+14}{x}$$

$$8x = x + 14$$

Solve for x.

$$7x = 14$$

$$x = 2$$

Check $x = 2$.

$$\log_2(2+14)-\log_2 2=3$$

$$4-1=3$$

The 2 works.

564. $x = -2$

Rewrite the expression on the left, using the law of logarithms involving the log of a product.

$$\log_7(5-x)(2-x)=\log_7 28$$

Now apply the equivalence $\log_b x = \log_b y \Leftrightarrow x = y$ to get

$$(5 - x)(2 - x) = 28.$$

Multiply on the left.

$$10 - 7x + x^2 = 28$$

Solve for x.

$$x^2 - 7x - 18 = 0$$

$$(x + 2)(x - 9) = 0$$

$x = -2$ or $x = 9$

Check $x = -2$ in the original equation.

$$\log_7(5+2) + \log_7(2+2) = \log_7 28$$

$$\log_7 7 + \log_7 4 = \log_7 28$$

$$\log_7(7 \cdot 4) = \log_7 28$$

It checks.

Now try $x = 9$.

$$\log_7(5-9) + \log_7(2-9) = \log_7 28$$

Throw this one out immediately. You can't have the log of a negative number.

565. **279,933**

Rewrite the expression on the left, using the law of logarithms involving the log of a quotient.

$$\log_6 \frac{x^2 - 9}{x - 3} = 7$$

Factor and reduce the fraction.

$$\log_6 \frac{\cancel{(x-3)}(x+3)}{\cancel{x-3}} = 7$$

$$\log_6(x+3) = 7$$

Now apply the equivalence $\log_b x = y \Leftrightarrow b^y = x$ to get

$$6^7 = x + 3.$$

Solve for x.

$$x = 6^7 - 3 = 279{,}933$$

566. $x = 3$

Rewrite the expression on the left, using the law of logarithms involving the log of a product.

$$\ln(x+1)(x-2) = \ln 4$$

Now apply the equivalence $\ln x = \ln y \Leftrightarrow x = y$ to get

$$(x + 1)(x - 2) = 4.$$

Multiply on the left.

$$x^2 - x - 2 = 4$$

Solve for x.

$$x^2 - x - 6 = 0$$

$$(x + 2)(x - 3) = 0$$

$x = -2$ or $x = 3$

Check $x = -2$ in the original equation.

$$\ln(-2+1) + \ln(-2-2) = \ln 4$$

Throw this one out immediately. You can't have the log of a negative number.

Now try $x = 3$.

$$\ln(3+1) + \ln(3-2) = \ln 4$$

$$\ln 4 + \ln 1 = \ln 4$$

$$\ln(4 \cdot 1) = \ln 4$$

It checks.

567. $x = 8$

Rewrite the expression on the left, using the law of logarithms involving the log of a product.

$$\ln(x-5)(x-7) = \ln 3$$

Now apply the equivalence $\ln x = \ln y \Leftrightarrow x = y$ to get

$$(x - 5)(x - 7) = 3.$$

Multiply on the left.

$$x^2 - 12x + 35 = 3$$

Solve for x.

$$x^2 - 12x + 32 = 0$$

$$(x - 4)(x - 8) = 0$$

$x = 4$ or $x = 8$

Check $x = 4$ in the original equation.

$$\ln(4-5) + \ln(4-7) = \ln 3$$

Throw this one out immediately. You can't have the log of a negative number.

Now try $x = 8$.

$$\ln(8-5) + \ln(8-7) = \ln 3$$

$$\ln 3 + \ln 1 = \ln 3$$

$$\ln(3 \cdot 1) = \ln 3$$

It checks.

568. **No solution**

Rewrite the expression on the left, using the law of logarithms involving the log of a product.

$$\ln x(x-4) = \ln(x-6)$$

Now apply the equivalence $\ln x = \ln y \Leftrightarrow x = y$ to get

$$x(x-4) = x-6.$$

Multiply on the left.

$$x^2 - 4x = x - 6$$

Solve for x.

$$x^2 - 5x + 6 = 0$$

$$(x-2)(x-3) = 0$$

$x = 2$ or $x = 3$

Check $x = 2$ in the original equation.

$$\ln 2 + \ln(2-4) = \ln(2-6)$$

Throw this one out immediately. You can't have the log of a negative number.

Now try $x = 3$.

$$\ln 3 + \ln(3-4) = \ln(3-6)$$

This one doesn't work, either! There is no solution.

569. $x = 3$

Rewrite the expression on the left, using the law of logarithms involving the log of a product.

$$\ln x(x+3) = \ln(2x+12)$$

Now apply the equivalence $\ln x = \ln y \Leftrightarrow x = y$ to get

$$x(x+3) = 2x + 12$$

Multiply on the left.

$$x^2 + 3x = 2x + 12$$

Solve for x.

$$x^2 + x - 12 = 0$$

$$(x+4)(x-3) = 0$$

$x = -4$ or $x = 3$

Check $x = -4$ in the original equation.

$$\ln(-4) + \ln(-4+3) = \ln(-8+12)$$

Throw this one out immediately. You can't have the log of a negative number.

Now try $x = 3$.

$$\ln(3)+\ln(3+3)=\ln(6+12)$$
$$\ln 3+\ln 6=\ln 18$$
$$\ln(3\cdot 6)=\ln 18$$

It checks.

570. $x = 7$

Rewrite the expression on the left, using the law of logarithms involving the log of a quotient.

$$\ln\frac{x+3}{x-2}=\ln 2$$

Now apply the equivalence $\ln x=\ln y \Leftrightarrow x=y$ to get $\frac{x+3}{x-2}=2$.

Solve for x by first multiplying each side by $x - 2$.

$$x+3=2(x-2)$$
$$x+3=2x-4$$
$$7=x$$

571. $y + 7 = (x - 3)^2$

First, add –2 to each side of the equation.

$$y-2=x^2-6x$$

Complete the square on the right by adding 9 to each side.

$$y-2+9=x^2-6x+9$$

Simplify on the left, and factor on the right.

$$y+7=(x-3)^2$$

572. $y - 3 = -2(x - 2)^2$

First, add 5 to each side of the equation.

$$y+5=-2x^2+8x$$

Next, factor the –2 out of the terms on the right.

$$y+5=-2(x^2-4x)$$

Complete the square in the parentheses by adding 4; then add –8 to the other side, because the 4 in the parentheses is being multiplied by the –2.

$$y+5-8=-2(x^2-4x+4)$$

Simplify on the left, and factor on the right.

$$y-3=-2(x-2)^2$$

573. $y + 27 = \frac{1}{2}(x - 8)^2$

First, add –10 to each side of the equation.

$$2y - 10 = x^2 - 16x$$

Complete the square on the right by adding 64 to each side.

$$2y - 10 + 64 = x^2 - 16x + 64$$

Simplify on the left, and factor on the right.

$$2y + 54 = (x - 8)^2$$

Now factor the 2 from the terms on the left.

$$2(y + 27) = (x - 8)^2$$

Divide each side of the equation by 2.

$$\frac{\cancel{2}(y+27)}{\cancel{2}} = \frac{(x-8)^2}{2}$$

$$y + 27 = \frac{1}{2}(x - 8)^2$$

574. $y - 11 = -(x + 2)^2$

First, add $-y$ to each side of the equation and do a "reversal" (symmetric property of equations) of the two sides.

$$-y + 7 = x^2 + 4x$$

Complete the square on the right by adding 4 to each side.

$$-y + 7 + 4 = x^2 + 4x + 4$$

Simplify on the left, and factor on the right.

$$-y + 11 = (x + 2)^2$$

Now multiply each side of the equation by –1.

$$y - 11 = -(x + 2)^2$$

575. $y - 44 = -4(x - 3)^2$

First, add $-y$ to each side of the equation and do a "reversal" (symmetric property of equations) of the two sides.

$$-y + 8 = 4x^2 - 24x$$

Factor the 4 out of the two terms on the right.

$$-y + 8 = 4(x^2 - 6x)$$

Complete the square in the parentheses by adding 9; then add 36 to the other side, because the 9 in the parentheses is being multiplied by the 4.

$$-y + 8 + 36 = 4(x^2 - 6x + 9)$$

Simplify on the left, and factor on the right.

$$-y + 44 = 4(x - 3)^2$$

Now multiply each side of the equation by –1.

$$y - 44 = -4(x - 3)^2$$

576. $\mathbf{x + 23 = (y - 5)^2}$

First, add –2 to each side of the equation.

$$x - 2 = y^2 - 10y$$

Complete the square on the right by adding 25 to each side.

$$x - 2 + 25 = y^2 - 10y + 25$$

Simplify on the left, and factor on the right.

$$x + 23 = (y - 5)^2$$

577. $\mathbf{x + 12 = 3(y + 2)^2}$

First, factor the 3 out of the terms on the right.

$$x = 3(y^2 + 4y)$$

Complete the square in the parentheses by adding 4; then add 12 to the other side, because the 4 in the parentheses is being multiplied by the 3.

$$x + 12 = 3(y^2 + 4y + 4)$$

Factor on the right.

$$x + 12 = 3(y + 2)^2$$

578. $\mathbf{x + 10 = 2\left(y - \frac{5}{2}\right)^2}$

First, add $-2x$ to each side of the equation and do a "reversal" (symmetric property of equations) of the two sides.

$$-2x + 5 = -4y^2 + 20y$$

Next, factor the –4 out of the terms on the right.

$$-2x + 5 = -4(y^2 - 5y)$$

Complete the square in the parentheses by adding $\frac{25}{4}$; then add –25 to the other side, because the $\frac{25}{4}$ in the parentheses is being multiplied by the –4.

$$-2x + 5 - 25 = -4\left(y^2 - 5y + \frac{25}{4}\right)$$

Simplify on the left and factor on the right.

$$-2x - 20 = -4\left(y - \frac{5}{2}\right)^2$$

Now divide both sides of the equation by –2.

$$x + 10 = 2\left(y - \frac{5}{2}\right)^2$$

579. $\mathbf{x = \frac{2}{3}y^2}$

Do a "reversal" (symmetric property of equations) of the two sides, and then divide both sides of the equation by 3.

$$x = \frac{2}{3}y^2$$

This is in the $x - h = a(y - k)^2$ form. Just insert 0's for the h and k if needed.

$$x - 0 = \frac{2}{3}(y - 0)^2$$

580. $\mathbf{y = \frac{9}{5}x^2}$

Do a "reversal" (symmetric property of equations) of the two sides, and then divide both sides of the equation by 5.

$$y = \frac{9}{5}x^2$$

This is in the $y - k = a(x - h)^2$ form. Just insert 0's for the h and k if needed.

$$y - 0 = \frac{9}{5}(x - 0)^2$$

581. **Vertex: (–2, 3); focus:** $\mathbf{\left(-2, \frac{49}{16}\right)}$**; directrix:** $\mathbf{y = \frac{47}{16}}$**; axis of symmetry:** $\mathbf{x = -2}$

Using the standard equation, the vertex is (–2, 3).

Using a variation of the standard equation, $4a(y - k) = (x - h)^2$, the equation is written $\frac{1}{4}(y-3) = (x+2)^2$, which makes $4a = \frac{1}{4}$ and $a = \frac{1}{16}$.

The focus occurs at $(h, k + a)$, or, in this case, $\left(-2, 3 + \frac{1}{16}\right) = \left(-2, \frac{49}{16}\right)$.

The directrix is $y = k - a$, or, in this case, $y = 3 - \frac{1}{16} = \frac{47}{16}$.

The axis of symmetry is $x = -2$.

582. **Vertex: (1, 2); focus:** $\mathbf{\left(1, \frac{23}{12}\right)}$**; directrix:** $\mathbf{y = \frac{25}{12}}$**; axis of symmetry:** $\mathbf{x = 1}$

Rewrite the equation in standard form: $y - 2 = -3(x - 1)^2$.

Using the standard equation, the vertex is (1, 2).

Using a variation of the standard equation, $4a(y - k) = (x - h)^2$, the equation is written $-\frac{1}{3}(y-2) = (x-1)^2$, which makes $4a = -\frac{1}{3}$ and $a = -\frac{1}{12}$.

The focus occurs at $(h, k + a)$, or, in this case, $\left(1, 2 - \frac{1}{12}\right) = \left(1, \frac{23}{12}\right)$.

The directrix is $y = k - a$, or, in this case, $y = 2 - \left(-\frac{1}{12}\right) = \frac{25}{12}$.

The axis of symmetry is $x = 1$.

583. **Vertex: (1, 4); focus: $\left(\frac{7}{8}, 4\right)$; directrix: $x = \frac{9}{8}$; axis of symmetry: $y = 4$**

Rewrite the equation in standard form: $x - 1 = -2(y - 4)^2$.

Using the standard equation, the vertex is (1, 4).

Using a variation of the standard equation, $4a(x - h) = (y - k)^2$, the equation is written $-\frac{1}{2}(x-1) = (y-4)^2$, which makes $4a = -\frac{1}{2}$ and $a = -\frac{1}{8}$.

The focus occurs at $(h + a, k)$, or, in this case, $\left(1 - \frac{1}{8}, 4\right) = \left(\frac{7}{8}, 4\right)$.

The directrix is $x = h - a$, or, in this case, $x = 1 - \left(-\frac{1}{8}\right) = \frac{9}{8}$.

The axis of symmetry is $y = 4$.

584. **Vertex: (–8, 0); focus: $\left(-\frac{15}{2}, 0\right)$; directrix: $x = -\frac{17}{2}$; axis of symmetry: $y = 0$**

Rewrite the equation in standard form: $x + 8 = \frac{1}{2}y^2$.

Using the standard equation, the vertex is (–8, 0).

Using a variation of the standard equation, $4a(x - h) = (y - k)^2$, the equation is written $2(x + 8) = y^2$, which makes $4a = 2$ and $a = \frac{1}{2}$.

The focus occurs at $(h + a, k)$, or, in this case, $\left(-8 + \frac{1}{2}, 0\right) = \left(-\frac{15}{2}, 0\right)$.

The directrix is $x = h - a$, or, in this case, $x = -8 - \frac{1}{2} = -\frac{17}{2}$.

The axis of symmetry is $y = 0$.

585. **Vertex: (0, 0); focus: $\left(0, \frac{1}{6}\right)$; directrix: $y = -\frac{1}{6}$; axis of symmetry: $x = 0$**

Rewrite the equation in standard form: $y = \frac{3}{2}x^2$.

Using the standard equation, the vertex is (0, 0).

Using a variation of the standard equation, $4a(y - k) = (x - h)^2$, the equation is written $\frac{2}{3}y = x^2$, which makes $4a = \frac{2}{3}$ and $a = \frac{1}{6}$.

The focus occurs at $(h, k + a)$, or, in this case, $\left(0, 0 + \frac{1}{6}\right) = \left(0, \frac{1}{6}\right)$.

The directrix is $y = k - a$, or, in this case, $y = 0 - \frac{1}{6} = -\frac{1}{6}$.

The axis of symmetry is $x = 0$.

586. Vertex: (–2, 5); opens downward through (1, 2)

The vertex is at (–2, 5), and the parabola opens downward, also going through the points (1, 2) and (–5, 2).

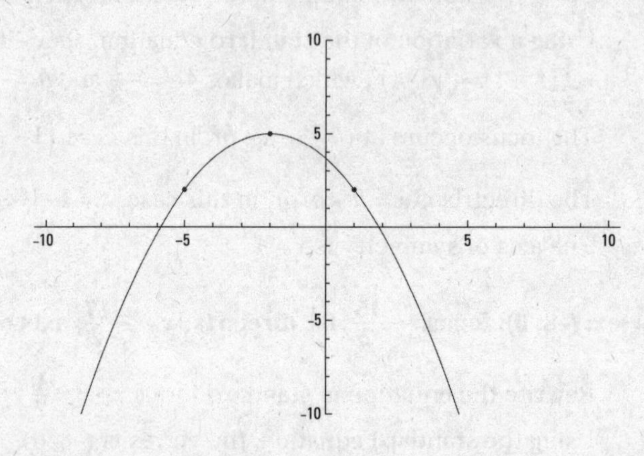

587. Vertex: (–3, 1); opens right through (1, 2)

The vertex is at (–3, 1), and the parabola opens to the right, also going through (1, 2) and (1, 0).

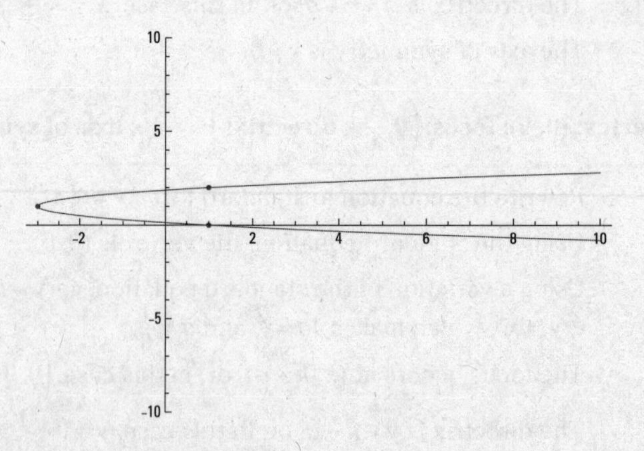

588. Vertex: (4, –15); opens upward through (8, 1)

First, put the equation in standard form.

$$y - 1 = x^2 - 8x$$

$$y - 1 + 16 = x^2 - 8x + 16$$

$$y + 15 = (x - 4)^2$$

The vertex is at (4, –15). The parabola opens upward and also goes through the points (0, 1) and (8, 1).

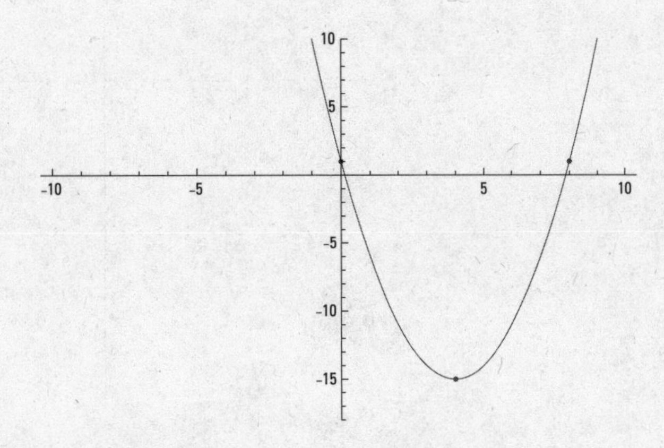

589. Vertex: (–7, 2); opens right through (–3, 4)

First, put the equation in standard form.

$$x + 3 = y^2 - 4y$$

$$x + 3 + 4 = y^2 - 4y + 4$$

$$x + 7 = (y - 2)^2$$

The vertex is at (–7, 2). The parabola opens to the right, also going through the points (–3, 0) and (–3, 4).

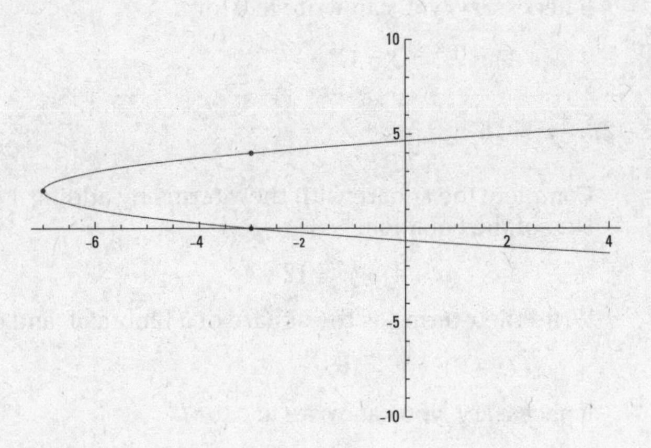

590. Vertex: (0, 0); opens downward through (5, –15)

First, put the equation in standard form.

$$y = -\frac{3}{5}x^2$$

The vertex is at (0, 0), and the parabola opens downward. It also goes through the points (5, –15) and (–5, –15).

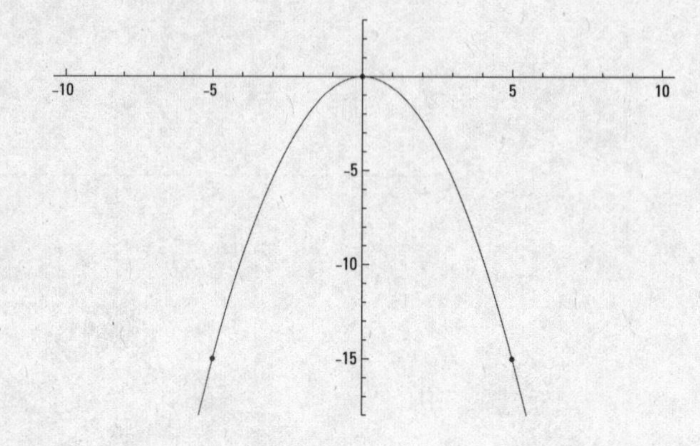

591. $x^2 + (y - 1)^2 = 9$

Complete the square with the *y*-terms by adding 1; you also need to add 1 to the other side of the equation.

$$x^2 + y^2 - 2y = 8$$

$$x^2 + (y^2 - 2y + 1) = 8 + 1$$

Write the *y*-terms as the square of a binomial, and simplify on the right.

$$x^2 + (y - 1)^2 = 9$$

If necessary, you can write in 0 for *h*.

$$(x - 0)^2 + (y - 1)^2 = 9$$

592. $(x + 2)^2 + y^2 = 16$

Complete the square with the *x*-terms by adding 4; you also need to add 4 to the other side of the equation.

$$(x^2 + 4x + 4) + y^2 = 12 + 4$$

Write the *x*-terms as the square of a binomial, and simplify on the right.

$$(x + 2)^2 + y^2 = 16$$

If necessary, you can write in 0 for *k*.

$$(x + 2)^2 + (y - 0)^2 = 16$$

593. $(x - 4)^2 + (y + 5)^2 = 64$

Rearrange the terms, pairing the *x*-terms and *y*-terms.

$$x^2 - 8x + y^2 + 10y = 23$$

Complete the square with the x-terms by adding 16, and complete the square with the y-terms by adding 25; you also need to add 16 + 25 = 41 to the other side of the equation.

$$(x^2 - 8x + 16) + (y^2 + 10y + 25) = 23 + 41$$

Write the x-terms and y-terms as squares of binomials, and simplify on the right.

$$(x - 4)^2 + (y + 5)^2 = 64$$

594. $(x + 6)^2 + (y + 1)^2 = 36$

Rearrange the terms, pairing the x-terms and y-terms, and add –1 to each side of the equation.

$$x^2 + 12x + y^2 + 2y = -1$$

Complete the square with the x-terms by adding 36, and complete the square with the y-terms by adding 1; you also need to add 36 + 1 = 37 to the other side of the equation.

$$(x^2 + 12x + 36) + (y^2 + 2y + 1) = -1 + 37$$

Write the x-terms and y-terms as squares of binomials, and simplify on the right.

$$(x + 6)^2 + (y + 1)^2 = 36$$

595. $(x+1)^2 + \left(y - \frac{1}{2}\right)^2 = \frac{9}{4}$

Rearrange the terms, pairing the x-terms and y-terms.

$$x^2 + 2x + y^2 - y = 1$$

Complete the square with the x-terms by adding 1, and complete the square with the y-terms by adding $\frac{1}{4}$; you also need to add $1 + \frac{1}{4} = \frac{5}{4}$ to the other side of the equation.

$$x^2 + 2x + 1 + y^2 - y + \frac{1}{4} = 1 + \frac{5}{4}$$

Write the x-terms and y-terms as squares of binomials, and simplify on the right.

$$(x+1)^2 + \left(y - \frac{1}{2}\right)^2 = \frac{9}{4}$$

596. $(x-3)^2 + \left(y + \frac{5}{2}\right)^2 = 16$

Rearrange the terms, pairing the x-terms and y-terms.

$$12x^2 - 72x + 12y^2 + 60y = 9$$

Next, divide each term by 12.

$$x^2 - 6x + y^2 + 5y = \frac{9}{12} = \frac{3}{4}$$

Complete the square with the x-terms by adding 9, and complete the square with the y-terms by adding $\frac{25}{4}$; you also need to add $9 + \frac{25}{4} = \frac{61}{4}$ to the other side of the equation.

$$x^2 - 6x + 9 + y^2 + 5y + \frac{25}{4} = \frac{3}{4} + \frac{61}{4}$$

Write the x-terms and y-terms as squares of binomials, and simplify on the right.

$$(x-3)^2 + \left(y + \frac{5}{2}\right)^2 = \frac{64}{4} = 16$$

597. $x^2 + (y-1)^2 = \frac{16}{5}$

First, divide each term by 5.

$$x^2 + y^2 - 2y = \frac{11}{5}$$

Complete the square with the y-terms by adding 1; you also need to add 1 to the other side of the equation.

$$x^2 + y^2 - 2y + 1 = \frac{11}{5} + 1$$

Write the y-terms as the square of a binomial, and simplify on the right.

$$x^2 + (y-1)^2 = \frac{16}{5}$$

598. $(x+2)^2 + y^2 = 9$

First, add y^2 to each side of the equation.

$$x^2 + 4x + y^2 = 5$$

Complete the square with the x-terms by adding 4; you also need to add 4 to the other side of the equation.

$$x^2 + 4x + 4 + y^2 = 5 + 4$$

Write the x-terms as the square of a binomial, and simplify on the right.

$$(x+2)^2 + y^2 = 9$$

599. $x^2 + (y-5)^2 = 36$

First, add x^2 and 11 to each side of the equation; then add $-10y$ to each side.

$$x^2 + y^2 = 10y + 11$$

$$x^2 + y^2 - 10y = 11$$

Complete the square with the y-terms by adding 25; you also need to add 25 to the other side of the equation.

$$x^2 + (y^2 - 10y + 25) = 11 + 25$$

Write the y-terms as the square of a binomial, and simplify on the right.

$$x^2 + (y-5)^2 = 36$$

600. $x^2 + y^2 = \frac{25}{9}$

First, add $9y^2$ to each side.

$$9x^2 + 9y^2 = 25$$

Divide each side of the equation by 9.

$$x^2 + y^2 = \frac{25}{9}$$

601. **Center: (4, –3), radius: 2**

Using the standard form $(x - h)^2 + (y - k)^2 = r^2$, the center (h, k) is at (4, –3), and the radius, r, is 2.

602. **Center: (–1, 0), radius: 5**

Using the standard form $(x - h)^2 + (y - k)^2 = r^2$, think of the equation as being $(x + 1)^2 + (y - 0)^2 = 25$.

The center (h, k) is at (–1, 0), and the radius, r, is 5.

603. **Center: $\left(-3, \frac{1}{4}\right)$, radius: $\frac{1}{4}$**

Using the standard form $(x - h)^2 + (y - k)^2 = r^2$, the center (h, k) is at $\left(-3, \frac{1}{4}\right)$ and the radius, r, is $\frac{1}{4}$.

604. **Center: (2, 4), radius: 5**

First, write the equation in the standard form:

$$(x - h)^2 + (y - k)^2 = r^2$$

$$x^2 - 4x + y^2 - 8y = 5$$

$$x^2 - 4x + 4 + y^2 - 8y + 16 = 5 + 4 + 16$$

$$(x - 2)^2 + (y - 4)^2 = 25$$

The center (h, k) is at (2, 4), and the radius, r, is 5.

605. **Center: (0, 0); radius: $\frac{1}{3}$**

First, write the equation in the standard form:

$$(x - h)^2 + (y - k)^2 = r^2$$

$$x^2 + y^2 = \frac{1}{9}$$

The center (h, k) is at (0, 0), and the radius, r, is $\frac{1}{3}$.

606. **Center (0, 0), goes through $\left(3, \sqrt{7}\right)$**

The center is at (0, 0), and the radius is 4. The circle also goes through the point $\left(3, \sqrt{7}\right)$, as one example.

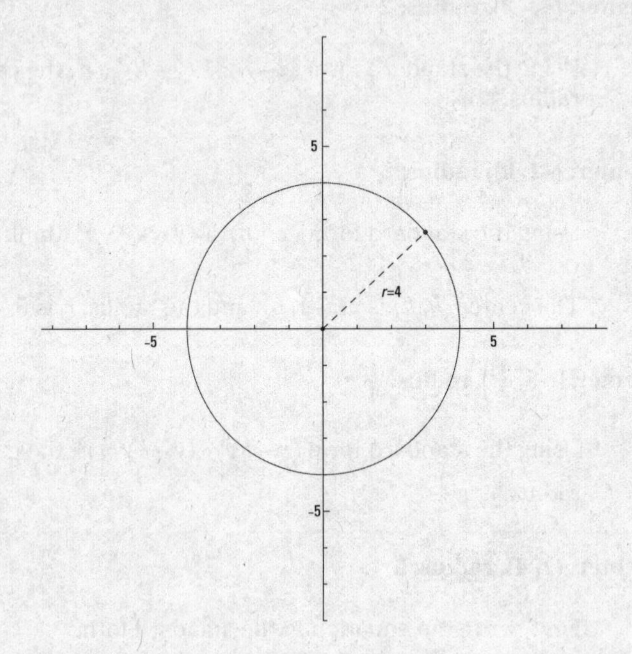

607. **Center (3, 1), goes through $\left(1, 1+\sqrt{5}\right)$**

The center is at (3, 1), and the radius is 3. The circle also goes through the point $\left(1, 1+\sqrt{5}\right)$, as one example.

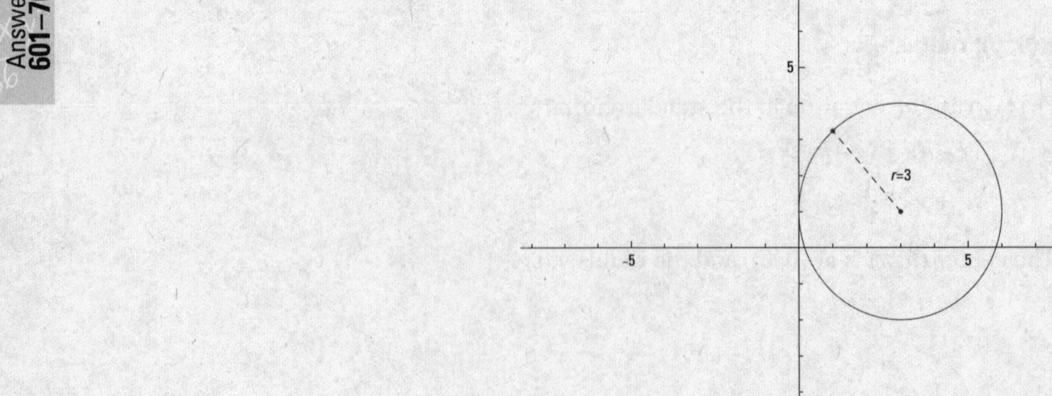

608. Center (–1, 0), goes through $\left(-\frac{1}{2}, \frac{\sqrt{3}}{2}\right)$

The center is at (–1, 0), and the radius is 1. The circle also goes through the point $\left(-\frac{1}{2}, \frac{\sqrt{3}}{2}\right)$, as one example.

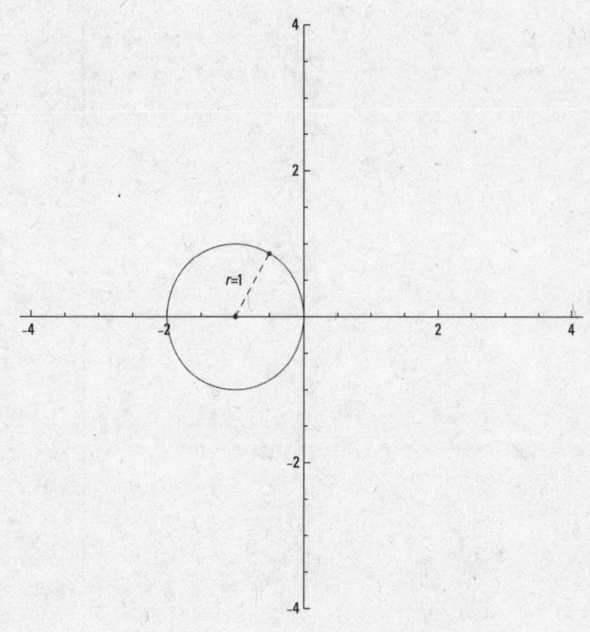

609. Center (2, 2), goes through $\left(1, 2+\sqrt{3}\right)$

The center is at (2, 2), and the radius is 2. The circle also goes through the point $\left(1, 2+\sqrt{3}\right)$, as one example.

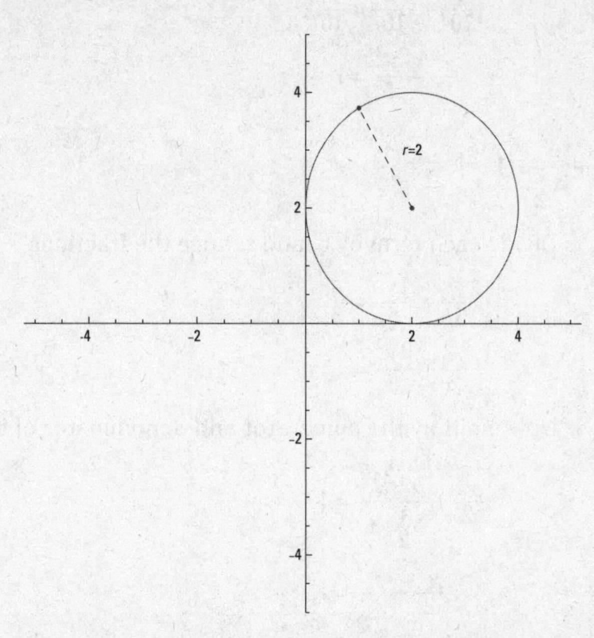

610. **Center (1, 0), goes through $\left(\frac{5}{4}, \frac{\sqrt{3}}{4}\right)$**

The center is at (1, 0), and the radius is $\frac{1}{2}$. The circle also goes through the point $\left(\frac{5}{4}, \frac{\sqrt{3}}{4}\right)$, as one example.

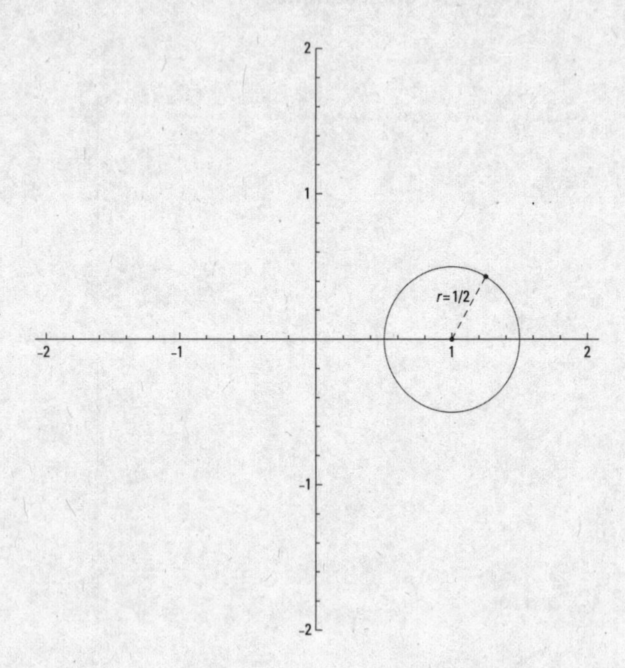

611. $\frac{x^2}{2} + \frac{y^2}{5} = 1$

Divide each term by 10, and reduce the fractions.

$$\frac{5x^2}{10} + \frac{2y^2}{10} = \frac{10}{10}$$

$$\frac{x^2}{2} + \frac{y^2}{5} = 1$$

612. $\frac{x^2}{2} + \frac{y^2}{3/2} = 1$

Divide each term by 6, and reduce the fractions.

$$\frac{3x^2}{6} + \frac{4y^2}{6} = \frac{6}{6}$$

$$\frac{x^2}{2} + \frac{2y^2}{3} = 1$$

Now multiply the numerator and denominator of the y^2 term by $\frac{1}{2}$.

$$\frac{x^2}{2} + \frac{\frac{1}{2} \cdot 2y^2}{\frac{1}{2} \cdot 3} = 1$$

$$\frac{x^2}{2} + \frac{y^2}{3/2} = 1$$

613. $\frac{(x-1)^2}{4}+\frac{(y+2)^2}{9}=1$

Group the x terms and y terms together.

$$9x^2 - 18x + 4y^2 + 16y = 11$$

Factor 9 from the two x terms and 4 from the two y terms.

$$9(x^2 - 2x) + 4(y^2 + 4y) = 11$$

Complete the squares within the parentheses by adding 1 and 4, respectively. You have to add 9(1) and 4(4) to the right side, because those two numbers in the parentheses are being multiplied, too.

$$9(x^2 - 2x + 1) + 4(y^2 + 4y + 4) = 11 + 9 + 16$$

Write the trinomials in the parentheses as the squares of binomials, and simplify on the right.

$$9(x - 1)^2 + 4(y + 2)^2 = 36$$

Now divide each term in the equation by 36 and reduce the fractions.

$$\frac{9(x-1)^2}{36}+\frac{4(y+2)^2}{36}=\frac{36}{36}$$

$$\frac{(x-1)^2}{4}+\frac{(y+2)^2}{9}=1$$

614. $\frac{(x+5)^2}{25}+\frac{(y-1)^2}{4}=1$

Group the x terms and y terms together; add –25 to each side of the equation.

$$4x^2 + 40x + 25y^2 - 50y = -25$$

Factor 4 from the two x terms and 25 from the two y terms.

$$4(x^2 + 10x) + 25(y^2 - 2y) = -25$$

Complete the squares within the parentheses by adding 25 and 1, respectively. You have to add 4(25) and 25(1) to the right side, because those two numbers in the parentheses are being multiplied, too.

$$4(x^2 + 10x + 25) + 25(y^2 - 2y + 1) = -25 + 100 + 25$$

Write the trinomials in the parentheses as the squares of binomials, and simplify on the right.

$$4(x + 5)^2 + 25(y - 1)^2 = 100$$

Now divide each term in the equation by 100 and reduce the fractions.

$$\frac{4(x+5)^2}{100}+\frac{25(y-1)^2}{100}=\frac{100}{100}$$

$$\frac{(x+5)^2}{25}+\frac{(y-1)^2}{4}=1$$

615. $\frac{(x+3)^2}{9}+\frac{y^2}{16}=1$

Group the x terms together.

$$16x^2 + 96x + 9y^2 = 0$$

Factor 16 from the two x terms.

$$16(x^2 + 6x) + 9y^2 = 0$$

Complete the square within the parentheses by adding 9. You have to add 16(9) to the right side, because the numbers in the parentheses are being multiplied, too.

$$16(x^2 + 6x + 9) + 9y^2 = 16(9)$$

Write the trinomial in the parentheses as the square of a binomial, and simplify on the right.

$$16(x + 3)^2 + 9y^2 = 144$$

Now divide each term in the equation by 144 and reduce the fractions.

$$\frac{16(x+3)^2}{144}+\frac{9y^2}{144}=\frac{144}{144}$$

$$\frac{(x+3)^2}{9}+\frac{y^2}{16}=1$$

616. $\frac{x^2}{32}+\frac{(y-2)^2}{4}=1$

Factor 8 from the two y terms.

$$x^2 + 8(y^2 - 4y) = 0$$

Complete the square within the parentheses by adding 4. You have to add 8(4) to the right side, because the numbers in the parentheses are being multiplied, too.

$$x^2 + 8(y^2 - 4y + 4) = 8(4)$$

Write the trinomial in the parentheses as the square of a binomial, and simplify on the right.

$$x^2 + 8(y - 2)^2 = 32$$

Now divide each term in the equation by 32 and reduce the fractions.

$$\frac{x^2}{32}+\frac{8(y-2)^2}{32}=\frac{32}{32}$$

$$\frac{x^2}{32}+\frac{(y-2)^2}{4}=1$$

617. $\frac{(x-1)^2}{10}+\frac{(y+3)^2}{20}=1$

Group the x terms and y terms together; add 9 to each side of the equation.

$$2x^2 - 4x + y^2 + 6y = 9$$

Factor 2 from the two x terms.

$$2(x^2 - 2x) + y^2 + 6y = 9$$

Complete the square within the parentheses by adding 1. You have to add 2(1) to the right side, because the number in the parentheses is being multiplied, too. Also, complete the square involving the y terms by adding 9; add 9 to the right side, too.

$$2(x^2 - 2x + 1) + (y^2 + 6y + 9) = 9 + 2 + 9$$

Write the trinomials as the squares of binomials, and simplify on the right.

$$2(x - 1)^2 + (y + 3)^2 = 20$$

Now divide each term in the equation by 20 and reduce the fractions.

$$\frac{2(x-1)^2}{20} + \frac{(y+3)^2}{20} = \frac{20}{20}$$

$$\frac{(x-1)^2}{10} + \frac{(y+3)^2}{20} = 1$$

618. $\mathbf{\frac{(x+5)^2}{54} + \frac{(y+5)^2}{18} = 1}$

Group the x terms and y terms together; add –46 to each side of the equation.

$$x^2 + 10x + 3y^2 + 30y = -46$$

Factor 3 from the two y terms.

$$x^2 + 10x + 3(y^2 + 10y) = -46$$

Complete the square within the parentheses by adding 25. You have to add 3(25) to the right side, because the number in the parentheses is being multiplied, too. Also, complete the square involving the x terms by adding 25; add 25 to the right side, too.

$$(x^2 + 10x + 25) + 3(y^2 + 10y + 25) = -46 + 25 + 75$$

Write the trinomials as the squares of binomials, and simplify on the right.

$$(x + 5)^2 + 3(y + 5)^2 = 54$$

Now divide each term in the equation by 54 and reduce the fractions.

$$\frac{(x+5)^2}{54} + \frac{3(y+5)^2}{54} = \frac{54}{54}$$

$$\frac{(x+5)^2}{54} + \frac{(y+5)^2}{18} = 1$$

619. $\mathbf{\frac{(x-1)^2}{a^2} + \frac{(y-1)^2}{b^2} = 1}$

Group the x terms and y terms together; add $-a^2$ and $-b^2$ to each side of the equation.

$$b^2x^2 - 2b^2x + a^2y^2 - 2a^2y = a^2b^2 - a^2 - b^2$$

Factor b^2 from the two x terms and a^2 from the two y terms.

$$b^2(x^2 - 2x) + a^2(y^2 - 2y) = a^2b^2 - a^2 - b^2$$

Complete the square within each pair of parentheses by adding 1. You have to add $b^2(1)$ and $a^2(1)$ to the right side, because the number in the parentheses is being multiplied, too.

$$b^2(x^2 - 2x + 1) + a^2(y^2 - 2y + 1) = a^2b^2 - a^2 - b^2 + b^2 + a^2$$

Write the trinomials as the squares of binomials, and simplify on the right.

$$b^2(x-1)^2 + a^2(y-1)^2 = a^2b^2$$

Now divide each term in the equation by a^2b^2 and reduce the fractions.

$$\frac{\cancel{b^2}(x-1)^2}{a^2\cancel{b^2}} + \frac{\cancel{a^2}(y-1)^2}{\cancel{a^2}b^2} = \frac{\cancel{a^2b^2}}{\cancel{a^2b^2}}$$

$$\frac{(x-1)^2}{a^2} + \frac{(y-1)^2}{b^2} = 1$$

620. $\mathbf{\frac{(x-h)^2}{a^2} + \frac{(y-k)^2}{b^2} = 1}$

Group the x terms and y terms together; add $-a^2$ and $-b^2$ to each side of the equation.

$$b^2x^2 - 2b^2hx + a^2y^2 - 2a^2ky = a^2b^2 - a^2k^2 - b^2h^2$$

Factor b^2 from the two x terms and a^2 from the two y terms.

$$b^2(x^2 - 2hx) + a^2(y^2 - 2ky) = a^2b^2 - a^2 - b^2$$

Complete the square within each pair of parentheses by adding 1. You have to add $b^2(h^2)$ and $a^2(k^2)$ to the right side, because the number in the parentheses is being multiplied, too.

$$b^2(x^2 - 2hx + h^2) + a^2(y^2 - 2ky + k^2) = a^2b^2 - a^2k^2 - b^2h^2 + b^2h^2 + a^2k^2$$

Write the trinomials as the squares of binomials, and simplify on the right.

$$b^2(x-h)^2 + a^2(y-k)^2 = a^2b^2$$

Now divide each term in the equation by a^2b^2 and reduce the fractions.

$$\frac{\cancel{b^2}(x-h)^2}{a^2\cancel{b^2}} + \frac{\cancel{a^2}(y-k)^2}{\cancel{a^2}b^2} = \frac{\cancel{a^2b^2}}{\cancel{a^2b^2}}$$

$$\frac{(x-h)^2}{a^2} + \frac{(y-k)^2}{b^2} = 1$$

You should recognize this as the standard form for the equation of an ellipse.

621. **Center: (0, 0); foci: (0, 3), (0, –3); major axis: (0, –5) to (0, 5); minor axis: (–4, 0) to (4, 0)**

The equation can be written $\frac{(x-0)^2}{16} + \frac{(y-0)^2}{25} = 1$, making h and k both 0. So the center is at (0, 0).

The foci are found with $c = \pm\sqrt{25-16} = \pm\sqrt{9} = \pm 3$, so they are at (0, 3) and (0, –3).

The major axis is vertical, and with $b = \pm\sqrt{25}$, it has endpoints at (0, –5) and (0, 5).

The minor axis is horizontal, and with $a = \pm\sqrt{16}$, its endpoints are at (–4, 0) and (4, 0).

622. **Center: (2, –3); foci: $\left(\sqrt{5}, 0\right), \left(-\sqrt{5}, 0\right)$; major axis: (–1, –3) to (5, –3); minor axis: (2, –5) and (2, –1)**

Using this standard form, the center is at (2, –3).

The foci are found with $c = \pm\sqrt{9-4} = \pm\sqrt{5}$, so they are $\sqrt{5}$ units from the center along the major axis, placing them at $\left(2-\sqrt{5}, -3\right)$ and $\left(2+\sqrt{5}, -3\right)$.

The major axis is horizontal with $a = \pm\sqrt{9}$ and so has endpoints at 3 units from the center at (–1, –3) and (5, –3).

The minor axis is vertical with $b = \pm\sqrt{4}$ and so has endpoints 2 units from the center at (2, –5) and (2, –1).

623. **Center: (0, 0); foci: $\left(0, \sqrt{21}\right), \left(0, -\sqrt{21}\right)$; major axis: (0, –5) to (0, 5); minor axis: (–2, 0) to (2, 0)**

First, write the equation in standard form.

$$\frac{x^2}{4} + \frac{y^2}{25} = 1$$

The equation can be written $\frac{(x-0)^2}{4} + \frac{(y-0)^2}{25} = 1$, making h and k both 0. So the center is at (0, 0).

The foci are found with $c = \pm\sqrt{25-4} = \pm\sqrt{21}$, so they are at $\left(0, \sqrt{21}\right)$ and $\left(0, -\sqrt{21}\right)$.

The major axis is vertical with $b = \pm\sqrt{25}$ and has endpoints at (0, –5) and (0, 5).

The minor axis is horizontal with $a = \pm\sqrt{4}$ and has endpoints at (–2, 0) and (2, 0).

624. **Center: (–3, 5); foci: $\left(-3, 5-2\sqrt{2}\right)$ and $\left(-3, 5+2\sqrt{2}\right)$; major axis: (–3, 2) and (–3, 8); minor axis: (–4, 5) and (–2, 5)**

The equation can be written $\frac{(x+3)^2}{1} + \frac{(y-5)^2}{9} = 1$.

The center is at (–3, 5).

The foci are found with $c = \pm\sqrt{9-1} = \pm\sqrt{8} = \pm 2\sqrt{2}$. Because they're $2\sqrt{2}$ units above and below the center, their coordinates are $\left(-3, 5-2\sqrt{2}\right)$ and $\left(-3, 5+2\sqrt{2}\right)$.

The major axis is vertical with $b = \pm\sqrt{9}$ so the endpoints are 3 units above and below the center at (–3, 2) and (–3, 8).

The minor axis is horizontal with $a = \pm\sqrt{1}$ and has endpoints to the left and right of the center at (–4, 5) and (–2, 5).

625. **Center: (1, –3); foci: $\left(1-4\sqrt{6}, -3\right)$ and $\left(1+4\sqrt{6}, -3\right)$; major axis: (–9, –3) and (11, –3); minor axis: (1, –5) and (1, –1)**

The center is at (1, –3).

The foci are found with $c = \pm\sqrt{100-4} = \pm\sqrt{96} = \pm 4\sqrt{6}$ so they are $4\sqrt{6}$ units on either side of the center at $\left(1-4\sqrt{6}, -3\right)$ and $\left(1+4\sqrt{6}, -3\right)$.

The major axis is horizontal with $a = \pm\sqrt{100}$ and has endpoints 10 units on either side of the center at (–9, –3) and (11, –3).

The minor axis is vertical with $b = \pm\sqrt{4}$ and has endpoints 2 units above and below the center at (1, –5) and (1, –1).

626. **Center: (0, 0); major axis: (0, –6) to (0, 6); minor axis: (–2, 0) to (2, 0)**

The center is at (0, 0). The major axis is vertical with $b = \pm\sqrt{36}$, so it runs above and below the center from (0, –6) to (0, 6). The minor axis is horizontal with $a = \pm\sqrt{4}$, so the endpoints are 2 units on either side of the center running from (–2, 0) to (2, 0).

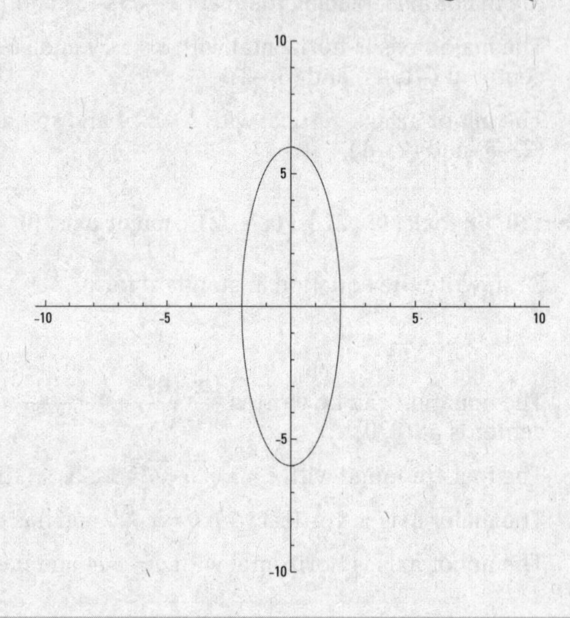

627. **Center: (0, 0); major axis: (–8, 0) to (8, 0); minor axis: (0, –3) to (0, 3)**

The center is at (0, 0). The major axis is horizontal with $a = \pm\sqrt{64}$, so it runs from (–8, 0) to (8, 0). The minor axis is vertical with $b = \pm\sqrt{9}$, so it runs from (0, –3) to (0, 3).

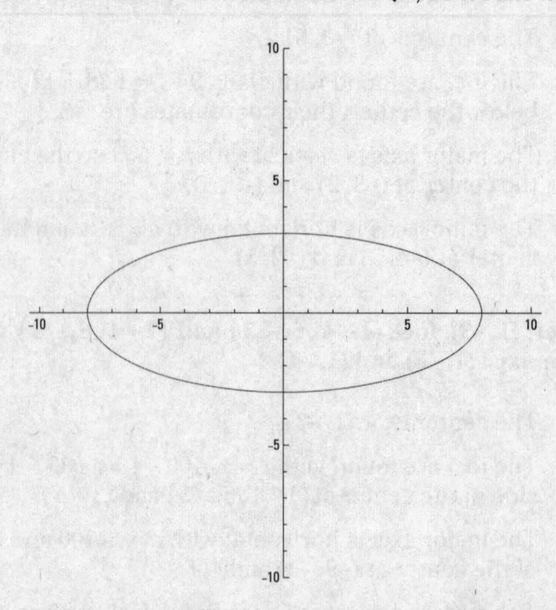

628. **Center: (3, –2); major axis: (3, –9) to (3, 5); minor axis: (–2, –2) to (8, –2)**

The center is at (3, –2). The major axis is vertical with $b = \pm\sqrt{49}$, so the endpoints are 7 units below and above the center at (3, –9) and (3, 5). The minor axis is horizontal with $a = \pm\sqrt{25}$, so the endpoints are 5 units on either side of the center at (–2, –2) and (8, –2).

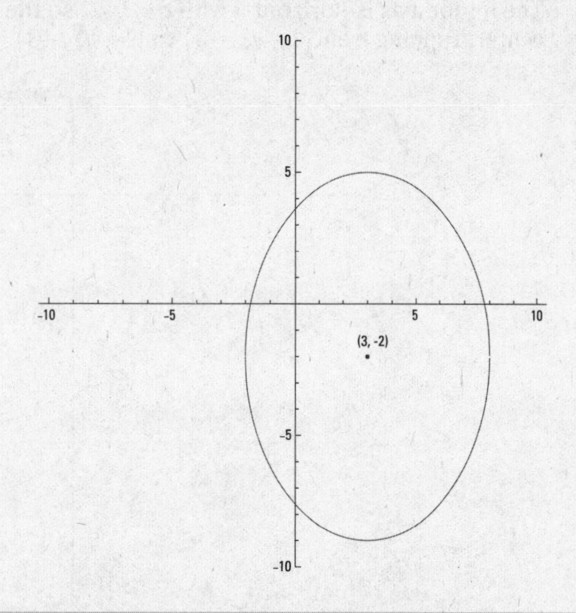

629. **Center: (0, 1); major axis: (–6, 1) to (6, 1); minor axis: (0, –4) to (0, 6)**

The center is at (0, 1). The major axis is horizontal with $a = \pm\sqrt{36}$, so the endpoints are 6 units on either side of the center running from (–6, 1) to (6, 1). The minor axis is vertical with $b = \pm\sqrt{25}$, so the endpoints are 5 units below and above the center running from (0, –4) to (0, 6).

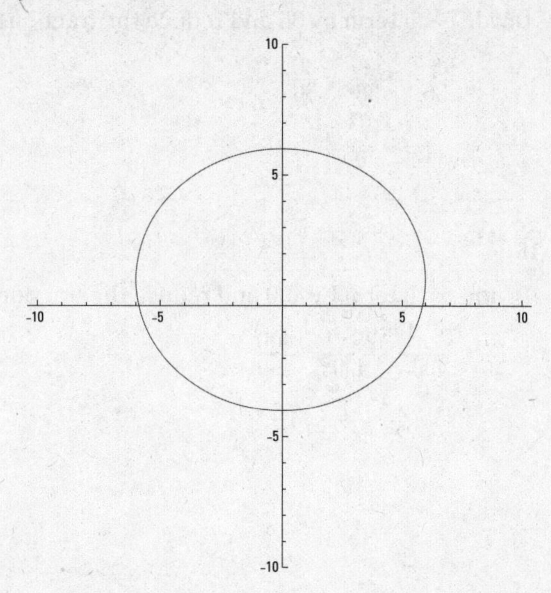

630. **Center: (1, –3); major axis: $\left(1,-3-2\sqrt{2}\right)$ to $\left(1,-3+2\sqrt{2}\right)$; minor axis: $\left(1-\sqrt{2},-3\right)$ to $\left(1+\sqrt{2},-3\right)$**

The center is at (1, –3). The major axis is vertical with $b=\pm\sqrt{8}=\pm2\sqrt{2}$. The endpoints are $2\sqrt{2}$ units below and above the center, so it runs from $\left(1,-3-2\sqrt{2}\right)$ to $\left(1,-3+2\sqrt{2}\right)$. The minor axis is horizontal with $a=\pm\sqrt{2}$, so the endpoints are $\sqrt{2}$ on either side of the center, running from $\left(1-\sqrt{2},-3\right)$ to $\left(1+\sqrt{2},-3\right)$.

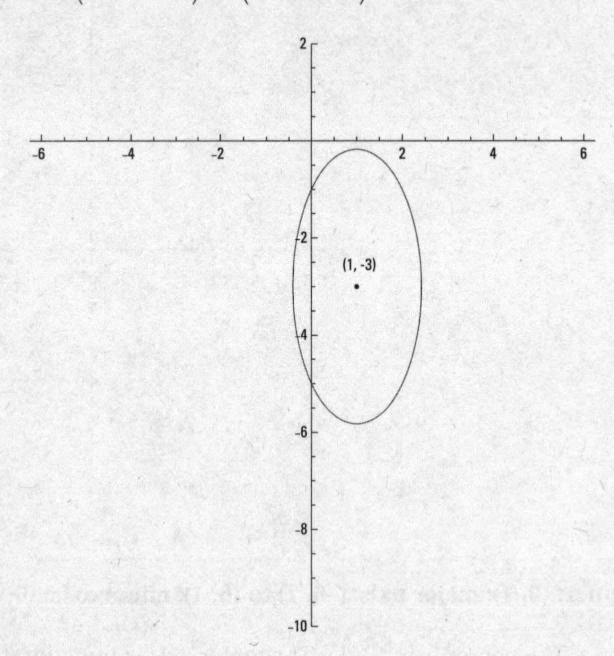

631. $\frac{x^2}{4}-\frac{y^2}{9}=1$

Divide each term by 36 and reduce the fractions.

$$\frac{9x^2}{36}-\frac{4y^2}{36}=\frac{36}{36}$$

$$\frac{x^2}{4}-\frac{y^2}{9}=1$$

632. $\frac{y^2}{25}-\frac{x^2}{16}=1$

Divide each term by 400 and reduce the fractions.

$$\frac{16y^2}{400}-\frac{25x^2}{400}=\frac{400}{400}$$

$$\frac{x^2}{25}-\frac{y^2}{16}=1$$

633. $\frac{(y+3)^2}{4} - \frac{(x-1)^2}{36} = 1$

Group the x terms and y terms together and add –176 to each side of the equation.

$$36y^2 + 216y - 4x^2 + 8x = -176$$

Factor 36 from the y terms and –4 from the x terms.

$$36(y^2 + 6y) - 4(x^2 - 2x) = -176$$

Complete the square of the y terms by adding 9, and complete the square of the x terms by adding 1. This means adding 36(9) and –4(1) to the right side.

$$36(y^2 + 6y + 9) - 4(x^2 - 2x + 1) = -176 + 36(9) - 4(1)$$

Write the trinomials in the parentheses as the squares of binomials, and simplify on the right.

$$36(y + 3)^2 - 4(x - 1)^2 = -176 + 324 - 4 = 144$$

Divide each term by 144 and reduce the fractions.

$$\frac{36(y+3)^2}{144} - \frac{4(x-1)^2}{144} = \frac{144}{144}$$

$$\frac{(y+3)^2}{4} - \frac{(x-1)^2}{36} = 1$$

634. $\frac{(x-1)^2}{4} - \frac{(y-5)^2}{9} = 1$

Group the x terms and y terms together.

$$9x^2 - 18x - 4y^2 + 40y = 127$$

Factor 9 from the x terms and –4 from the y terms.

$$9(x^2 - 2x) - 4(y^2 - 10y) = 127$$

Complete the square of the x terms by adding 1, and complete the square of the y terms by adding 25. This means adding 9(1) and –4(25) to the right side.

$$9(x^2 - 2x + 1) - 4(y^2 - 10y + 25) = 127 + 9(1) - 4(25)$$

Write the trinomials in the parentheses as the squares of binomials, and simplify on the right.

$$9(x - 1)^2 - 4(y - 5)^2 = 36$$

Now divide each term by 36 and reduce the fractions.

$$\frac{9(x-1)^2}{36} - \frac{4(y-5)^2}{36} = \frac{36}{36}$$

$$\frac{(x-1)^2}{4} - \frac{(y-5)^2}{9} = 1$$

635. $\frac{(x-2)^2}{16}-\frac{(y-1)^2}{4}=1$

Group the x terms and y terms together.

$$4x^2 - 16x - 16y^2 - 32y = 64$$

Factor 4 from the x terms and –16 from the y terms.

$$4(x^2 - 4x) - 16(y^2 - 2y) = 64$$

Complete the square of the x terms by adding 4, and complete the square of the y terms by adding 1. This means adding 4(4) and –16(1) to the right side.

$$4(x^2 - 4x + 4) - 16(y^2 - 2y + 1) = 64 + 4(4) - 16(1)$$

Write the trinomials in the parentheses as the squares of binomials, and simplify on the right.

$$4(x-2)^2 - 16(y-1)^2 = 64$$

Now divide each term by 64 and reduce the fractions.

$$\frac{{}^1\cancel{4}(x-2)^2}{{}^{16}\cancel{64}}-\frac{{}^1\cancel{16}(y-1)^2}{{}^4\cancel{64}}=\frac{{}^1\cancel{64}}{{}^1\cancel{64}}$$

$$\frac{(x-2)^2}{16}-\frac{(y-1)^2}{4}=1$$

636. $\frac{(x+3)^2}{25}-\frac{(y+2)^2}{4}=1$

Group the x terms and y terms together.

$$4x^2 + 24x - 25y^2 - 100y = 164$$

Factor 4 from the x terms and –25 from the y terms.

$$4(x^2 + 6x) - 25(y^2 + 4y) = 164$$

Complete the square of the x terms by adding 9, and complete the square of the y terms by adding 4. This means adding 4(9) and –25(4) to the right side.

$$4(x^2 + 6x + 9) - 25(y^2 + 4y + 4) = 164 + 4(9) - 25(4)$$

Write the trinomials in the parentheses as the squares of binomials, and simplify on the right.

$$4(x+3)^2 - 25(y+2)^2 = 100$$

Divide each term by 100 and reduce the fractions.

$$\frac{4(x+3)^2}{100}-\frac{25(y+2)^2}{100}=\frac{100}{100}$$

$$\frac{(x+3)^2}{25}-\frac{(y+2)^2}{4}=1$$

637. $\frac{(x-5)^2}{16}-\frac{(y+3)^2}{48}=1$

Group the x terms and y terms together; also, add –18 to each side of the equation.

$$3x^2 - 30x - y^2 - 6y = -18$$

Factor 3 from the x terms and –1 from the y terms.

$$3(x^2 - 10x) - (y^2 + 6y) = -18$$

Complete the square of the x terms by adding 25, and complete the square of the y terms by adding 9. This means adding 3(25) and –1(9) to the right side.

$$3(x^2 - 10x + 25) - (y^2 + 6y + 9) = -18 + 3(25) - 1(9)$$

Write the trinomials in the parentheses as the squares of binomials, and simplify on the right.

$$3(x-5)^2 - (y+3)^2 = 48$$

Divide each term by 48 and reduce the fractions.

$$\frac{3(x-5)^2}{48} - \frac{(y+3)^2}{48} = \frac{{}^1\cancel{48}}{{}^1\cancel{48}}$$

$$\frac{(x-5)^2}{16} - \frac{(y+3)^2}{48} = 1$$

638. $\frac{(y-1)^2}{9} - \frac{(x+4)^2}{1} = 1$

Group the x terms and y terms together.

$$y^2 - 2y - 9x^2 - 72x = 152$$

Factor –9 from the x terms.

$$y^2 - 2y - 9(x^2 + 8x) = 152$$

Complete the square of the y terms by adding 1, and complete the square of the x terms by adding 16. This means adding 1 and –9(16) to the right side.

$$(y^2 - 2y + 1) - 9(x^2 + 8x + 16) = 152 + 1 - 9(16)$$

Write the trinomials as the squares of binomials, and simplify on the right.

$$(y-1)^2 - 9(x+4)^2 = 9$$

Now divide each term by 9 and reduce the fractions.

$$\frac{(y-1)^2}{9} - \frac{9(x+4)^2}{9} = \frac{9}{9}$$

$$\frac{(y-1)^2}{9} - \frac{(x+4)^2}{1} = 1$$

639. $\frac{(x+2)^2}{1/4} - \frac{y^2}{1/9} = 1$

Group the x terms together and add –15 to each side of the equation.

$$4x^2 + 16x - 9y^2 = -15$$

Factor 4 from the x terms.

$$4(x^2 + 4x) - 9y^2 = -15$$

Complete the square within the parentheses by adding 4; also add 4(4) to the right side.

$$4(x^2 + 4x + 4) - 9y^2 = -15 + 4(4)$$

Write the trinomial as the square of a binomial, and simplify on the right.

$4(x+2)^2 - 9y^2 = 1$

Multiply the first term by $\frac{1/4}{1/4}$ and the second term by $\frac{1/9}{1/9}$.

$$\frac{1/4}{1/4} \cdot \frac{4(x+2)^2}{1} - \frac{1/9}{1/9} \cdot \frac{9y^2}{1} = 1$$

$$\frac{(x+2)^2}{1/4} - \frac{y^2}{1/9} = 1$$

640. $\frac{(x-2)^2}{1} - \frac{(y+1)^2}{4} = 1$

Group the x terms and y terms together; add –11 to each side of the equation.

$4x^2 - 16x - y^2 - 2y = -11$

Factor 4 from the x terms and –1 from the y terms.

$4(x^2 - 4x) - 1(y^2 + 2y) = -11$

Complete the square of the x terms by adding 4, and complete the square of the y terms by adding 1. This means adding 4(4) and –1(1) to the right side.

$4(x^2 - 4x + 4) - 1(y^2 + 2y + 1) = -11 + 4(4) - 1(1)$

Write the trinomials as the squares of binomials, and simplify on the right.

$4(x-2)^2 - 1(y+1)^2 = 4$

Now divide each term by 4 and reduce the fractions.

$$\frac{4(x-2)^2}{4} - \frac{(y+1)^2}{4} = \frac{4}{4}$$

$$\frac{(x-2)^2}{1} - \frac{(y+1)^2}{4} = 1$$

641. **Center: (–1, –3); foci: (–1, 7), (–1, –13); asymptotes:** $y = \frac{4}{3}x - \frac{5}{3}$, $y = -\frac{4}{3}x - \frac{13}{3}$

The center is at (–1, –3).

Using $c = \pm\sqrt{a^2 + b^2}$, the foci are at (–1, 7) and (–1, –13).

To find the equations of the asymptotes, change the 1 on the right of the equation to 0, add the second fraction to each side, and find the square root of each side.

$$\frac{(y+3)^2}{64} - \frac{(x+1)^2}{36} = 0$$

$$\frac{(y+3)^2}{64} = \frac{(x+1)^2}{36}$$

$$\sqrt{\frac{(y+3)^2}{64}} = \pm\sqrt{\frac{(x+1)^2}{36}}$$

$$\frac{y+3}{8} = \pm\frac{x+1}{6}$$

Taking the positive and negative versions of the fraction on the right separately and simplifying, the equations of the asymptotes are $y=\frac{4}{3}x-\frac{5}{3}$ and $y=-\frac{4}{3}x-\frac{13}{3}$.

642. **Center: (0, 3); foci: (–5, 3), (5, 3); asymptotes: $y=\frac{3}{4}x+3$, $y=-\frac{3}{4}x+3$**

The center is at (0, 3).

Using $c=\pm\sqrt{a^2+b^2}$, the foci are at (–5, 3) and (5, 3).

To find the equations of the asymptotes, change the 1 on the right of the equation to 0, add the second fraction to each side, and find the square root of each side.

$$\frac{x^2}{16}-\frac{(y-3)^2}{9}=0$$

$$\frac{x^2}{16}=\frac{(y-3)^2}{9}$$

$$\sqrt{\frac{x^2}{16}}=\pm\sqrt{\frac{(y-3)^2}{9}}$$

$$\frac{x}{4}=\pm\frac{y-3}{3}$$

Taking the positive and negative versions of the fraction on the right separately and simplifying, the equations of the asymptotes are $y=\frac{3}{4}x+3$ and $y=-\frac{3}{4}x+3$.

643. **Center: (3, –1); foci: (–12, –1), (18, –1); asymptotes: $y=-\frac{4}{3}x+3$, $y=\frac{4}{3}x-5$**

The center is at (3, –1).

Using $c=\pm\sqrt{a^2+b^2}$, the foci are at (–12, –1) and (18, –1).

To find the equations of the asymptotes, change the 1 on the right of the equation to 0, add the second fraction to each side, and find the square root of each side.

$$\frac{(x-3)^2}{81}-\frac{(y+1)^2}{144}=0$$

$$\frac{(x-3)^2}{81}=\frac{(y+1)^2}{144}$$

$$\sqrt{\frac{(x-3)^2}{81}}=\pm\sqrt{\frac{(y+1)^2}{144}}$$

$$\frac{x-3}{9}=\pm\frac{y+1}{12}$$

Taking the positive and negative versions of the fraction on the right separately and simplifying, the equations of the asymptotes are $y=-\frac{4}{3}x+3$ and $y=\frac{4}{3}x-5$.

644. **Center: (0, 0); foci: $\left(0, \sqrt{17}\right), \left(0, -\sqrt{17}\right)$; asymptotes: $y = \frac{1}{4}x, y = -\frac{1}{4}x$**

First, write the equation in the standard form.

$$\frac{y^2}{1} - \frac{x^2}{16} = 1$$

The center is at (0, 0).

Using $c = \pm\sqrt{a^2 + b^2}$, the foci are at $\left(0, \sqrt{17}\right)$ and $\left(0, -\sqrt{17}\right)$.

To find the equations of the asymptotes, change the 1 on the right of the equation to 0, add the second fraction to each side, and find the square root of each side.

$$\frac{y^2}{1} - \frac{x^2}{16} = 0$$

$$\frac{y^2}{1} = \frac{x^2}{16}$$

$$\sqrt{\frac{y^2}{1}} = \pm\sqrt{\frac{x^2}{16}}$$

$$\frac{y}{1} = \pm\frac{x}{4}$$

Taking the positive and negative versions of the fraction on the right separately and simplifying, the equations of the asymptotes are $y = \frac{1}{4}x$ and $y = -\frac{1}{4}x$.

645. **Center: (0, 0); foci: $\left(\sqrt{45}, 0\right), \left(-\sqrt{45}, 0\right)$; asymptotes: $y = \frac{1}{2}x, y = -\frac{1}{2}x$**

First, write the equation in the standard form.

$$\frac{x^2}{36} - \frac{y^2}{9} = 1$$

The center is at (0, 0).

Using $c = \pm\sqrt{a^2 + b^2}$, the foci are at $\left(\sqrt{45}, 0\right)$ and $\left(-\sqrt{45}, 0\right)$.

$$\frac{x^2}{36} - \frac{y^2}{9} = 0$$

$$\frac{x^2}{36} = \frac{y^2}{9}$$

$$\sqrt{\frac{x^2}{36}} = \pm\sqrt{\frac{y^2}{9}}$$

$$\frac{x}{6} = \pm\frac{y}{3}$$

The equations of the asymptotes are $y = \frac{1}{2}x$ and $y = -\frac{1}{2}x$.

646. **Center: (0, 0); asymptotes: $y = \frac{7}{2}x, y = -\frac{7}{2}x$; opens left and right**

The center is at (0, 0).

The asymptotes are $y = \frac{7}{2}x$ and $y = -\frac{7}{2}x$.

647. Center: (0, 0); asymptotes: $y = 4x$, $y = -4x$; opens up and down

The center is at (0, 0).

The asymptotes are $y = 4x$ and $y = -4x$.

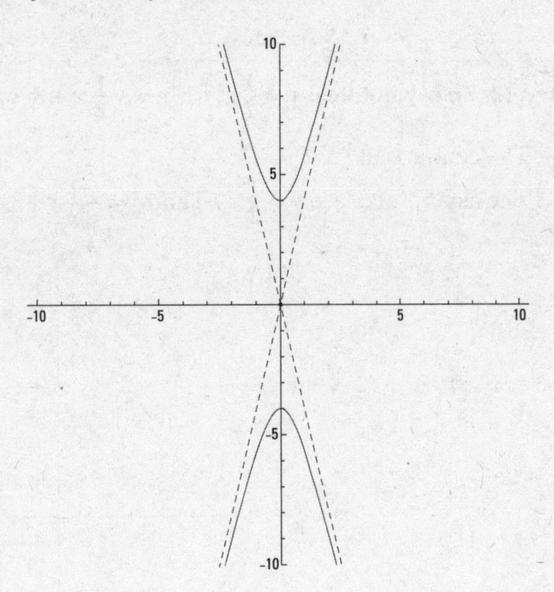

648. **Center: (1, –1); asymptotes: $y = \frac{2}{5}x - \frac{7}{5}$, $y = -\frac{2}{5}x - \frac{3}{5}$; opens left and right**

The center is at (1, –1).

The asymptotes are $y = \frac{2}{5}x - \frac{7}{5}$ and $y = -\frac{2}{5}x - \frac{3}{5}$.

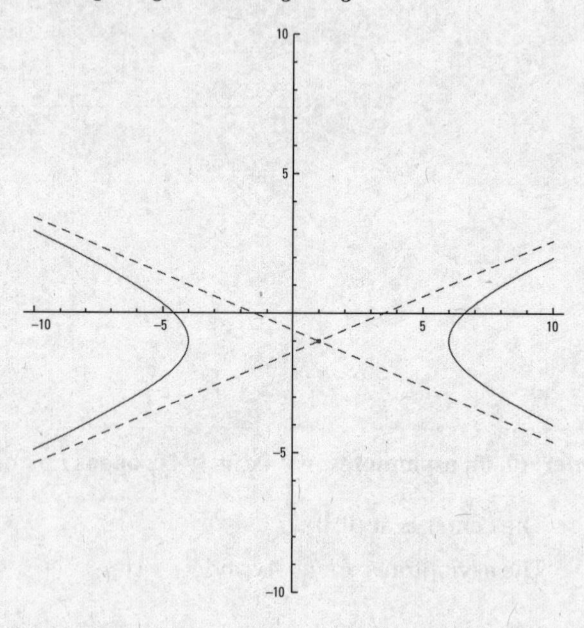

649. **Center: (2, 2); asymptotes: $y = \frac{1}{2}x + 1$, $y = -\frac{1}{2}x + 3$; opens up and down**

The center is at (2, 2).

The asymptotes are $y = \frac{1}{2}x + 1$ and $y = -\frac{1}{2}x + 3$.

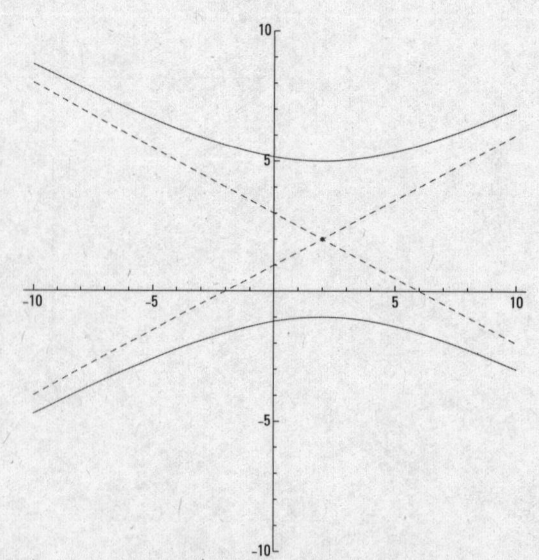

650. Center: (0, 0); asymptotes: $y = 5x$, $y = -5x$; opens left and right

Rewrite the equation in standard form.

$$\frac{x^2}{1} - \frac{y^2}{25} = 1$$

The center is at (0, 0).

The asymptotes are $y = 5x$ and $y = -5x$.

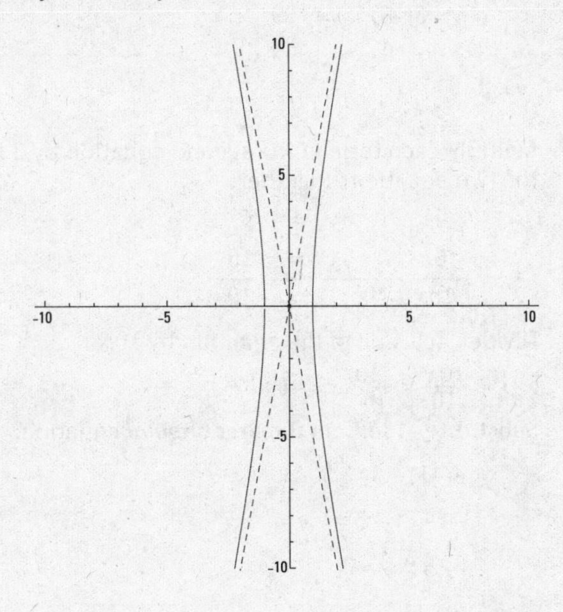

651. $x = 7, y = 17$

Because the equations are both set equal to y, use substitution to write:

$$2x + 3 = 3x - 4$$

Add $-3x$ and -3 to each side of the equation:

$$\begin{array}{rcrcrcr} 2x & + & 3 & = & 3x & - & 4 \\ -3x & & -3 & & -3x & & -3 \\ \hline -x & & & = & & - & 7 \end{array}$$

Now, multiply each side of the equation by -1.

$$-1(-x) = -1(-7),\ x = 7$$

Substitute 7 for x in the first *original* equation:

$$y = 2(7) + 3 = 17$$

652. $x = -6, y = -29$

Because the equations are both set equal to y, use substitution to write:

$$5x + 1 = 7x + 13$$

Add –7x and –1 to each side of the equation:

$$\begin{array}{rcrcrcr} 5x & + & 1 & = & 7x & + & 13 \\ -7x & & -1 & & -7x & & -1 \\ \hline -2x & & & = & & + & 12 \end{array}$$

Now, divide each side of the equation by –2.

$$\frac{-2x}{-2} = \frac{12}{-2}, x = -6$$

Substitute –6 for x in the first *original* equation:

$$y = 5(-6) + 1 = -29$$

653. $x = -1, y = 3$

Multiply each term in the second equation by 3 to create opposite y terms. Then add the two equations together.

$$\begin{array}{rcrcr} 4x & + & 3y & = & 5 \\ 6x & - & 3y & = & -15 \\ \hline 10x & & & = & -10 \end{array}$$

Divide each side of the equation by 10.

$$\frac{10x}{10} = \frac{-10}{10}, x = -1$$

Substitute –1 for x in the first *original* equation:

$$4(-1) + 3y = 5$$

$$-4 + 3y = 5$$

$$3y = 9, y = 3$$

654. $x = -2, y = -4$

Multiply each term in the first equation by 2 and each term in the second equation by –3 to create opposite x terms. Then add the two equations together.

$$\begin{array}{rcrcr} 6x & - & 8y & = & 20 \\ -6x & + & 15y & = & -48 \\ \hline & + & 7y & = & -28 \end{array}$$

Divide each side of the equation by 7.

$$\frac{7y}{7} = \frac{-28}{7}, y = -4$$

Substitute –4 for y in the first *original* equation:

$$3x - 4(-4) = 10$$

$$3x + 16 = 10$$

$$3x = -6, x = -2$$

655. $x = \frac{1}{2}, y = 3$

Add the two equations together:

$$\begin{array}{rcrcr} 4x & + & y & = & 5 \\ 2x & - & y & = & -2 \\ \hline 6x & & & = & 3 \end{array}$$

Divide each side of the equation by 6.

$$\frac{6x}{6} = \frac{3}{6}, x = \frac{1}{2}$$

Substitute $\frac{1}{2}$ for x in the first *original* equation:

$$4\left(\frac{1}{2}\right) + y = 5$$

$$2 + y = 5$$

$$y = 3$$

656. $x = -2, y = \frac{1}{3}$

Multiply each term in the second equation by –3 to create opposite y terms. Then add the two equations together.

$$\begin{array}{rcrcr} 2x & + & 9y & = & -1 \\ -9x & - & 9y & = & 15 \\ \hline -7x & & & = & 14 \end{array}$$

Divide each side of the equation by –7.

$$\frac{-7x}{-7} = \frac{14}{-7}, x = -2$$

Substitute –2 for x in the first *original* equation:

$$2(-2) + 9y = -1$$

$$-4 + 9y = -1$$

$$9y = 3, y = \frac{1}{3}$$

657. $x = k, y = k + 5$

Because the first equation is already solved for y, replace the y in the second equation with that equivalence.

$$2x - 2(x + 5) + 10 = 0$$

$$2x - 2x - 10 + 10 = 0$$

Simplifying, the equation becomes $0 = 0$.

The fact that $0 = 0$ is always true tells you that the equations are two different versions of the same equation — their graphs are the same line. One way to write the solution is to assign x to be the arbitrary value k; then $y = k + 5$.

658. $\mathbf{x = 2 - 3k, y = k}$

Because the second equation is already solved for x, replace the x in the first equation with that equivalence.

$$6y = 4 - 2(2 - 3y)$$

$$6y = 4 - 4 + 6y$$

$$6y = 6y$$

The fact that $6y = 6y$ is always true tells you that the equations are two different versions of the same equation — their graphs are the same line. One way to write the solution is to assign y to be the arbitrary value k; then $x = 2 - 3k$.

659. **No solution**

Because the second equation is already solved for y, replace the y in the first equation with that equivalence.

$$2x + 4 - 2x = 6$$

$$4 = 6$$

This is not a true statement. There is no solution; the graphs of the lines are parallel to one another.

660. **No solution**

Multiply each term in the first equation by –2. Then add the two equations together.

$$\begin{array}{rcrcr} -6x & - & 4y & = & -8 \\ 6x & + & 4y & = & 5 \\ \hline & & 0 & = & -3 \end{array}$$

The result is a false statement. There is no solution; the graphs of the lines are parallel to one another.

661. **Interchange rows 1 and 2.**

The operation $R_1 \leftrightarrow R_2$ tells you to interchange rows 1 and 2 and not change any of the terms in either equation. This change in the rows does not affect the solution of the system of equations.

$$\begin{cases} x + y - 2z = 11 \\ 2x - 3y - z = 12 \\ 3x - 2y + z = 3 \end{cases}$$

662. **Multiply row 2 by –3 and replace the row with the products.**

The operation $-3R_2 \rightarrow R_2$ tells you to multiply each term in row 2 by –3 and then replace all the terms in row 2 with those products. This multiplication does not affect the solution of the system of equations.

$$\begin{cases} 3x+3y-2z=4 \\ -3x+6y-3z=-21 \\ 2x+4y+z=1 \end{cases}$$

663. **Multiply row 1 by –3, add the products to row 2, and replace row 2 with the sums.**

The operation $-3R_1+R_2 \rightarrow R_2$ involves several steps. First, multiply each term in row 1 by –3; then add each of the new terms to the corresponding terms in row 2. Then replace row 2 with all those sums. This multiplication and addition does not affect the solution of the system of equations.

$$\begin{cases} x-3y+2z=7 \\ 10y-7z=-17 \\ 2x-y+z=1 \end{cases}$$

664. **Multiply row 2 by –3, add the products to row 1, and replace row 1 with the sums.**

The operation $-3R_2+R_1 \rightarrow R_1$ involves several steps. First, multiply each term in row 2 by –3; then add each of the new terms to the corresponding terms in row 1. Then replace row 1 with all those sums. This multiplication and addition does not affect the solution of the system of equations.

$$\begin{cases} x \quad +5z=13 \\ y-z=-2 \\ 2y+3z=4 \end{cases}$$

665. **Multiply row 2 by –7, add the products to row 1, and replace row 1 with the sums.**

The operation $-7R_2+R_1 \rightarrow R_1$ involves several steps. First multiply each term in row 2 by –7; then add each of the new terms to the corresponding terms in row 1. Then replace row 1 with all those sums. This multiplication and addition does not affect the solution of the system of equations.

$$\begin{cases} x \quad +12z=-22 \\ y-z=4 \\ z=1 \end{cases}$$

666. $x = -1, y = 1, z = -1$

$$\begin{matrix} -1R_1 + R_2 \to R_2 \\ -2R_1 + R_3 \to R_3 \end{matrix} : \begin{cases} x + y + 2z = -2 \\ -3y + 1z = -4 \\ -7y - 3z = -4 \end{cases}$$

$$-\frac{1}{3}R_2 \to R_2 : \begin{cases} x + y + 2z = -2 \\ y - \frac{1}{3}z = \frac{4}{3} \\ -7y - 3z = -4 \end{cases}$$

$$7R_2 + R_3 \to R_3 : \begin{cases} x + y + 2z = -2 \\ y - \frac{1}{3}z = \frac{4}{3} \\ -\frac{16}{3}z = \frac{16}{3} \end{cases}$$

$$-\frac{3}{16}R_3 \to R_3 : \begin{cases} x + y + 2z = -2 \\ y - \frac{1}{3}z = \frac{4}{3} \\ 1z = -1 \end{cases}$$

The variable $z = -1$. Replace the z in the second equation with -1 and solve for y:

$$y - \frac{1}{3}(-1) = \frac{4}{3}$$
$$y + \frac{1}{3} = \frac{4}{3}, \; y = 1$$

And now, with $z = -1$ and $y = 1$, substitute those values into the first equation to solve for x.

$$x + 1 + 2(-1) = -2$$
$$x - 1 = -2, \; x = -1$$

667. $x = 1, y = 0, z = -8$

$$\begin{matrix} -2R_1 + R_2 \to R_2 \\ -3R_1 + R_3 \to R_3 \end{matrix} : \begin{cases} x + y - 4z = 33 \\ 2y + 7z = -56 \\ -4y + 13z = -104 \end{cases}$$

$$\frac{1}{2}R_2 \to R_2 : \begin{cases} x + y - 4z = 33 \\ y + \frac{7}{2}z = -28 \\ -4y + 13z = -104 \end{cases}$$

$$4R_2 + R_3 \to R_3 : \begin{cases} x + y - 4z = 33 \\ y + \frac{7}{2}z = -28 \\ 27z = -216 \end{cases}$$

$$\frac{1}{27}R_3 \to R_3 : \begin{cases} x + y - 4z = 33 \\ y + \frac{7}{2}z = -28 \\ z = -8 \end{cases}$$

The variable $z = -8$. Replace the z in the second equation with -8 and solve for y:

$$y + \frac{7}{2}(-8) = -28$$

$$y - 28 = -28,\ y = 0$$

And now, with $z = -8$ and $y = 0$, substitute those values into the first equation to solve for x.

$$x + 0 - 4(-8) = 33$$

$$x + 32 = 33,\ x = 1$$

668. $\mathbf{x = 2, y = 3, z = -1}$

$$R_1 \leftrightarrow R_2 : \begin{cases} x + 3y + z = 10 \\ 2x - y - 4z = 5 \\ x + 2y - 3z = 11 \end{cases}$$

$$\begin{matrix} -2R_1 + R_2 \to R_2 \\ -1R_1 + R_3 \to R_3 \end{matrix} : \begin{cases} x + 3y + z = 10 \\ -7y - 6z = -15 \\ -y - 4z = 1 \end{cases}$$

$$-\frac{1}{7}R_2 \to R_2 : \begin{cases} x + 3y + z = 10 \\ y + \frac{6}{7}z = \frac{15}{7} \\ -y - 4z = 1 \end{cases}$$

$$R_2 + R_3 \to R_3 : \begin{cases} x + 3y + z = 10 \\ y + \frac{6}{7}z = \frac{15}{7} \\ -\frac{22}{7}z = \frac{22}{7} \end{cases}$$

$$-\frac{7}{22}R_3 \to R_3 : \begin{cases} x + 3y + z = 10 \\ y + \frac{6}{7}z = \frac{15}{7} \\ z = -1 \end{cases}$$

The variable $z = -1$. Replace the z in the second equation with -1 and solve for y:

$$y + \frac{6}{7}(-1) = \frac{15}{7}$$

$$y - \frac{6}{7} = \frac{15}{7},\ y = 3$$

And now, with $z = -1$ and $y = 3$, substitute those values into the first equation to solve for x.

$$2x - 3 - 4(-1) = 5$$

$$2x - 3 + 4 = 5$$

$$2x + 1 = 5, 2x = 4, x = 2$$

669. $x = 0, y = 1, z = 1$

$$R_1 \leftrightarrow R_3 : \begin{cases} x + y + z = 2 \\ 2x + 3y - z = 2 \\ 3x - 2y + z = -1 \end{cases}$$

$$\begin{matrix} -2R_1 + R_2 \rightarrow R_2 \\ -3R_1 + R_3 \rightarrow R_3 \end{matrix} : \begin{cases} x + y + z = 2 \\ 1y - 3z = -2 \\ -5y - 2z = -7 \end{cases}$$

$$5R_2 + R_3 \rightarrow R_3 : \begin{cases} x + y + z = 2 \\ 1y - 3z = -2 \\ -17z = -17 \end{cases}$$

$$-\frac{1}{17}R_3 \rightarrow R_3 : \begin{cases} x + y + z = 2 \\ 1y - 3z = -2 \\ z = 1 \end{cases}$$

The variable $z = 1$. Replace the z in the second equation with 1 and solve for y:

$$y - 3(1) = -2$$

$$y - 3 = -2, y = 1$$

And now, with $z = 1$ and $y = 1$, substitute those values into the first equation to solve for x.

$$x + 1 + 1 = 2, x = 0$$

670. $x = 3, y = 0, z = -3$

$$R_1 \leftrightarrow R_2 : \begin{cases} x - 4y - 2z = 9 \\ 2x - 3y + z = 3 \\ 3x + y + z = 6 \end{cases}$$

$$\begin{matrix} -2R_1 + R_2 \rightarrow R_2 \\ -3R_1 + R_3 \rightarrow R_3 \end{matrix} : \begin{cases} x - 4y - 2z = 9 \\ 5y + 5z = -15 \\ 13y + 7z = -21 \end{cases}$$

$$\frac{1}{5}R_2 \to R_2 : \begin{cases} x - 4y - 2z = 9 \\ y + z = -3 \\ 13y + 7z = -21 \end{cases}$$

$$-13R_2 + R_3 \to R_3 : \begin{cases} x - 4y - 2z = 9 \\ y + z = -3 \\ -6z = 18 \end{cases}$$

$$-\frac{1}{6}R_3 \to R_3 : \begin{cases} x - 4y - 2z = 9 \\ y + z = -3 \\ z = -3 \end{cases}$$

The variable $z = -3$. Replace the z in the second equation with 1 and solve for y:

$y + (-3) = -3, y = 0$

And now, with $z = -3$ and $y = 0$, substitute those values into the first equation to solve for x.

$2x - 3(0) + (-3) = 3, 2x = 6, x = 3$

671. $\mathbf{x = 2, y = 1, z = \frac{1}{2}}$

$$\begin{matrix} -2R_1 + R_2 \to R_2 \\ -3R_1 + R_3 \to R_3 \end{matrix} : \begin{cases} x - 2y + 2z = 1 \\ 5y - 8z = 1 \\ 5y \qquad = 5 \end{cases}$$

$$\frac{1}{5}R_3 \to R_3 : \begin{cases} x - 2y + 2z = 1 \\ 5y - 8z = 1 \\ y \qquad = 1 \end{cases}$$

$$R_2 \leftrightarrow R_3 : \begin{cases} x - 2y + 2z = 1 \\ y \qquad = 1 \\ 5y - 8z = 1 \end{cases}$$

$$-5R_2 + R_3 \to R_3 : \begin{cases} x - 2y + 2z = 1 \\ y \qquad = 1 \\ -8z = -4 \end{cases}$$

$$-\frac{1}{8}R_3 \to R_3 : \begin{cases} x - 2y + 2z = 1 \\ y \qquad = 1 \\ z = \frac{1}{2} \end{cases}$$

The variable $z = \frac{1}{2}$ and $y = 1$. Replace the y and z in the first equation and solve for x:

$x - 2(1) + 2\left(\frac{1}{2}\right) = 1$

$x - 2 + 1 = 1,\ x - 1 = 1,\ x = 2$

672. $x = 1, y = \frac{1}{3}, z = -1$

$$R_1 \leftrightarrow R_3: \begin{cases} x + 6y - z = 4 \\ 2x - 3y - z = 2 \\ 3x + 3y + z = 3 \end{cases}$$

$$\begin{matrix} -2R_1 + R_2 \to R_2 \\ -3R_1 + R_3 \to R_3 \end{matrix} : \begin{cases} x + 6y - z = 4 \\ -15y + z = -6 \\ -15y + 4z = -9 \end{cases}$$

$$-\frac{1}{15}R_2 \to R_2: \begin{cases} x + 6y - z = 4 \\ y - \frac{1}{15}z = \frac{2}{5} \\ -15y + 4z = -9 \end{cases}$$

$$15R_2 + R_3 \to R_3: \begin{cases} x + 6y - z = 4 \\ y - \frac{1}{15}z = \frac{2}{5} \\ 3z = -3 \end{cases}$$

$$\frac{1}{3}R_3 \to R_3: \begin{cases} x + 6y - z = 4 \\ y - \frac{1}{15}z = \frac{2}{5} \\ z = -1 \end{cases}$$

The variable $z = -1$. Replace the z with -1 in the second equation and solve for y:

$$y - \frac{1}{15}(-1) = \frac{2}{5}$$

$$y + \frac{1}{15} = \frac{2}{5}, y = \frac{6}{15} - \frac{1}{15} = \frac{5}{15} = \frac{1}{3}$$

And now, with $z = -1$ and $y = \frac{1}{3}$, substitute those values into the first equation to solve for x.

$$x + 6\left(\frac{1}{3}\right) - (-1) = 4$$

$$x + 2 + 1 = 4,\ x + 3 = 4,\ x = 1$$

673. $x = 0, y = 0, z = 2$

$$R_1 \leftrightarrow R_2: \begin{cases} x - 2y + 3z = 6 \\ 3x + y - z = -2 \\ 4x + 4y + z = 2 \end{cases}$$

$$\begin{matrix} -3R_1 + R_2 \to R_2 \\ -4R_1 + R_3 \to R_3 \end{matrix} : \begin{cases} x - 2y + 3z = 6 \\ 7y - 10z = -20 \\ 12y - 11z = -22 \end{cases}$$

$$\frac{1}{7}R_2 \to R_2 : \begin{cases} x - 2y + 3z = 6 \\ y - \frac{10}{7}z = -\frac{20}{7} \\ 12y - 11z = -22 \end{cases}$$

$$-12R_2 + R_3 \to R_3 : \begin{cases} x - 2y + 3z = 6 \\ y - \frac{10}{7}z = -\frac{20}{7} \\ \frac{43}{7}z = \frac{86}{7} \end{cases}$$

$$\frac{7}{43}R_3 \to R_3 : \begin{cases} x - 2y + 3z = 6 \\ y - \frac{10}{7}z = -\frac{20}{7} \\ z = 2 \end{cases}$$

The variable $z = 2$. Replace the z with 2 in the second equation and solve for y:

$$y - \frac{10}{7}(2) = -\frac{20}{7}$$

$$y - \frac{20}{7} = -\frac{20}{7}, y = 0$$

And now, with $z = 2$ and $y = 0$, substitute those values into the first equation to solve for x.

$$x - 2(0) + 3(2) = 6, x + 6 = 6, x = 0$$

674. $\boldsymbol{x = \frac{5}{8} + \frac{1}{8}k, y = \frac{1}{8} + \frac{5}{8}k, z = k}$

$$R_1 \leftrightarrow R_2 : \begin{cases} x + 3y - 2z = 1 \\ 3x + y - z = 2 \\ 5x + 7y - 5z = 4 \end{cases}$$

$$\begin{matrix} -3R_1 + R_2 \to R_2 \\ -5R_1 + R_3 \to R_3 \end{matrix} : \begin{cases} x + 3y - 2z = 1 \\ -8y + 5z = -1 \\ -8y + 5z = -1 \end{cases}$$

$$-\frac{1}{8}R_2 \to R_2 : \begin{cases} x + 3y - 2z = 1 \\ y - \frac{5}{8}z = \frac{1}{8} \\ -8y + 5z = -1 \end{cases}$$

$$8R_2 + R_3 \to R_3 : \begin{cases} x + 3y - 2z = 1 \\ y - \frac{5}{8}z = \frac{1}{8} \\ 0 = 0 \end{cases}$$

The equation 0 = 0 indicates that there is no unique solution; there are an infinite number of solutions. Choosing a parameter to represent z, let $k = z$. Substitute k for z in the second equation to solve for y.

$$y - \frac{5}{8}(k) = \frac{1}{8}$$
$$y = \frac{1}{8} + \frac{5}{8}k$$

And now, letting $z = k$ and $y = \frac{1}{8} + \frac{5}{8}k$, substitute those values into the first equation to solve for x.

$$x + 3\left(\frac{1}{8} + \frac{5}{8}k\right) - 2(k) = 1$$
$$x + \frac{3}{8} + \frac{15}{8}k - 2k = 1$$
$$x - \frac{1}{8}k = \frac{5}{8}$$
$$x = \frac{5}{8} + \frac{1}{8}k$$

The complete solution is

$$x = \frac{5}{8} + \frac{1}{8}k,\ y = \frac{1}{8} + \frac{5}{8}k,\ z = k.$$

For an example of finding a solution, let $k = 3$. This gives you the solution $x = 1$, $y = 2$, $z = 3$.

675. $\boldsymbol{x = 2 - \frac{3}{2}k,\ y = \frac{7}{2} - \frac{5}{4}k,\ z = k}$

$$\begin{matrix} -2R_1 + R_2 \to R_2 \\ -1R_1 + R_3 \to R_3 \end{matrix} : \begin{cases} x + 2y + 4z = 9 \\ -4y - 5z = -14 \\ -4y - 5z = -14 \end{cases}$$

$$-\frac{1}{4}R_2 \to R_2 : \begin{cases} x + 2y + 4z = 9 \\ y + \frac{5}{4}z = \frac{14}{4} \\ -4y - 5z = -14 \end{cases}$$

$$4R_2 + R_3 \to R_3 : \begin{cases} x + 2y + 4z = 9 \\ y + \frac{5}{4}z = \frac{14}{4} \\ 0 = 0 \end{cases}$$

The equation 0 = 0 indicates that there is no unique solution; there are an infinite number of solutions. Choosing a parameter to represent z, let $k = z$. Substitute k for z in the second equation to solve for y.

$$y + \frac{5}{4}(k) = \frac{14}{4}$$
$$y = \frac{14}{4} - \frac{5}{4}k = \frac{7}{2} - \frac{5}{4}k$$

And now, letting $z = k$ and $y = \frac{7}{2} - \frac{5}{4}k$, substitute those values into the first equation to solve for x.

$$x + 2\left(\frac{7}{2} - \frac{5}{4}k\right) + 4(k) = 9$$

$$x + 7 - \frac{5}{2}k + 4k = 9$$

$$x + \frac{3}{2}k = 2$$

$$x = 2 - \frac{3}{2}k$$

The complete solution is

$$x = 2 - \frac{3}{2}k,\ y = \frac{7}{2} - \frac{5}{4}k,\ z = k.$$

For an example of finding a solution, let $k = -2$. This gives you the solution $x = 21$, $y = 1$, $z = -2$.

676. $x = 1, y = 2, z = 3$

$$\begin{matrix} -2R_1 + R_2 \to R_2 \\ -1R_1 + R_3 \to R_3 \end{matrix} : \begin{cases} x - 2y + z = 0 \\ y = 2 \\ 3y - 3z = -3 \end{cases}$$

$$\begin{matrix} 2R_2 + R_1 \to R_1 \\ -3R_2 + R_3 \to R_3 \end{matrix} : \begin{cases} x + z = 4 \\ y = 2 \\ -3z = -9 \end{cases}$$

$$-\frac{1}{3}R_3 \to R_3 : \begin{cases} x + z = 4 \\ y = 2 \\ z = 3 \end{cases}$$

$$-1R_3 + R_1 \to R_1 : \begin{cases} x = 1 \\ y = 2 \\ z = 3 \end{cases}$$

$x = 1, y = 2, z = 3$

677. $x = -2, y = 0, z = 1$

$$R_1 \leftrightarrow R_2 : \begin{cases} x + y - z = -3 \\ 2x - 3y + 7z = 3 \\ 3x + 2y + z = -5 \end{cases}$$

$$\begin{matrix} -2R_1 + R_2 \to R_2 \\ -3R_1 + R_3 \to R_3 \end{matrix} : \begin{cases} x + y - z = -3 \\ -5y + 9z = 9 \\ -y + 4z = 4 \end{cases}$$

$$R_2 \leftrightarrow R_3: \begin{cases} x+y-z=-3 \\ -y+4z=4 \\ -5y+9z=9 \end{cases}$$

$$-1R_2 \to R_2: \begin{cases} x+y-z=-3 \\ y-4z=-4 \\ -5y+9z=9 \end{cases}$$

$$\begin{matrix} -1R_2+R_1 \to R_1 \\ 5R_2+R_3 \to R_3 \end{matrix}: \begin{cases} x \quad +3z=1 \\ y-4z=-4 \\ -11z=-11 \end{cases}$$

$$-\frac{1}{11}R_3 \to R_3: \begin{cases} x \quad +3z=1 \\ y-4z=-4 \\ z=1 \end{cases}$$

$$\begin{matrix} -3R_3+R_1 \to R_1 \\ 4R_3+R_2 \to R_2 \end{matrix}: \begin{cases} x \quad =-2 \\ y \quad =0 \\ z=1 \end{cases}$$

$x = -2, y = 0, z = 1$

678. $x = 4, y = 1, z = -2$

$$\begin{matrix} -1R_1+R_2 \to R_2 \\ -2R_1+R_3 \to R_3 \end{matrix}: \begin{cases} x-y-2z=7 \\ y+5z=-9 \\ 3y+4z=-5 \end{cases}$$

$$\begin{matrix} R_2+R_1 \to R_1 \\ -3R_2+R_3 \to R_3 \end{matrix}: \begin{cases} x \quad +3z=-2 \\ y+5z=-9 \\ -11z=22 \end{cases}$$

$$-\frac{1}{11}R_3 \to R_3: \begin{cases} x \quad +3z=-2 \\ y+5z=-9 \\ z=-2 \end{cases}$$

$$\begin{matrix} -3R_3+R_1 \to R_1 \\ -5R_3+R_2 \to R_2 \end{matrix}: \begin{cases} x \quad =4 \\ y \quad =1 \\ z=-2 \end{cases}$$

$x = 4, y = 1, z = -2$

679. $x = -1, y = 2, z = 3$

$$R_1 \leftrightarrow R_3 : \begin{cases} x \quad -2z = -7 \\ 4y + z = 11 \\ 2x + y - 3z = -9 \end{cases}$$

$$-2R_1 + R_3 \rightarrow R_3 : \begin{cases} x \quad -2z = -7 \\ 4y + z = 11 \\ y + z = 5 \end{cases}$$

$$R_2 \leftrightarrow R_3 : \begin{cases} x \quad -2z = -7 \\ y + z = 5 \\ 4y + z = 11 \end{cases}$$

$$-4R_2 + R_3 \rightarrow R_3 : \begin{cases} x \quad -2z = -7 \\ y + z = 5 \\ -3z = -9 \end{cases}$$

$$-\frac{1}{3}R_3 \rightarrow R_3 : \begin{cases} x \quad -2z = -7 \\ y + z = 5 \\ z = 3 \end{cases}$$

$$\begin{matrix} 2R_3 + R_1 \rightarrow R_1 \\ -1R_3 + R_2 \rightarrow R_2 \end{matrix} : \begin{cases} x \quad = -1 \\ y = 2 \\ z = 3 \end{cases}$$

$x = -1, y = 2, z = 3$

680. $x = \frac{1}{7}k + \frac{10}{7}, y = \frac{3}{7}k + \frac{23}{7}, z = k$

$$R_1 \leftrightarrow R_2 : \begin{cases} x + 2y - z = 8 \\ 5x + 3y - 2z = 17 \\ 3x - y \quad = 1 \end{cases}$$

$$\begin{matrix} -5R_1 + R_2 \rightarrow R_2 \\ -3R_1 + R_3 \rightarrow R_3 \end{matrix} : \begin{cases} x + 2y - z = 8 \\ -7y + 3z = -23 \\ -7y + 3z = -23 \end{cases}$$

$$-\frac{1}{7}R_2 \rightarrow R_2 : \begin{cases} x + 2y - z = 8 \\ y - \frac{3}{7}z = \frac{23}{7} \\ -7y + 3z = -23 \end{cases}$$

$$7R_2 + R_3 \to R_3 : \begin{cases} x + 2y - z = 8 \\ y - \frac{3}{7}z = \frac{23}{7} \\ 0 + 0 = 0 \end{cases}$$

$$-2R_3 + R_1 \to R_1 : \begin{cases} x \quad - \frac{1}{7}z = \frac{10}{7} \\ y - \frac{3}{7}z = \frac{23}{7} \\ 0 + 0 = 0 \end{cases}$$

The system does not have a single, unique solution. Let k represent z, and solve for x and y in terms of k.

$$x = \frac{1}{7}k + \frac{10}{7},\ y = \frac{3}{7}k + \frac{23}{7},\ z = k$$

To find one of the solutions, pick a value for k. For example, if $k = -3$, then $x = 1$, $y = 2$, $z = -3$.

681. $\boldsymbol{x = 1, y = 2, z = -1}$

$$\begin{matrix} -2R_1 + R_2 \to R_2 \\ -R_1 + R_3 \to R_3 \end{matrix} : \begin{cases} x + 3y + z = 6 \\ -5y - 3z = -7 \\ -4y - 4z = -4 \end{cases}$$

$$-\frac{1}{4}R_3 \to R_3 : \begin{cases} x + 3y + z = 6 \\ -5y - 3z = -7 \\ y \quad + z = 1 \end{cases}$$

$$5R_3 + R_2 \to R_2 : \begin{cases} x + 3y + z = 6 \\ 2z = -2 \\ y + \quad z = 1 \end{cases}$$

The second equation reads $2z = -2$.

Divide both sides of the equation by -2 to get $z = -1$.

Replace the z in the third equation with -1 and solve for y:

$$y + (-1) = 1,\ y = 2$$

And now, with $z = -1$ and $y = 2$, substitute those values into the first equation to solve for x.

$$x + 3(2) + (-1) = 6,\ x + 6 - 1 = 6,\ x = 1$$

682. $\boldsymbol{x = -2, y = 3, z = 4}$

$$\begin{matrix} -2R_2 + R_1 \to R_1 \\ -1R_2 + R_3 \to R_3 \end{matrix} : \begin{cases} 3y - 7z = -19 \\ x - y + 4z = 11 \\ 3y - 3z = -3 \end{cases}$$

$$\frac{1}{3}R_3 \rightarrow R_3: \begin{cases} 3y - 7z = -19 \\ x - y + 4z = 11 \\ y - z = -1 \end{cases}$$

$$-3R_3 + R_1 \rightarrow R_1: \begin{cases} -4z = -16 \\ x - y + 4z = 11 \\ y - z = -1 \end{cases}$$

The first equation reads $-4z = -16$.

Divide both sides of the equation by -4 to get $z = 4$.

Replace the z in the third equation with 4 and solve for y:

$y(-4) = -1, y = 3$

And now, with $z = 4$ and $y = 3$, substitute those values into the second equation to solve for x.

$x - 3 + 4(4) = 11, x + 13 = 11, x = -2$

683. $x = 2, y = 0, z = 1$

$$\begin{matrix} -3R_1 + R_2 \rightarrow R_2 \\ -4R_1 + R_3 \rightarrow R_3 \end{matrix} : \begin{cases} x - y + 2z = 4 \\ 5y - 5z = -5 \\ y - 9z = -9 \end{cases}$$

$$\frac{1}{5}R_2 \rightarrow R_2: \begin{cases} x - y + 2z = 4 \\ y - z = -1 \\ y - 9z = -9 \end{cases}$$

$$-1R_2 + R_3 \rightarrow R_3: \begin{cases} x - y + 2z = 4 \\ y - z = -1 \\ -8z = -8 \end{cases}$$

The third equation reads $-8z = -8$.

Divide both sides of the equation by -8 to get $z = 1$.

Replace the z in the second equation with 1 and solve for y:

$y - 1 = -1, y = 0$

And now, with $z = 1$ and $y = 0$, substitute those values into the first equation to solve for x.

$x - 0 + 2(1) = 4, x + 2 = 4, x = 2.$

684. $x = 0, y = 2, z = -4$

$$\begin{matrix}-3R_3 + R_1 \to R_1 \\ -4R_3 + R_2 \to R_2\end{matrix} : \begin{cases} -8y - 4z = 0 \\ -13y - 7z = 2 \\ x + 3y + z = 2 \end{cases}$$

$$-\frac{1}{8}R_1 \to R_1 : \begin{cases} y + \frac{1}{2}z = 0 \\ -13y - 7z = 2 \\ x + 3y + z = 2 \end{cases}$$

$$13R_1 + R_2 \to R_2 : \begin{cases} y + \frac{1}{2}z = 0 \\ -\frac{1}{2}z = 2 \\ x + 3y + z = 2 \end{cases}$$

The second equation reads $-\frac{1}{2}z = 2$.

Multiply both sides of the equation by –2 to get $z = -4$.

Replace the z in the first equation with –4 and solve for y:

$$y + \frac{1}{2}(-4) = 0,\ y - 2 = 0,\ y = 2$$

And now, with $z = -4$ and $y = 2$, substitute those values into the third equation to solve for x.

$$x + 3(2) + (-4) = 2,\ x + 2 = 2,\ x = 0$$

685. $x = -1, y = -1, z = -2$

$$\begin{matrix}-2R_1 + R_2 \to R_2 \\ -3R_1 + R_3 \to R_3\end{matrix} : \begin{cases} x - 3y - z = 4 \\ 7y - 2z = -3 \\ 8y + 2z = -12 \end{cases}$$

$$\frac{1}{7}R_2 \to R_2 : \begin{cases} x - 3y - z = 4 \\ y - \frac{2}{7}z = -\frac{3}{7} \\ 8y + 2z = -12 \end{cases}$$

$$-8R_2 + R_3 \to R_3 : \begin{cases} x - 3y - z = 4 \\ y - \frac{2}{7}z = -\frac{3}{7} \\ \frac{30}{7}z = -\frac{60}{7} \end{cases}$$

The third equation reads $\frac{30}{7}z = -\frac{60}{7}$.

Multiply both sides of the equation by $\frac{7}{30}$ to get $z = -2$.

Replace the z in the second equation with –2 and solve for y:

$$y - \frac{2}{7}(-2) = -\frac{3}{7},\ y + \frac{4}{7} = -\frac{3}{7},\ y = -1$$

And now, with $z = -2$ and $y = -1$, substitute those values into the first equation to solve for x.

$$x - 3(-1) - (-2) = 4,\ x + 5 = 4,\ x = -1$$

686. $\mathbf{x = 3,\ y = -3,\ z = 2}$

$$\begin{matrix} -1R_1 + R_2 \rightarrow R_2 \\ -1R_1 + R_3 \rightarrow R_3 \end{matrix} : \begin{cases} x + y + z = 2 \\ -2y \quad\ = 6 \\ -2y - 2z = 2 \end{cases}$$

The second equation reads $-2y = 6$.

Divide both sides of the equation by –2 to get $y = -3$.

Replace the y in the third equation with –3 and solve for z:

$$-2(-3) - 2z = 2,\ -2z = -4,\ z = 2$$

And now, with $z = 2$ and $y = -3$, substitute those values into the first equation to solve for x.

$$x + (-3) + 2 = 2,\ x - 1 = 2,\ x = 3$$

687. $\mathbf{x = \frac{2}{3} - \frac{1}{3}k,\ y = \frac{11}{3} + \frac{5}{3}k,\ z = k}$

$$\begin{matrix} -2R_1 + R_2 \rightarrow R_2 \\ -3R_1 + R_3 \rightarrow R_3 \end{matrix} : \begin{cases} x + 2y - 3z = 8 \\ -3y + 5z = -11 \\ -3y + 5z = -11 \end{cases}$$

$$-1R_2 + R_3 \rightarrow R_3 : \begin{cases} x + 2y - 3z = 8 \\ -3y + 5z = -11 \\ 0 + 0 = 0 \end{cases}$$

The system does not have a single, unique solution. Let k represent z, and solve for y in terms of k.

$$-3y = -5k - 11$$

$$y = \frac{5}{3}k + \frac{11}{3}$$

Then solve for x:

$$x + 2\left(\frac{5}{3}k + \frac{11}{3}\right) - 3k = 8$$

$$x + \frac{10}{3}k + \frac{22}{3} - \frac{9}{3}k = 8$$

$$x + \frac{1}{3}k = 8 - \frac{22}{3}$$

$$x = \frac{2}{3} - \frac{1}{3}k$$

To find one of the solutions, pick a value for k. For example, if $k = z = -1$, then $x = \frac{2}{3} - \frac{1}{3}(-1) = 1$ and $y = \frac{11}{3} + \frac{5}{3}(-1) = 2$.

688. $x = \frac{19}{5} + \frac{3}{5}k,\ y = -\frac{21}{10} - \frac{7}{10}k,\ z = k$

$$\begin{array}{l} -4R_2 + R_1 \to R_1 \\ -6R_2 + R_3 \to R_3 \end{array} : \begin{cases} 10y + 7z = -21 \\ x - 2y - 2z = 8 \\ 10y + 7z = -21 \end{cases}$$

$$-1R_1 + R_3 \to R_3 : \begin{cases} 10y + 7z = -21 \\ x - 2y - 2z = 8 \\ 0 + 0 = 0 \end{cases}$$

The system does not have a single, unique solution. Let k represent z, and solve for y in terms of k.

$$10y = -7k - 21$$
$$y = -\frac{7}{10}k - \frac{21}{10}$$

Then solve for x:

$$x - 2\left(-\frac{7}{10}k - \frac{21}{10}\right) - 2k = 8$$
$$x + \frac{7}{5}k + \frac{21}{5} - \frac{10}{5}k = 8$$
$$x - \frac{3}{5}k = 8 - \frac{21}{5}$$
$$x = \frac{19}{5} + \frac{3}{5}k$$

To find one of the solutions, pick a value for k. For example, if $k = z = -3$, then $x = \frac{19}{5} + \frac{3}{5}(-3) = 2$ and $y = -\frac{7}{10}(-3) - \frac{21}{10} = 0$.

689. $x = 4,\ y = -2,\ z = 1$

$$-2R_1 + R_3 \to R_3 : \begin{cases} x + z = 5 \\ y - 2z = -4 \\ -y - 2z = 0 \end{cases}$$

$$R_2 + R_3 \to R_3 : \begin{cases} x + z = 5 \\ y - 2z = -4 \\ -4z = -4 \end{cases}$$

The third equation reads $-4z = -4$, so $z = 1$.

Replace the z in the first equation with 1 and solve for x:

$x + 1 = 5, x = 4$

And now, with $z = 1$, substitute that value into the second equation to solve for y:

$y - 2(1) = -4, y = -2.$

690. $x = 0, y = 3, z = 2$

$$-2R_1 + R_2 \to R_2 : \begin{cases} x + y = 3 \\ -2y - z = -8 \\ 3y - z = 7 \end{cases}$$

$$-1R_2 + R_3 \to R_3 : \begin{cases} x + y = 3 \\ -2y - z = -8 \\ 5y = 15 \end{cases}$$

The third equation reads $5y = 15$, so $y = 3$.

Replace the y in the first equation with 3 and solve for x:

$x + 3 = 3, x = 0$

And now, with $y = 3$, substitute that value into the second equation to solve for z:

$-2(3) - z = -8, -z = -2, z = 2$

691. $\frac{-2}{x} + \frac{3}{x-2}$

The fraction is the sum of $\frac{A}{x} + \frac{B}{x-2}$.

Set the original fraction equal to the sum.

$$\frac{x+4}{x(x-2)} = \frac{A}{x} + \frac{B}{x-2}$$

Now add the two fractions on the right together.

$$\begin{aligned} \frac{x+4}{x(x-2)} &= \frac{A}{x} \cdot \frac{x-2}{x-2} + \frac{B}{x-2} \cdot \frac{x}{x} \\ &= \frac{A(x-2)}{x(x-2)} + \frac{Bx}{x(x-2)} \\ &= \frac{Ax - 2A + Bx}{x(x-2)} \end{aligned}$$

Let the numerator on the left equal the numerator on the right.

$x + 4 = Ax - 2A + Bx$

Factor on the right.

$x + 4 = x(A + B) - 2A$

Write a system of equations involving the coefficients of the x terms and the constants.

$$\begin{cases} 1 = A + B \\ 4 = -2A \end{cases}$$

From the second equation, you find that $A = -2$. Substitute that into the first equation:

$1 = -2 + B$, so $B = 3$

The decomposition is $\frac{x+4}{x(x-2)} = \frac{-2}{x} + \frac{3}{x-2}$.

692. $\frac{2}{x+1}+\frac{-1}{x-3}$

The fraction is the sum of $\frac{A}{x+1}+\frac{B}{x-3}$.

Set the original fraction equal to the sum.

$$\frac{x-7}{(x+1)(x-3)}=\frac{A}{x+1}+\frac{B}{x-3}$$

Now add the two fractions on the right together.

$$\begin{aligned}\frac{x-7}{(x+1)(x-3)}&=\frac{A}{x+1}\cdot\frac{x-3}{x-3}+\frac{B}{x-3}\cdot\frac{x+1}{x+1}\\&=\frac{A(x-3)}{(x+1)(x-3)}+\frac{B(x+1)}{(x+1)(x-3)}\\&=\frac{Ax-3A+Bx+B}{(x+1)(x-3)}\end{aligned}$$

Let the numerator on the left equal the numerator on the right.

$$x-7=Ax-3A+Bx+B$$

Factor on the right.

$$x-7=x(A+B)-3A+B$$

Write a system of equations involving the coefficients of the x terms and the constants.

$$\begin{cases}1=A+B\\-7=-3A+B\end{cases}$$

Solve the system:

$$-1R_1+R_2\rightarrow R_2\begin{cases}1=A+B\\-8=-4A\end{cases}$$

From the second equation, you find that $A=2$. Substitute that into the first equation:

$1=2+B$, so $B=-1$.

The decomposition is $\frac{x-7}{(x+1)(x-3)}=\frac{2}{x+1}+\frac{-1}{x-3}$.

693. $\frac{-2}{x}+\frac{2}{x-3}$

The fraction is the sum of $\frac{A}{x}+\frac{B}{x-3}$.

Set the original fraction equal to the sum.

$$\frac{6}{x(x-3)}=\frac{A}{x}+\frac{B}{x-3}$$

Now add the two fractions on the right together.

$$\frac{6}{x(x-3)} = \frac{A}{x}\cdot\frac{x-3}{x-3} + \frac{B}{x-3}\cdot\frac{x}{x}$$
$$= \frac{A(x-3)}{x(x-3)} + \frac{Bx}{x(x-3)}$$
$$= \frac{Ax-3A+Bx}{x(x-3)}$$

Let the numerator on the left equal the numerator on the right.

$6 = Ax - 3A + Bx$

Factor on the right.

$6 = x(A + B) - 3A$

Write a system of equations involving the coefficients of the x terms and the constants.

$$\begin{cases} 0 = A + B \\ 6 = -3A \end{cases}$$

From the second equation, you find that $A = -2$. Substitute that into the first equation:

$0 = -2 + B$, so $B = 2$

The decomposition is $\frac{6}{x(x-3)} = \frac{-2}{x} + \frac{2}{x-3}$.

694. $\mathbf{\frac{1}{x} + \frac{-4}{x+1} + \frac{2}{x-3}}$

The fraction is the sum of $\frac{A}{x} + \frac{B}{x+1} + \frac{C}{x-3}$.

Set the original fraction equal to the sum.

$$\frac{-x^2+12x-3}{x(x+1)(x-3)} = \frac{A}{x} + \frac{B}{x+1} + \frac{C}{x-3}$$

Now add the three fractions on the right together.

$$\frac{-x^2+12x-3}{x(x+1)(x-3)} = \frac{A}{x}\cdot\frac{(x+1)(x-3)}{(x+1)(x-3)} + \frac{B}{x+1}\cdot\frac{x(x-3)}{x(x-3)} + \frac{C}{x-3}\cdot\frac{x(x+1)}{x(x+1)}$$
$$= \frac{A(x+1)(x-3)}{x(x+1)(x-3)} + \frac{Bx(x-3)}{x(x+1)(x-3)} + \frac{Cx(x+1)}{x(x+1)(x-3)}$$
$$= \frac{A(x^2-2x-3)+B(x^2-3x)+C(x^2+x)}{x(x+1)(x-3)}$$
$$= \frac{Ax^2-2Ax-3A+Bx^2-3Bx+Cx^2+Cx}{x(x+1)(x-3)}$$

Let the numerator on the left equal the numerator on the right.

$-x^2 + 12x - 3$

$= Ax^2 - 2Ax - 3A + Bx^2 - 3Bx + Cx^2 + Cx$

Factor on the right.

$$-x^2 + 12x - 3$$
$$= x^2(A + B + C) + x(-2A - 3B + C) - 3A$$

Write a system of equations involving the coefficients of the x terms and the constants.

$$\begin{cases} -1 = A + B + C \\ 12 = -2A - 3B + C \\ -3 = -3A \end{cases}$$

From the third equation, you find that $A = 1$. Substituting that into the first and second equations, a new system of two equations in two unknowns is created.

$$\begin{cases} -1 = 1 + B + C \\ 12 = -2(1) - 3B + C \end{cases}$$

$$\begin{cases} -2 = B + C \\ 14 = -3B + C \end{cases}$$

Solve the new system.

$$-1R_1 + R_2 \to R_2 \begin{cases} -2 = B + C \\ 16 = -4B \end{cases}$$

From the second equation, you find that $B = -4$. Substitute that into the first equation:

$$-2 = -4 + C, \text{ so } C = 2$$

The decomposition is:

$$\frac{-x^2 + 12x - 3}{x(x+1)(x-3)} = \frac{1}{x} + \frac{-4}{x+1} + \frac{2}{x-3}$$

695. $\frac{5}{x-1} + \frac{-1}{x+2} + \frac{-3}{x-5}$

The fraction is the sum of $\frac{A}{x-1} + \frac{B}{x+2} + \frac{C}{x-5}$.

Set the original fraction equal to the sum.

$$\frac{x^2 - 12x - 49}{(x-1)(x+2)(x-5)} = \frac{A}{x-1} + \frac{B}{x+2} + \frac{C}{x-5}$$

Now add the three fractions on the right together.

$$\frac{x^2 - 12x - 49}{(x-1)(x+2)(x-5)}$$
$$= \frac{A}{x-1} \cdot \frac{(x+2)(x-5)}{(x+2)(x-5)} + \frac{B}{x+2} \cdot \frac{(x-1)(x-5)}{(x-1)(x-5)} + \frac{C}{x-5} \cdot \frac{(x-1)(x+2)}{(x-1)(x+2)}$$
$$= \frac{A(x+2)(x-5)}{(x-1)(x+2)(x-5)} + \frac{B(x-1)(x-5)}{(x-1)(x+2)(x-5)} + \frac{C(x-1)(x+2)}{(x-1)(x+2)(x-5)}$$
$$= \frac{A(x^2 - 3x - 10) + B(x^2 - 6x + 5) + C(x^2 + x - 2)}{(x-1)(x+2)(x-5)}$$
$$= \frac{Ax^2 - 3Ax - 10A + Bx^2 - 6Bx + 5B + Cx^2 + Cx - 2C}{(x-1)(x+2)(x-5)}$$

Let the numerator on the left equal the numerator on the right.

$x^2 - 12x - 49$

$= Ax^2 - 3Ax - 10A + Bx^2 - 6Bx + 5B + Cx^2 + Cx - 2C$

Factor on the right.

$-x^2 + 12x - 3$

$= x^2(A + B + C) + x(-3A - 6B + C) - 10A + 5B - 2C$

Write a system of equations involving the coefficients of the x terms and the constants.

$$\begin{cases} 1 = A + B + C \\ -12 = -3A - 6B + C \\ -49 = -10A + 5B - 2C \end{cases}$$

Solve the system:

$$\begin{matrix} 3R_1 + R_2 \to R_2 \\ 10R_1 + R_3 \to R_3 \end{matrix} : \begin{cases} 1 = A + B + C \\ -9 = \quad -3B + 4C \\ -39 = \quad 15B + 8C \end{cases}$$

$$5R_2 + R_3 \to R_3 : \begin{cases} 1 = A + B + C \\ -9 = \quad -3B + 4C \\ -84 = \quad 28C \end{cases}$$

From the third equation, you find that $C = -3$. Substitute that into the second equation:

$-9 = -3B + 4(-3)$

$-9 = -3B - 12$

$3 = -3B$, so $B = -1$.

Then, solve for A in the first equation:

$1 = A + (-1) + -3, A = 5$

The decomposition is $\frac{x^2 - 12x - 49}{(x-1)(x+2)(x-5)} = \frac{5}{x-1} + \frac{-1}{x+2} + \frac{-3}{x-5}$.

Answers 601–700

696. **(3, 8), (–1, 4)**

The y-values are equal at the points of intersection, so set their equations equal to one another and solve for x.

$2x^2 - 3x - 1 = x + 5$

$2x^2 - 4x - 6 = 0$

$2(x^2 - 2x - 3) = 0$

$(x - 3)(x + 1) = 0$

$x = 3$ or $x = -1$

Substituting those values into the equation for the line:

When $x = 3$, $y = 3 + 5 = 8$.

When $x = -1$, $y = -1 + 5 = 4$.

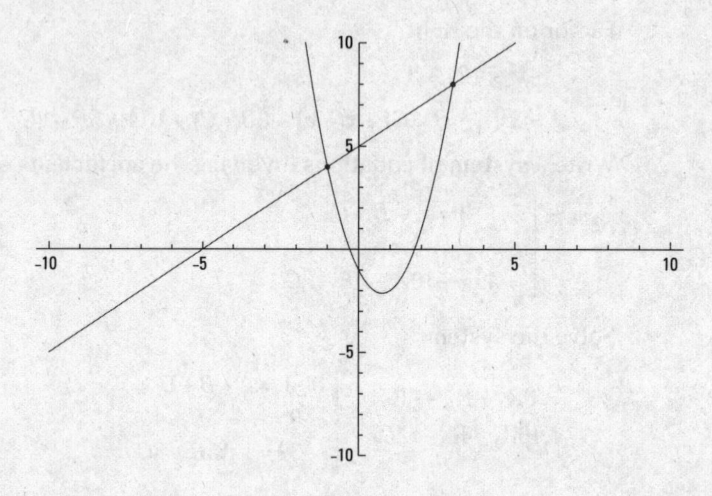

697. **(3, –9), (1, –7)**

The y-values are equal at the points of intersection, so set their equations equal to one another and solve for x.

$$x^2 - 5x - 3 = -x - 6$$

$$x^2 - 4x + 3 = 0$$

$$(x - 3)(x - 1) = 0$$

$$x = 3 \text{ or } x = 1$$

Substituting those values into the equation for the line:

When $x = 3$, $y = -3 - 6 = -9$.

When $x = 1$, $y = -1 - 6 = -7$.

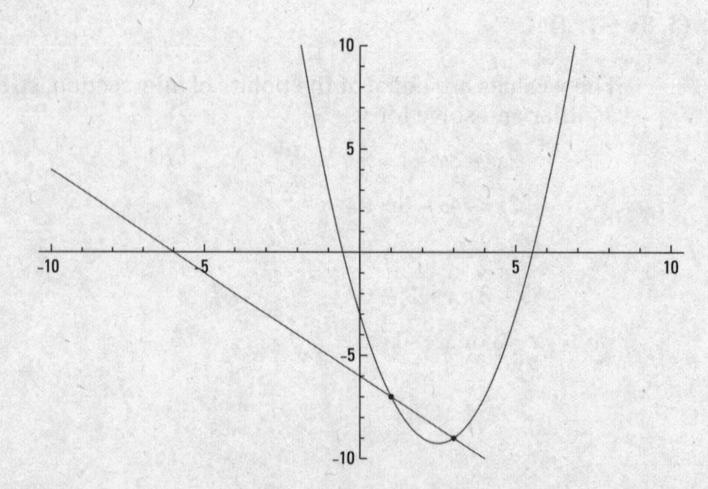

698. (2, 1), (–2, –7)

The y-values are equal at the points of intersection, so set their equations equal to one another and solve for x.

$$x^2 + 2x - 7 = 2x - 3$$
$$x^2 - 4 = 0$$
$$(x - 2)(x + 2) = 0$$
$$x = 2 \text{ or } x = -2$$

Substituting those values into the equation for the line:

When $x = 2$, $y = 2(2) - 3 = 4 - 3 = 1$.

When $x = -2$, $y = 2(-2) - 3 = -4 - 3 = -7$.

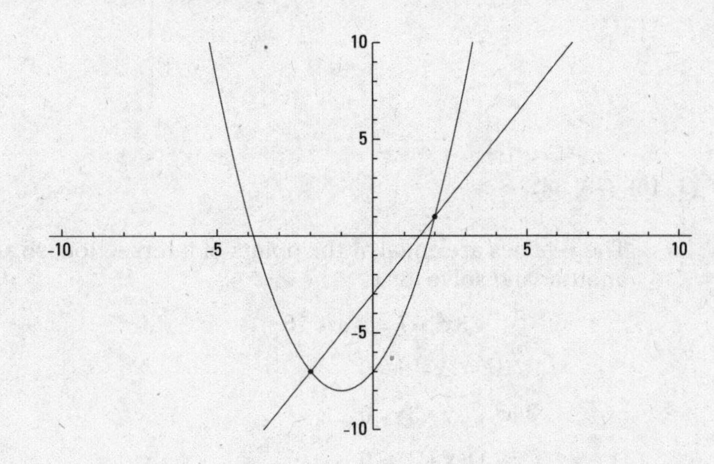

699. (4, –15), (–1, 5)

The y-values are equal at the points of intersection, so set their equations equal to one another and solve for x.

$$-x^2 - x + 5 = -4x + 1$$
$$x^2 - 3x - 4 = 0$$
$$(x - 4)(x + 1) = 0$$
$$x = 4 \text{ or } x = -1$$

Substituting those values into the equation for the line:

When $x = 4$, $y = -4(4) + 1 = -16 + 1 = -15$.

When $x = -1$, $y = -4(-1) + 1 = 4 + 1 = 5$.

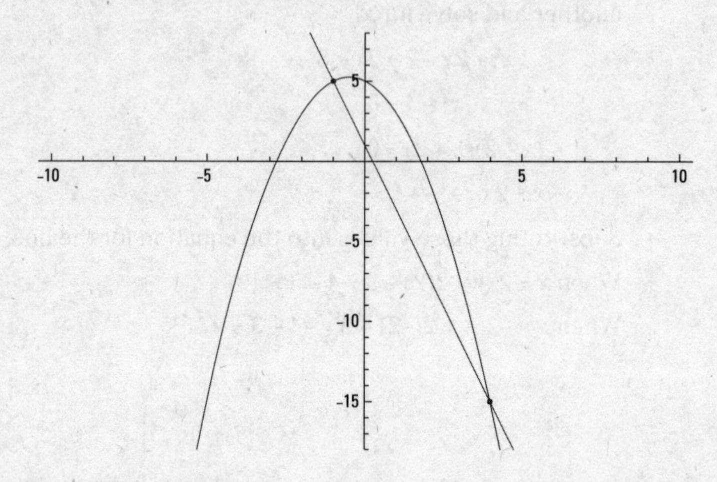

700. **(1, 10), (–3, 34)**

The y-values are equal at the points of intersection, so set their equations equal to one another and solve for x.

$$3x^2 + 7 = -6x + 16$$
$$3x^2 + 6x - 9 = 0$$
$$3(x^2 + 2x - 3) = 0$$
$$3(x - 1)(x + 3) = 0$$
$$x = 1 \text{ or } x = -3$$

Substituting those values into the equation for the line:

When $x = 1$, $y = -6(1) + 16 = -6 + 16 = 10$.

When $x = -3$, $y = -6(-3) + 16 = 18 + 16 = 34$.

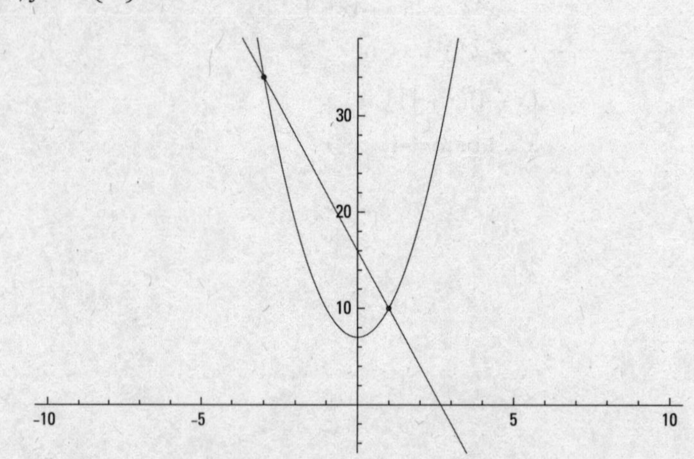

701. **(4, –28), (–1, –8)**

The y-values are equal at the points of intersection, so set their equations equal to one another and solve for x.

$$-2x^2 + 2x - 4 = -4x - 12$$

$$2x^2 - 6x - 8 = 0$$

$$2(x^2 - 3x - 4) = 0$$

$$2(x - 4)(x + 1) = 0$$

$$x = 4 \text{ or } x = -1$$

Substituting those values into the equation for the line:

When $x = 4$, $y = -4(4) - 12 = -16 - 12 = -28$.

When $x = -1$, $y = -4(-1) - 12 = 4 - 12 = -8$.

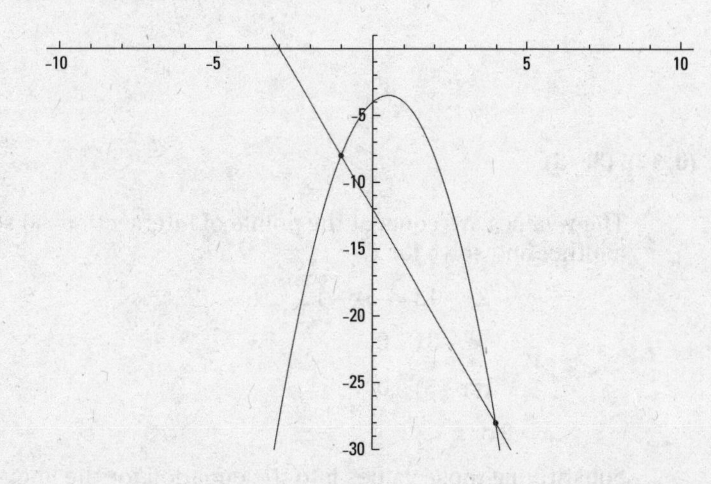

702. **(4, 6), (1, 3)**

The y-values are equal at the points of intersection, so set their equations equal to one another and solve for x.

$$-x^2 + 6x - 2 = x + 2$$

$$x^2 - 5x + 4 = 0$$

$$(x - 4)(x - 1) = 0$$

$$x = 4 \text{ or } x = 1$$

Substituting those values into the equation for the line:

When $x = 4$, $y = 4 + 2 = 6$.

When $x = 1$, $y = 1 + 2 = 3$.

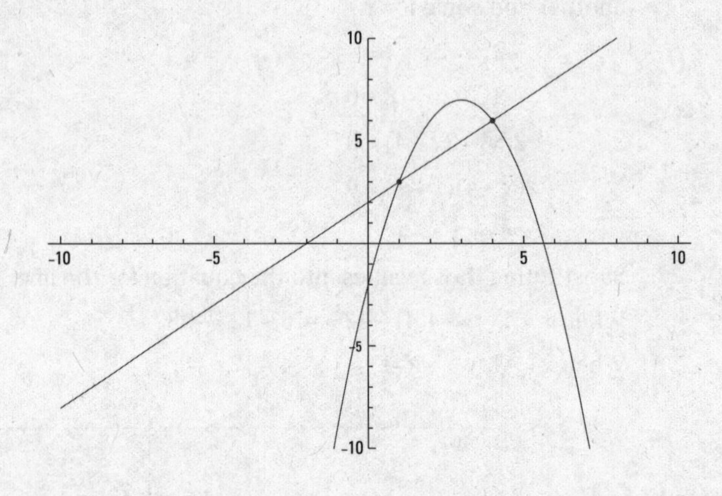

703. (0, 12), (3, –3)

The y-values are equal at the points of intersection, so set their equations equal to one another and solve for x.

$$-x^2 - 2x + 12 = -5x + 12$$

$$x^2 - 3x = 0$$

$$x(x - 3) = 0$$

$$x = 0 \text{ or } x = 3$$

Substituting those values into the equation for the line:

When $x = 0$, $y = -5(0) + 12 = 12$.

When $x = 3$, $y = -5(3) + 12 = -15 + 12 = -3$.

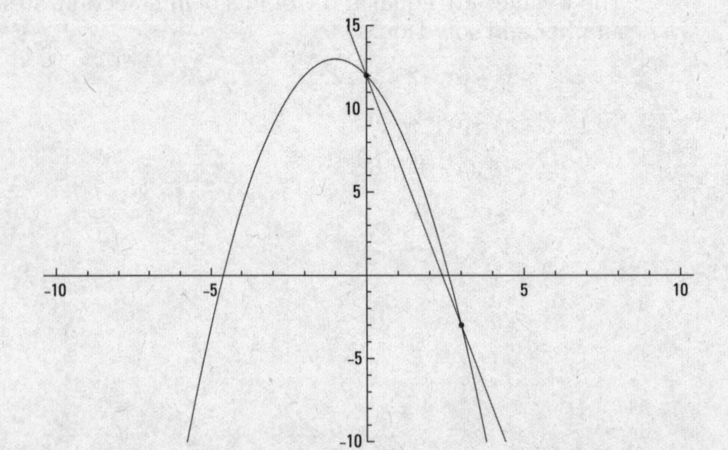

704. **(1, –4), (–3, 60)**

The *y*-values are equal at the points of intersection, so set their equations equal to one another and solve for *x*.

$$4x^2 - 8x = -16x + 12$$

$$4x^2 + 8x - 12 = 0$$

$$4(x^2 + 2x - 3) = 0$$

$$4(x - 1)(x + 3) = 0$$

$$x = 1 \text{ or } x = -3$$

Substituting those values into the equation for the line:

When $x = 1$, $y = -16(1) + 12 = -16 + 12 = -4$.

When $x = -3$, $y = -16(-3) + 12 = 48 + 12 = 60$.

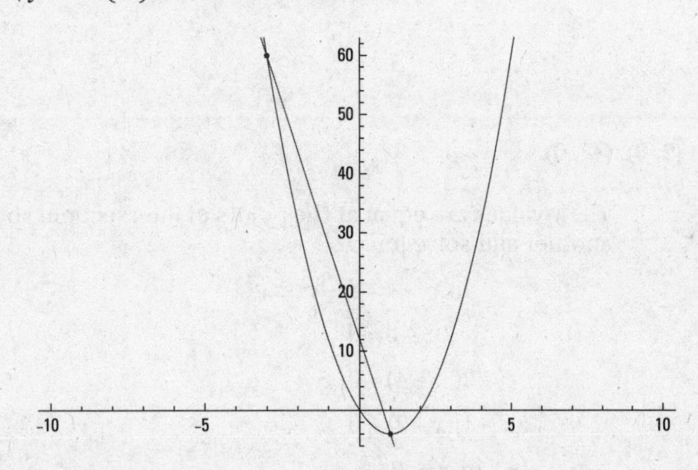

705. **(0, –5), (4, –37)**

The *y*-values are equal at the points of intersection, so set their equations equal to one another and solve for *x*.

$$-3x^2 + 4x - 5 = -8x - 5$$

$$3x^2 - 12x = 0$$

$$3x(x - 4) = 0$$

$$x = 0 \text{ or } x = 4$$

Substituting those values into the equation for the line:

When $x = 0$, $y = -8(0) - 5 = -5$.

When $x = 4$, $y = -8(4) - 5 = -32 - 5 = -37$.

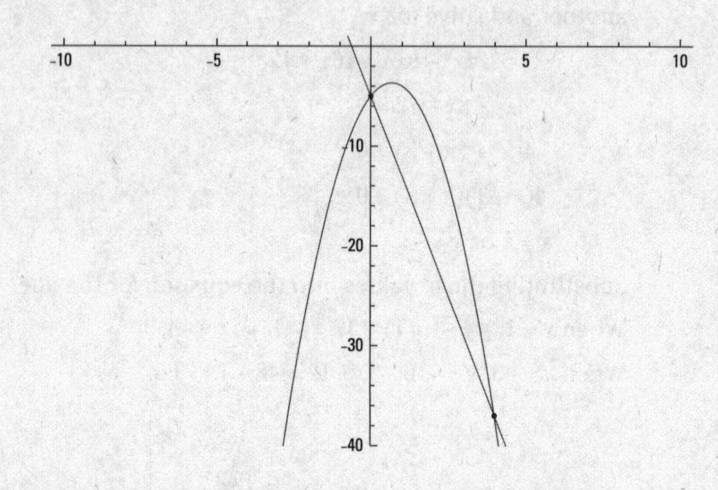

706. (2, 0), (–2, 0)

The y-values are equal at the points of intersection, so set their equations equal to one another and solve for x.

$$x^2 - 4 = 4 - x^2$$
$$2x^2 - 8 = 0$$
$$2(x^2 - 4) = 0$$
$$2(x - 2)(x + 2) = 0$$
$$x = 2 \text{ or } x = -2$$

Substituting those values into the first equation:

When $x = 2$, $y = (2)^2 - 4 = 4 - 4 = 0$.

When $x = -2$, $y = (-2)^2 - 4 = 4 - 4 = 0$.

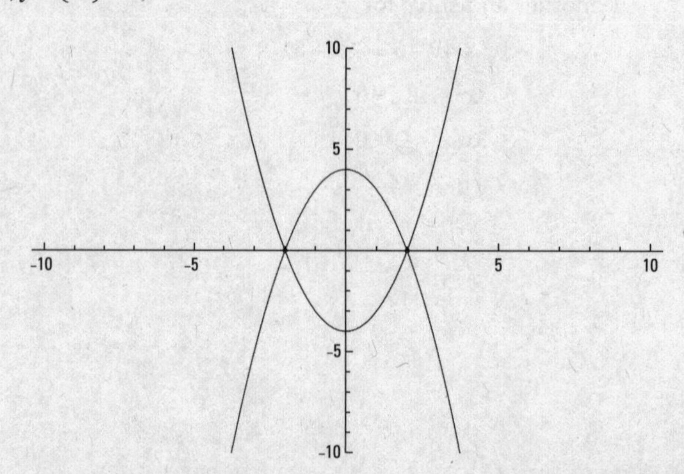

707. **(0, 2), (5, −13)**

The *y*-values are equal at the points of intersection, so set their equations equal to one another and solve for *x*.

$$x^2 - 8x + 2 = -2x^2 + 7x + 2$$
$$3x^2 - 15x = 0$$
$$3x(x - 5) = 0$$
$$x = 0 \text{ or } x = 5$$

Substituting those values into the first equation:

When $x = 0$, $y = (0)^2 - 8(0) + 2 = 2$.

When $x = 5$, $y = (5)^2 - 8(5) + 2 = 25 - 40 + 2 = -13$.

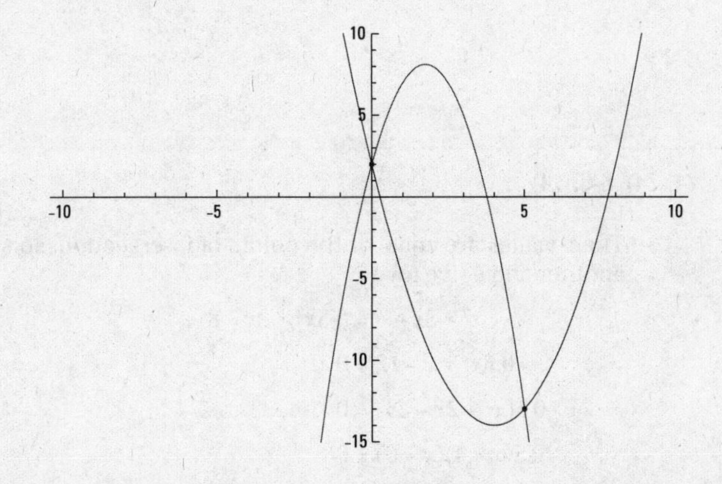

708. **(5, −20), (−1, 16)**

The *y*-values are equal at the points of intersection, so set their equations equal to one another and solve for *x*.

$$x^2 - 10x + 5 = -1.5x^2 + 17.5$$
$$2.5x^2 - 10x - 12.5 = 0$$
$$2.5(x^2 - 4x - 5) = 0$$
$$2.5(x - 5)(x + 1) = 0$$
$$x = 5 \text{ or } x = -1$$

Substituting those values into the first equation:

When $x = 5$, $y = (5)^2 - 10(5) + 5 = 25 - 50 + 5 = -20$.

When $x = -1$, $y = (-1)^2 - 10(-1) + 5 = 1 + 10 + 5 = 16$.

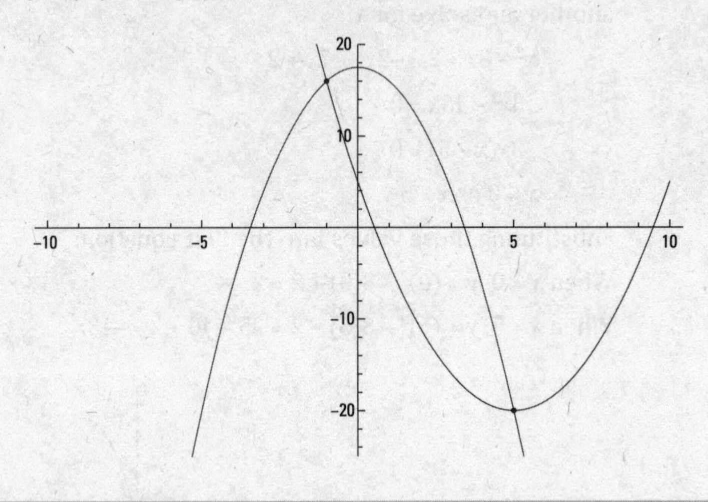

709. (4, 24), (–6, 14)

The *y*-values are equal at the points of intersection, so set their equations equal to one another and solve for *x*.

$$x^2 + 3x - 4 = 0.5x^2 + 2x + 8$$
$$0.5x^2 + x - 12 = 0$$
$$0.5(x^2 + 2x - 24) = 0$$
$$0.5(x - 4)(x + 6) = 0$$
$$x = 4 \text{ or } x = -6$$

Substituting those values into the first equation:

When $x = 4$, $y = (4)^2 + 3(4) - 4 = 16 + 12 - 4 = 24$.

When $x = -6$, $y = (-6)^2 + 3(-6) - 4 = 36 - 18 - 4 = 14$.

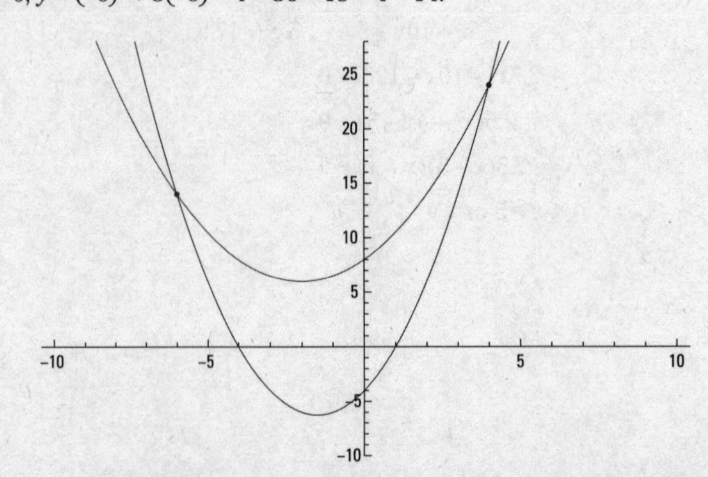

710. **(3, 16), (–4, –12)**

The y-values are equal at the points of intersection, so set their equations equal to one another and solve for x.

$$x^2 + 5x - 8 = -0.25x^2 + 3.75x + 7$$
$$1.25x^2 + 1.25x - 15 = 0$$
$$1.25(x^2 + x - 12) = 0$$
$$1.25(x - 3)(x + 4) = 0$$
$$x = 3 \text{ or } x = -4$$

Substituting those values into the first equation:

When $x = 3$, $y = (3)^2 + 5(3) - 8 = 9 + 15 - 8 = 16$.

When $x = -4$, $y = (-4)^2 + 5(-4) - 8 = 16 - 20 - 8 = -12$.

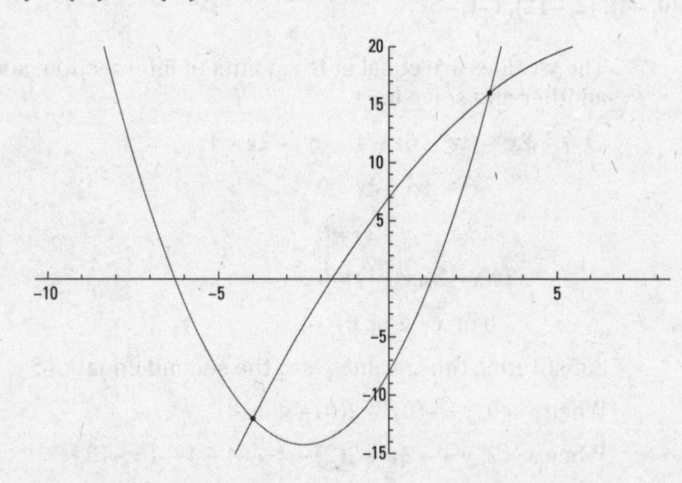

711. **(0, –1), (2, –9), (3, –1)**

The y-values are equal at the points of intersection, so set their equations equal to one another and solve for x.

$$x^3 - x^2 - 6x - 1 = 4x^2 - 12x - 1$$
$$x^3 - 5x^2 + 6x = 0$$
$$x(x^2 - 5x + 6) = 0$$
$$x(x - 2)(x - 3) = 0$$
$$x = 0 \text{ or } x = 2 \text{ or } x = 3$$

Substituting those values into the second equation:

When $x = 0$, $y = 4(0)^2 - 12(0) - 1 = -1$.

When $x = 2$, $y = 4(2)^2 - 12(2) - 1 = 16 - 24 - 1 = -9$.

When $x = 3$, $y = 4(3)^2 - 12(3) - 1 = 36 - 36 - 1 = -1$.

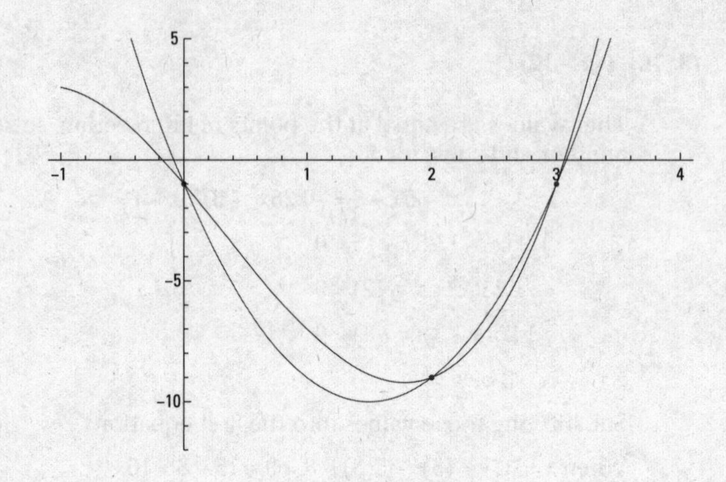

712. **(0, –4), (2, –12), (–1, –3)**

The *y*-values are equal at the points of intersection, so set their equations equal to one another and solve for *x*.

$$2x^3 - 3x^2 - 6x - 4 = -x^2 - 2x - 4$$

$$2x^3 - 2x^2 - 4x = 0$$

$$2x(x^2 - x - 2) = 0$$

$$2x(x - 2)(x + 1) = 0$$

$$x = 0 \text{ or } x = 2 \text{ or } x = -1$$

Substituting those values into the second equation:

When $x = 0$, $y = -(0)^2 - 2(0) - 4 = -4$.

When $x = 2$, $y = -(2)^2 - 2(2) - 4 = -4 - 4 - 4 = -12$.

When $x = -1$, $y = -(-1)^2 - 2(-1) - 4 = -1 + 2 - 4 = -3$.

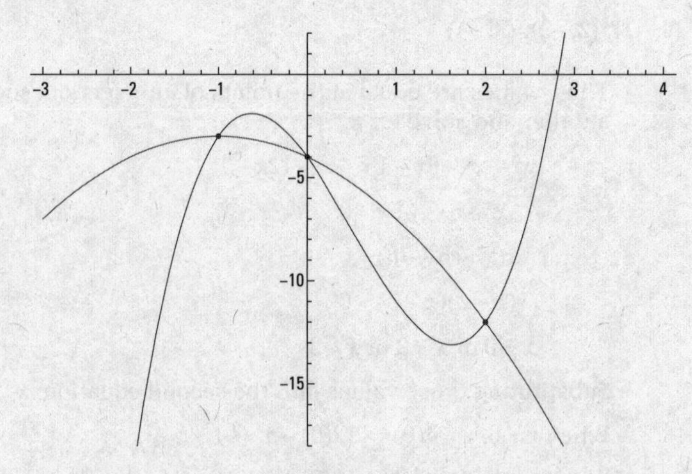

713. (3, –3), (2, 1), (5, –41)

The y-values are equal at the points of intersection, so set their equations equal to one another and solve for x.

$$-x^3 + 5x^2 - 10x + 9 = -5x^2 + 21x - 21$$

$$x^3 - 10x^2 + 31x - 30 = 0$$

This won't factor by grouping, so use synthetic division and try $x = 3$.

$$\begin{array}{r|rrrr} 3 & 1 & -10 & 31 & -30 \\ & & 3 & -21 & 30 \\ \hline & 1 & -7 & 10 & 0 \end{array}$$

The number 3 is a root, so you can write the equation as:

$$(x - 3)(x^2 - 7x + 10) = 0$$

$$(x - 3)(x - 2)(x - 5) = 0$$

$$x = 3 \text{ or } x = 2 \text{ or } x = 5$$

Substituting those values into the second equation:

When $x = 3$, $y = -5(3)^2 + 21(3) - 21 = -45 + 63 - 21 = -3$.

When $x = 2$, $y = -5(2)^2 + 21(2) - 21 = -20 + 42 - 21 = 1$.

When $x = 5$, $y = -5(5)^2 + 21(5) - 21 = -125 + 105 - 21 = -41$.

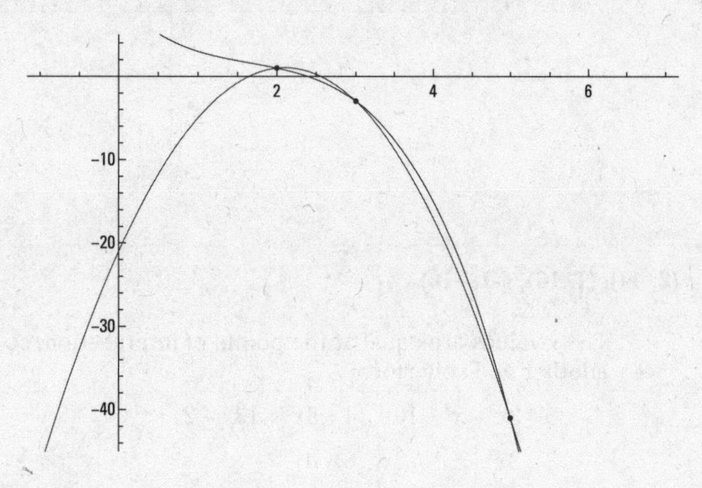

714. (1, 11), (4, –13), (–2, 17)

The y-values are equal at the points of intersection, so set their equations equal to one another and solve for x.

$$-x^3 + 2x^2 + 3x + 7 = -x^2 - 3x + 15$$

$$x^3 - 3x^2 - 6x + 8 = 0$$

This won't factor by grouping, so use synthetic division and try $x = 1$.

$$\begin{array}{r|rrrr} 1 & 1 & -3 & -6 & 8 \\ & & 1 & -2 & -8 \\ \hline & 1 & -2 & -8 & 0 \end{array}$$

The number 1 is a root, so you can write the equation as:

$$(x-1)(x^2-2x-8)=0$$

$$(x-1)(x-4)(x+2)=0$$

$$x=1 \text{ or } x=4 \text{ or } x=-2$$

Substituting those values into the second equation:

When $x = 1$, $y = -(1)^2 - 3(1) + 15 = -1 - 3 + 15 = 11$.

When $x = 4$, $y = -(4)^2 - 3(4) + 15 = -16 - 12 + 15 = -13$.

When $x = -2$, $y = -(-2)^2 - 3(-2) + 15 = -4 + 6 + 15 = 17$.

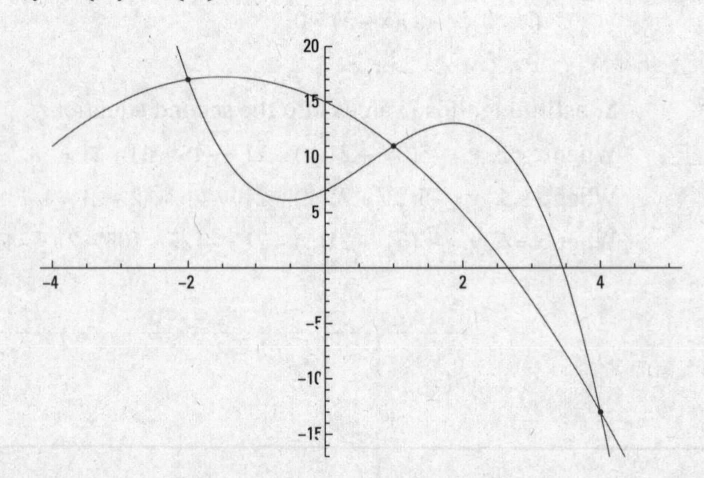

715. (2, 44), (1, 16), (–1, –10)

The y-values are equal at the points of intersection, so set their equations equal to one another and solve for x.

$$3x^3 - x^2 + 10x + 4 = 5x^2 + 13x - 2$$

$$3x^3 - 6x^2 - 3x + 6 = 0$$

This factors by grouping:

$$3x^2(x-2) - 3(x-2) = 0$$

$$(x-2)(3x^2-3) = 0$$

$$3(x-2)(x-1)(x+1) = 0$$

$$x = 2 \text{ or } x = 1 \text{ or } x = -1$$

Substituting those values into the second equation:

When $x = 2$, $y = 5(2)^2 + 13(2) - 2 = 20 + 26 - 2 = 44$.

When $x = 1$, $y = 5(1)^2 + 13(1) - 2 = 5 + 13 - 2 = 16$.

When $x = -1$, $y = 5(-1)^2 + 13(-1) - 2 = 5 - 13 - 2 = -10$.

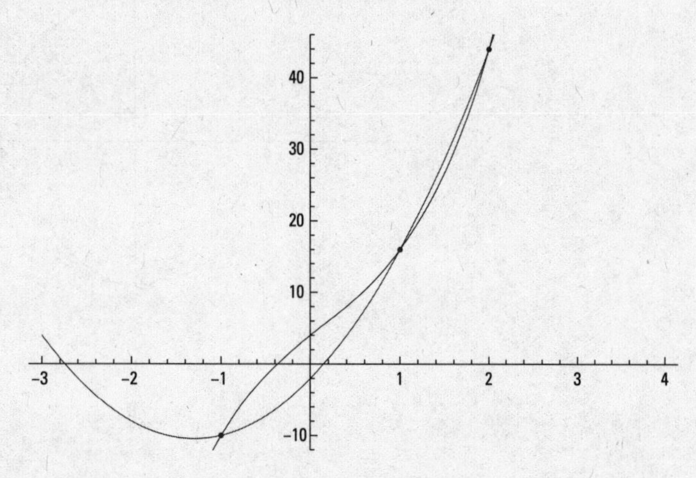

716. **(0, –5), (5, 0), (–5, 0)**

Because the second equation is solved for y, substitute its x-equivalence into the first equation.

$$x^2 + (0.2x^2 - 5)^2 = 25$$

Now solve for x.

$$\begin{aligned} x^2 + 0.04x^4 - 2x^2 + 25 &= 25 \\ 0.04x^4 - x^2 &= 0 \\ x^2(0.04x^2 - 1) &= 0 \end{aligned}$$

From the first factor, $x^2 = 0$, giving you $x = 0$.

From the second factor,

$$\begin{aligned} &0.04x^2 - 1 = 0 \\ &0.04x^2 = 1 \\ &x^2 = \frac{1}{0.04} = 25 \\ &x = \pm 5 \end{aligned}$$

Substituting those x values into the second equation:

When $x = 0$, $y = 0.2(0)^2 - 5 = -5$.

When $x = 5$, $y = 0.2(5)^2 - 5 = 0$.

When $x = -5$, $y = 0.2(-5)^2 - 5 = 0$.

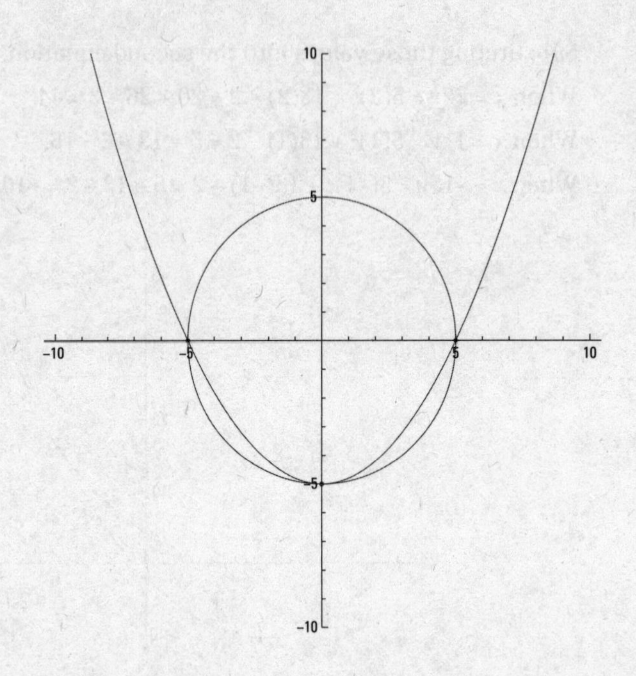

717. **(0, –8), (7, –1)**

Because the second equation is solved for y, substitute its x-equivalence into the first equation.

$$(x-3)^2+(x-8+4)^2=25$$
$$(x-3)^2+(x-4)^2=25$$

Now solve for x.

$$x^2-6x+9+x^2-8x+16=25$$
$$2x^2-14x+25=25$$
$$2x^2-14x=0$$
$$2x(x-7)=0$$
$$x=0 \text{ or } x=7$$

Substituting those x values into the second equation:

When $x=0$, $y=0-8=-8$.

When $x=7$, $y=7-8=-1$.

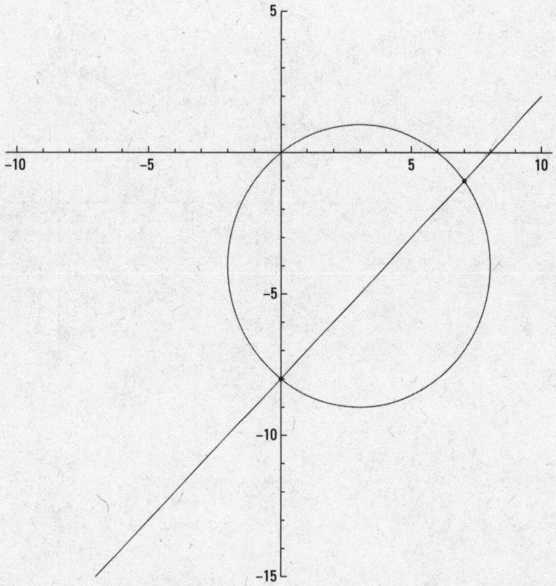

718. **(6, –8), (–6, –8), (8, 6), (–8, 6)**

Because the second equation is solved for y, substitute its x-equivalence into the first equation.

$$x^2 + (0.5x^2 - 26)^2 = 100$$

Now solve for x.

$$x^2 + 0.25x^4 - 26x^2 + 676 = 100$$
$$0.25x^4 - 25x^2 + 576 = 0$$
$$0.25(x^4 - 100x + 2304) = 0$$

The quadratic-like equation factors.

$$0.25(x^2 - 36)(x^2 - 64) = 0$$
$$0.25(x - 6)(x + 6)(x - 8)(x + 8) = 0$$
$$x = 6 \text{ or } x = -6 \text{ or } x = 8 \text{ or } x = -8$$

Substituting those x values into the second equation:

When $x = 6$, $y = 0.5(6)^2 - 26 = 18 - 26 = -8$.

When $x = -6$, $y = 0.5(-6)^2 - 26 = 18 - 26 = -8$.

When $x = 8$, $y = 0.5(8)^2 - 26 = 32 - 26 = 6$.

When $x = -8$, $y = 0.5(-8)^2 - 26 = 32 - 26 = 6$.

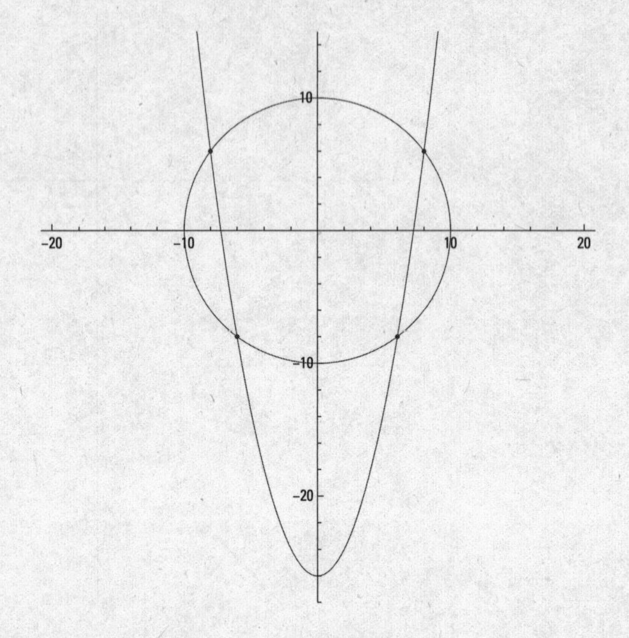

719. **(3, –2.4), (–3, –2.4), (4, 1.8), (–4, 1.8)**

Because the second equation is solved for *y*, substitute its *x*-equivalence into the first equation.

$$\frac{x^2}{25}+\frac{\left(0.6x^2-7.8\right)^2}{9}=1$$

Now multiply each term by the lowest common denominator of 225 and solve for *x*.

$$9x^2+25(0.6x^2-7.8)^2=225$$

$$9x^2+25(0.36x^4-9.36x^2+60.84)=225$$

$$9x^2+9x^4-234x^2+1521=225$$

$$9x^4-225x^2+1296=0$$

Factor.

$$9(x^4-25x^2+144)=0$$

$$9(x^2-9)(x^2-16)=0$$

$$9(x-3)(x+3)(x-4)(x+4)=0$$

$$x=3 \text{ or } x=-3 \text{ or } x=4 \text{ or } x=-4.$$

Substituting those x values into the second equation:

When $x = 3$, $y = 0.6(3)^2 - 7.8 = 5.4 - 7.8 = -2.4$.

When $x = -3$, $y = 0.6(-3)^2 - 26 = 5.4 - 7.8 = -2.4$.

When $x = 4$, $y = 0.6(4)^2 - 7.8 = 9.6 - 7.8 = 1.8$.

When $x = -4$, $y = 0.6(-4)^2 - 26 = 9.6 - 7.8 = 1.8$.

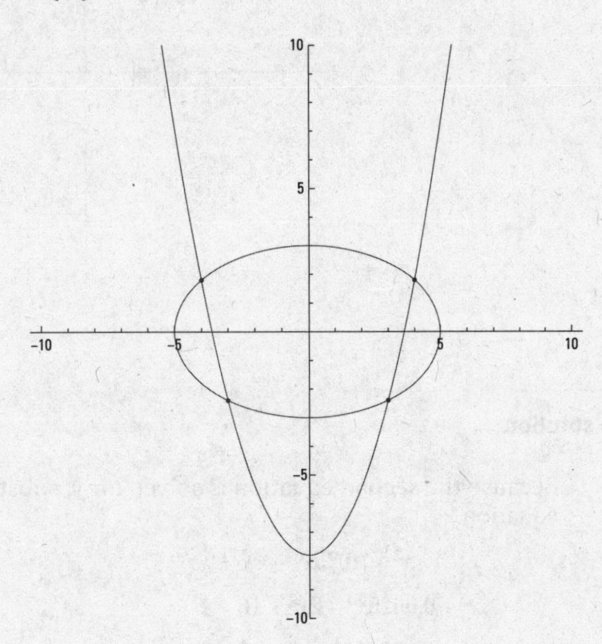

720. **(0, –2)**

Because the second equation is solved for y, substitute its x-equivalence into the first equation.

$$x^2 + (-x^2 - 2)^2 = 4$$
$$x^2 + x^4 + 4x^2 + 4 = 4$$
$$x^4 + 5x^2 = 0$$
$$x^2(x^2 + 5) = 0$$

From the first factor, $x^2 = 0$, giving you $x = 0$.

From the second factor, $x^2 + 5 = 0$, there is no real solution.

Substituting $x = 0$ into the second equation,

$$y = -0^2 - 2 = -2.$$

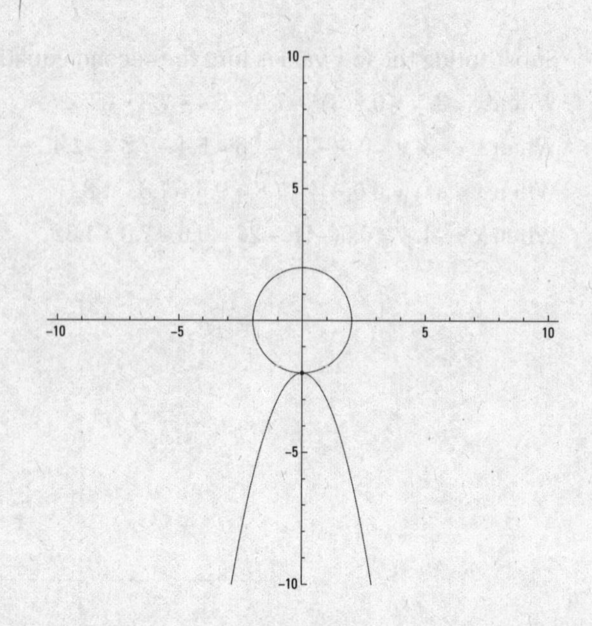

721. No solution

Because the second equation is solved for *y*, substitute its *x*-equivalence into the first equation.

$$x^2 + (0.25x^2 - 4)^2 = 1$$
$$x^2 + 0.0625x^4 - 2x^2 + 16 = 1$$
$$0.0625x^4 - x^2 + 15 = 0$$
$$0.0625(x^4 - 16x^2 + 240) = 0$$

The trinomial doesn't factor, and using the quadratic formula gives you just imaginary solutions. There are no points of intersection of the two curves.

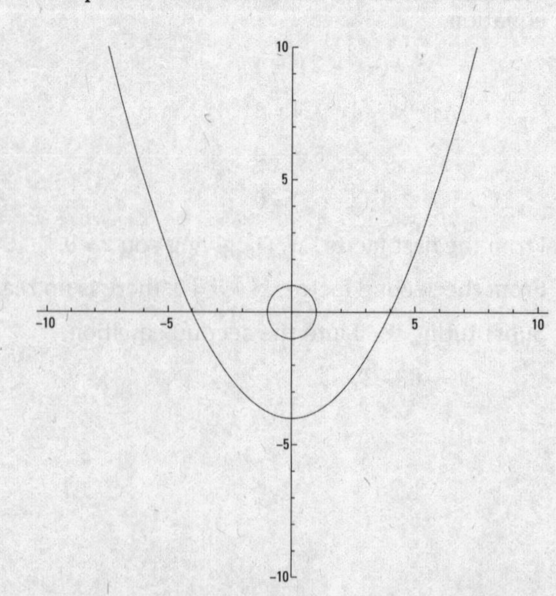

722. (4, 3), (4, –3)

Subtract the terms in the first equation from the corresponding terms in the second equation.

$$(x-8)^2 + y^2 = 25$$
$$-x^2 - y^2 = -25$$
$$(x-8)^2 - x^2 = 0$$
$$x^2 - 16x + 64 - x^2 = 0$$
$$-16x + 64 = 0$$
$$64 = 16x$$
$$x = 4$$

Substituting that value back into the first equation,

$$4^2 + y^2 = 25$$
$$y^2 = 25 - 16 = 9$$
$$y = 3 \text{ or } y = -3$$

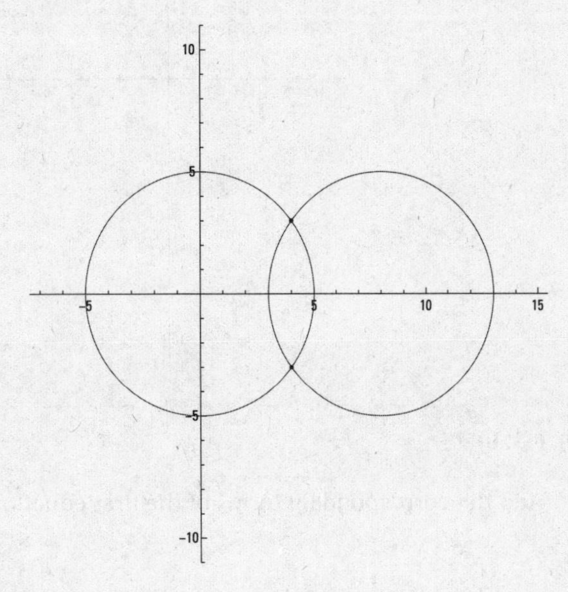

723. $\left(\frac{7}{8}, \frac{5}{8}\sqrt{39}\right), \left(\frac{7}{8}, -\frac{5}{8}\sqrt{39}\right)$

Subtract the terms in the first equation from the corresponding terms in the second equation.

$$(x-4)^2 + y^2 = 25$$
$$-x^2 - y^2 = -16$$
$$(x-4)^2 - x^2 = 9$$
$$x^2 - 8x + 16 - x^2 = 9$$
$$-8x = -7$$
$$x = \frac{7}{8}$$

Substituting that value back into the first equation,

$$\left(\frac{7}{8}\right)^2 + y^2 = 16$$

$$\frac{49}{64} + y^2 = 16$$

$$y^2 = 16 - \frac{49}{64} = \frac{975}{64}$$

$$y = \pm\sqrt{\frac{975}{64}} = \pm\sqrt{\frac{25 \cdot 39}{64}}$$

$$= \pm\sqrt{\frac{25}{64}}\sqrt{39} = \pm\frac{5}{8}\sqrt{39}$$

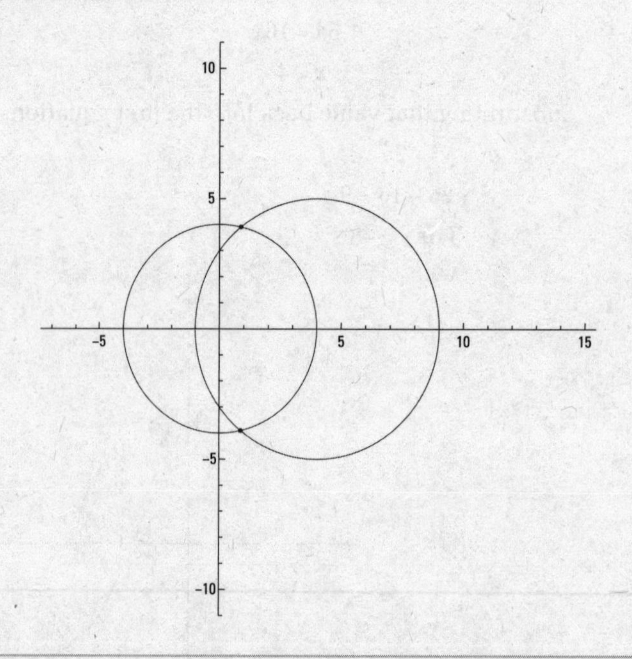

724. (1, 0), (–1, 0)

Add the corresponding terms of the first equation to those in the second equation.

$$x^2 + y^2 = 1$$

$$x^2 - y^2 = 1$$

$$2x^2 = 2$$

$$x^2 = 1$$

$$x = 1 \text{ or } x = -1$$

Substituting those values back into the first equation:

When $x = 1$, $1^2 + y^2 = 1$, $y^2 = 0$, $y = 0$.

When $x = -1$, $(-1)^2 + y^2 = 1$, $y^2 = 0$, $y = 0$.

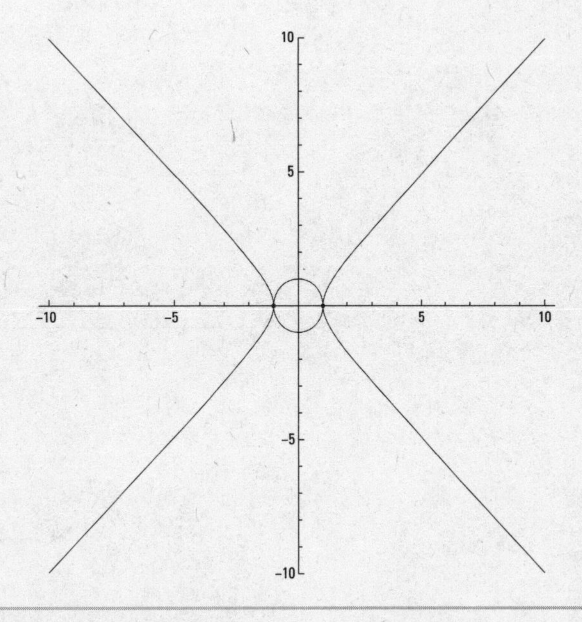

725. **(4, 0), (−4, 0), (4.96, −3.66), (−4.96, −3.66)**

Multiply the terms in the first equation by −25 and add the corresponding terms to those in the second equation.

$$25x^2 - 16y^2 = 400$$
$$-25x^2 - 25(y+3)^2 = -625$$
$$-16y^2 - 25(y+3)^2 = -225$$
$$-16y^2 - 25y^2 - 150y - 225 = -225$$
$$-41y^2 - 150y = 0$$
$$-y(41y + 150) = 0$$
$$y = 0 \text{ or } y = -\frac{150}{41}$$

Substituting those values back into the second equation,

when $y = 0$, $25x^2 - 16(0)^2 = 400$, $25x^2 = 400$, $x^2 = 16$, $x = 4$ or $x = -4$.

When $y = -\frac{150}{41}$, $25x^2 - 16\left(-\frac{150}{41}\right)^2 = 400$

$$25x^2 - 16\left(\frac{22{,}500}{1{,}681}\right) = 400$$
$$25x^2 = 400 + \frac{360{,}000}{1{,}681}$$
$$x^2 = 16 + \frac{14{,}400}{1{,}681} = \frac{41{,}296}{1{,}681}$$
$$x = \pm\sqrt{\frac{41{,}296}{1{,}681}} = \pm\sqrt{\frac{16 \cdot 2{,}581}{41^2}}$$
$$= \pm\sqrt{\frac{16}{41^2}}\sqrt{2{,}581} = \pm\frac{4}{41}\sqrt{2{,}581} \approx \pm 4.96$$

The approximate value of y for this x is −3.66.

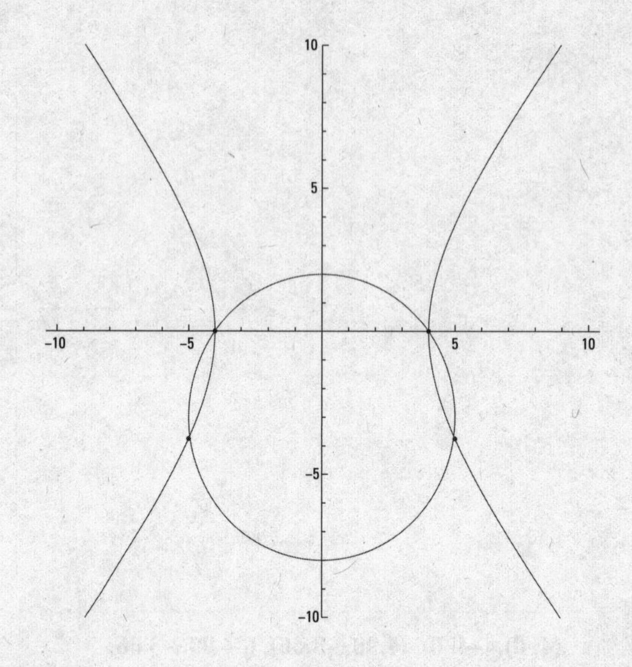

726. (0, 1)

The y-values are equal at the points of intersection, so set their equations equal to one another and solve for x.

$$2^x = 2^{-x}$$

When the bases are the same, the exponents are equal.

$$x = -x$$

$$2x = 0, x = 0$$

When $x = 0$, $y = 2^0 = 1$.

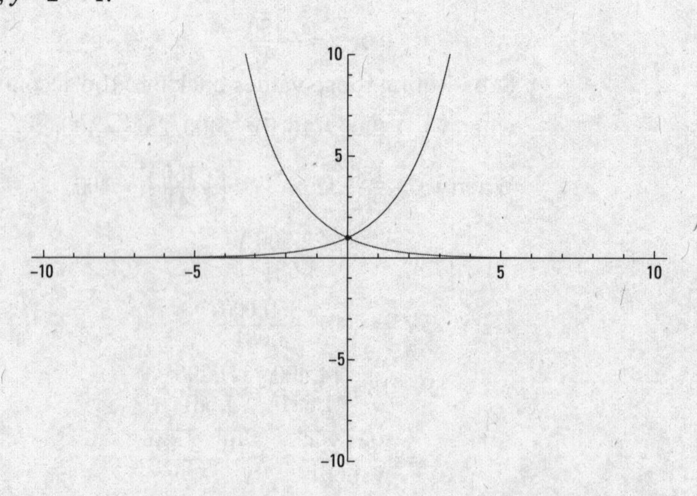

727. **(0, 1)**

The *y*-values are equal at the points of intersection, so set their equations equal to one another and solve for *x*.

$$3^x = 2 - 3^x$$

Add 3^x to each side of the equation.

$$2 \cdot 3^x = 2$$

Divide each side by 2.

$$3^x = 1, x = 0$$

When $x = 0$, $y = 3^0 = 1$.

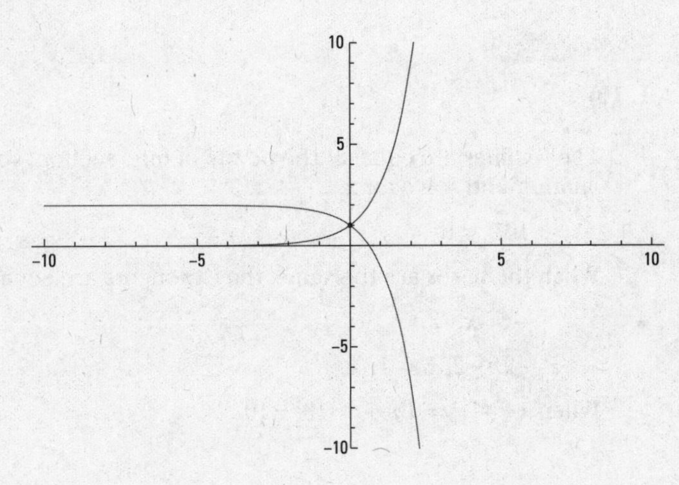

728. **(1, 5)**

The *y*-values are equal at the points of intersection, so set their equations equal to one another and solve for *x*.

$$5^x = 10 - 5^x$$

Add 5^x to each side of the equation.

$$2 \cdot 5^x = 10$$

Divide each side by 2.

$$5^x = 5, x = 1$$

When $x = 1$, $y = 5^1 = 5$.

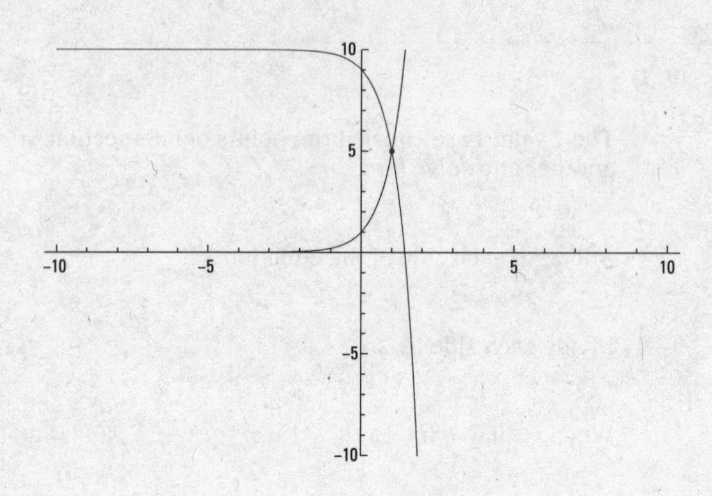

729. **(–1, 10)**

The y-values are equal at the points of intersection, so set their equations equal to one another and solve for x.

$$10^{-x} = 10^{x+2}$$

When the bases are the same, the exponents are equal.

$$-x = x + 2$$

$$-2x = 2, x = -1$$

When $x = -1$, $y = 10^{-(-1)} = 10^1 = 10$.

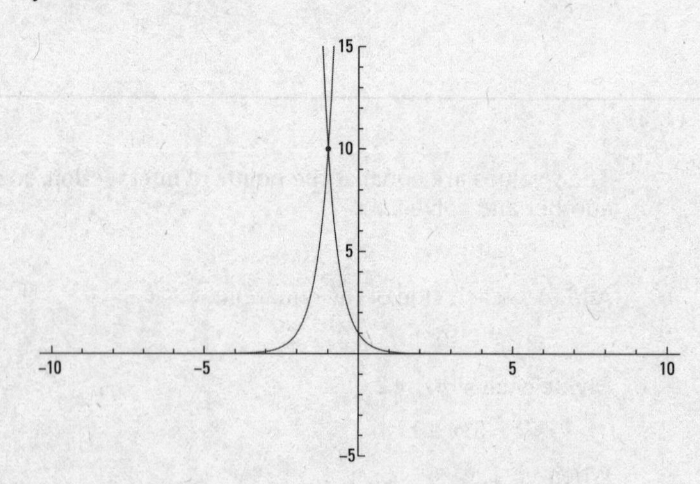

730. **(0, 6)**

The y-values are equal at the points of intersection, so set their equations equal to one another and solve for x.

$$6^{x+1} = 6^{-x+1}$$

When the bases are the same, the exponents are equal.

$$x + 1 = -x + 1$$

$$2x = 0, x = 0$$

When $x = 0$, $y = 6^{0+1} = 6$.

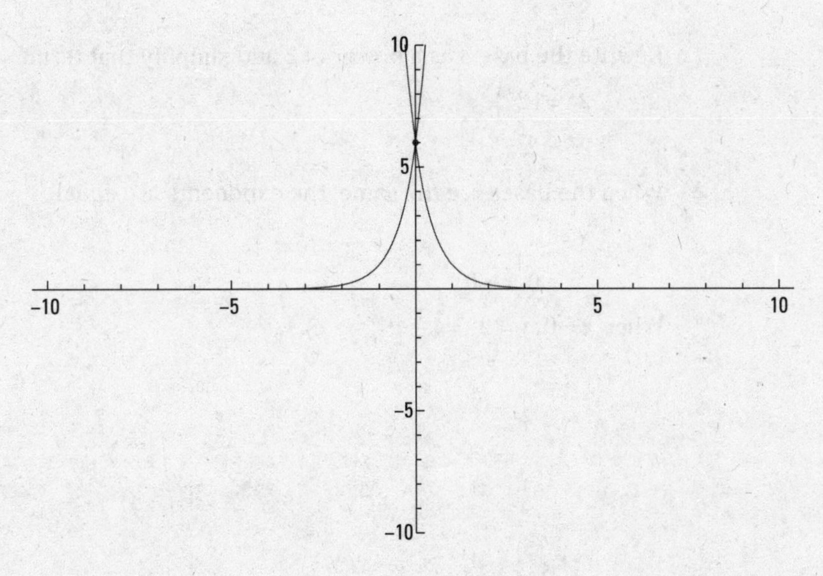

731. (4, 4)

The y-values are equal at the points of intersection, so set their equations equal to one another and solve for x.

$$4^{x-3} = 4^{5-x}$$

When the bases are the same, the exponents are equal.

$$x - 3 = 5 - x$$

$$2x = 8, x = 4$$

When $x = 4$, $y = 4^{4-3} = 4^1 = 4$.

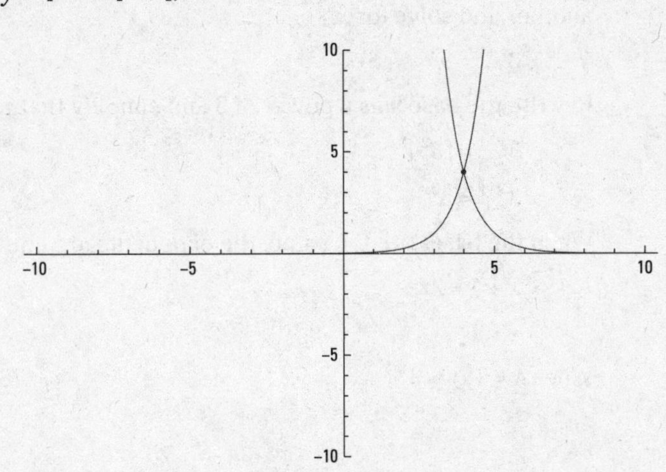

732. **(0, 1)**

The y-values are equal at the points of intersection, so set their equations equal to one another and solve for x.

$$2^x = 8^x$$

Rewrite the base 8 as a power of 2 and simplify that term.

$$2^x = \left(2^3\right)^x$$

$$2^x = 2^{3x}$$

When the bases are the same, the exponents are equal.

$$x = 3x$$

$$2x = 0, x = 0$$

When $x = 0$, $y = 2^0 = 1$.

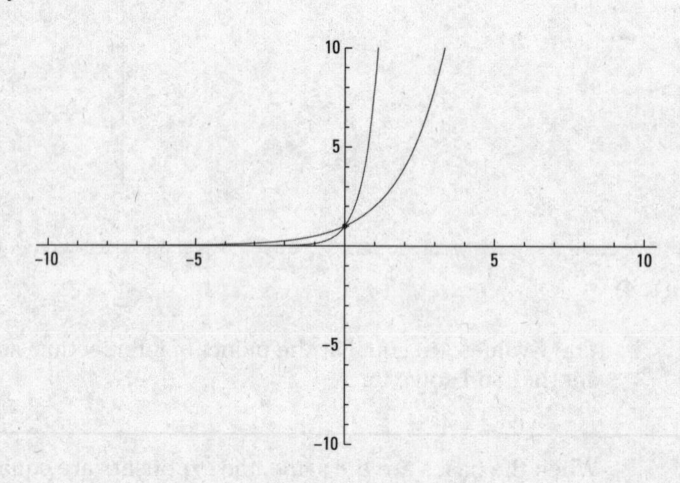

733. **(1, 9)**

The y-values are equal at the points of intersection, so set their equations equal to one another and solve for x.

$$3^{x+1} = 9^x$$

Rewrite the base 9 as a power of 3 and simplify that term.

$$3^{x+1} = \left(3^2\right)^x$$

$$3^{x+1} = 3^{2x}$$

When the bases are the same, the exponents are equal.

$$x + 1 = 2x$$

$$x = 1$$

When $x = 1$, $y = 3^{1+1} = 3^2 = 9$.

734. **(3, 1)**

The *y*-values are equal at the points of intersection, so set their equations equal to one another and solve for *x*.

$$4^{x-3} = 2^{x-3}$$

Rewrite the base 4 as a power of 2 and simplify that term.

$$\left(2^2\right)^{x-3} = 2^{x-3}$$

$$2^{2x-6} = 2^{x-3}$$

When the bases are the same, the exponents are equal.

$$2x - 6 = x - 3$$

$$x = 3$$

When $x = 3$, $y = 2^{3-3} = 2^0 = 1$.

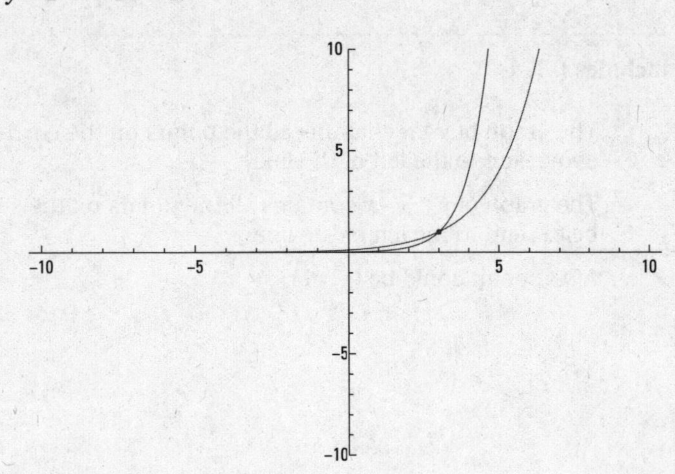

735. (1, 1)

The *y*-values are equal at the points of intersection, so set their equations equal to one another and solve for *x*.

$$25^{1-x} = 125^{x-1}$$

Rewrite the bases as powers of 5 and simplify those terms.

$$\left(5^2\right)^{1-x} = \left(5^3\right)^{x-1}$$
$$5^{2-2x} = 5^{3x-3}$$

When the bases are the same, the exponents are equal.

$$2 - 2x = 3x - 3$$
$$5x = 5, x = 1$$

When $x = 1$, $y = 25^{1-1} = 25^0 = 1$.

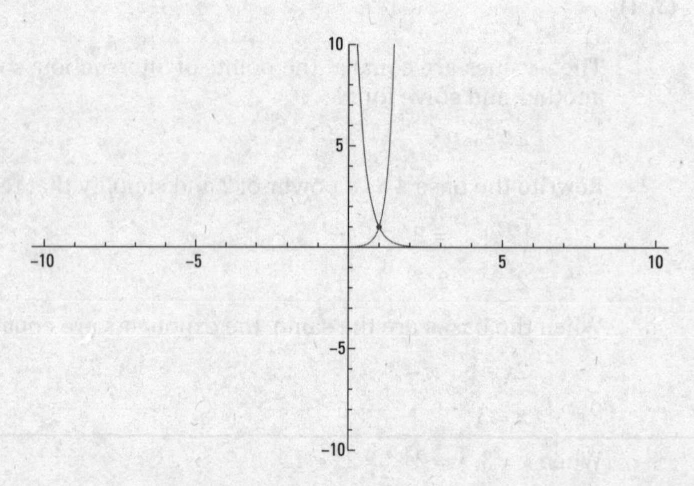

736. Includes (–1, 1)

The graph of $y \geq x$ contains all the points on the corresponding line and everything above and to the left of the line.

The graph of $y \leq 3 - x$ contains all the points on the corresponding line and everything below and to the left of the line.

A test point could be (–1, 1).

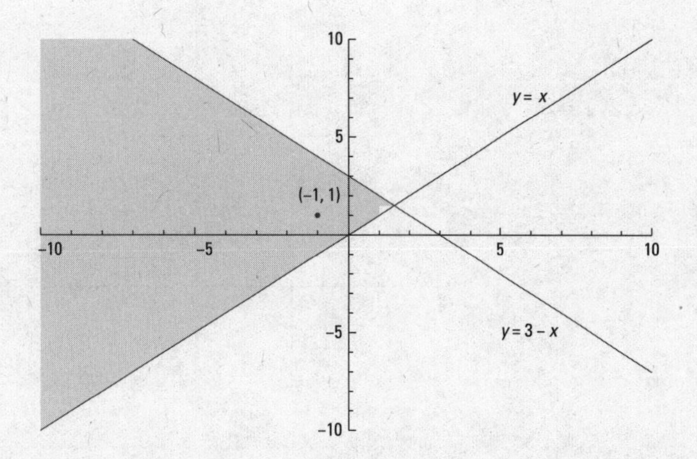

737. Includes (–1, 1)

The graph of $y \leq 5$ contains all the points on the corresponding horizontal line and everything below it.

The graph of $y \geq x + 1$ contains all the points on the corresponding line and above and to the left of the line.

A test point could be (–1, 1).

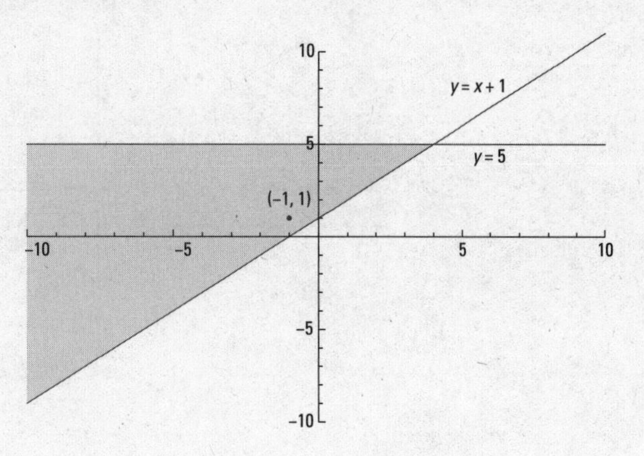

738. Includes (4, 0)

The graph of $x + y \leq 10$ contains all the points on the corresponding line and everything to the left of and below it.

The graph of $x \geq 3$ contains all the points on the corresponding vertical line and everything to the right of the line.

A test point could be (4, 0).

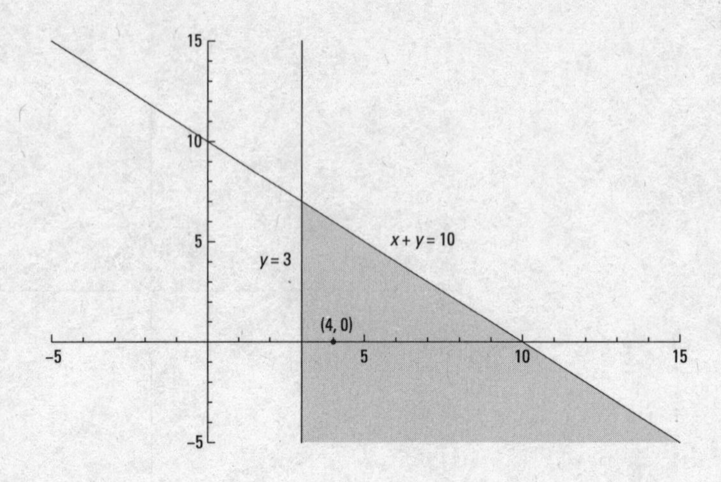

739. Includes (0, 0)

The graph of $x + y \leq 1$ contains all the points on the corresponding line and everything to the left of and below it.

The graph of $y \geq -2$ contains all the points on the corresponding horizontal line and everything above the line.

A test point could be (0, 0).

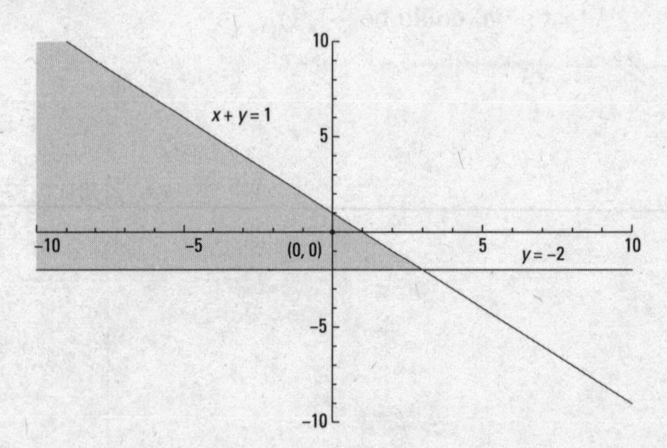

740. Includes (0, 1)

The graph of $y \leq 4$ contains all the points on the corresponding horizontal line and everything below it.

The graph of $y \geq x^2$ contains all the points on the corresponding parabola and everything above (inside) it.

A test point could be (0, 1).

741. Includes (1, 1)

The graph of $y \leq 4x - x^2$ contains all the points on the corresponding parabola and everything below it.

The graph of $y \geq 0$ contains all the points on the corresponding horizontal line (x-axis) and everything above it.

A test point could be (1, 1).

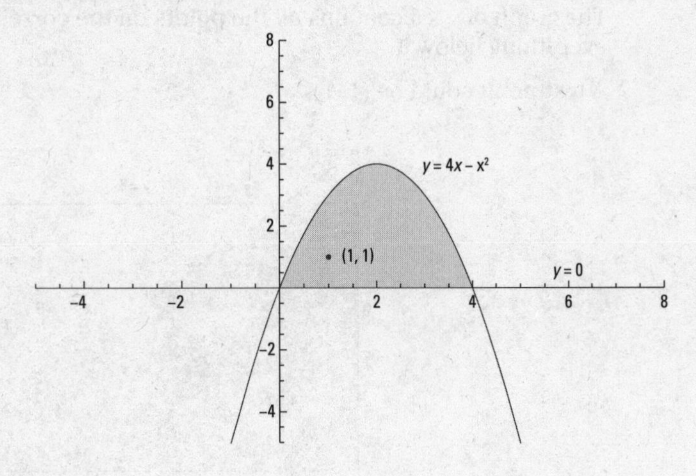

742. Includes (0, 2)

The graph of $y \leq 4$ contains all the points on the corresponding horizontal line and everything below it.

The graph of $y \geq 2^x$ contains all the points on the corresponding exponential curve and everything above it.

A test point could be (0, 2).

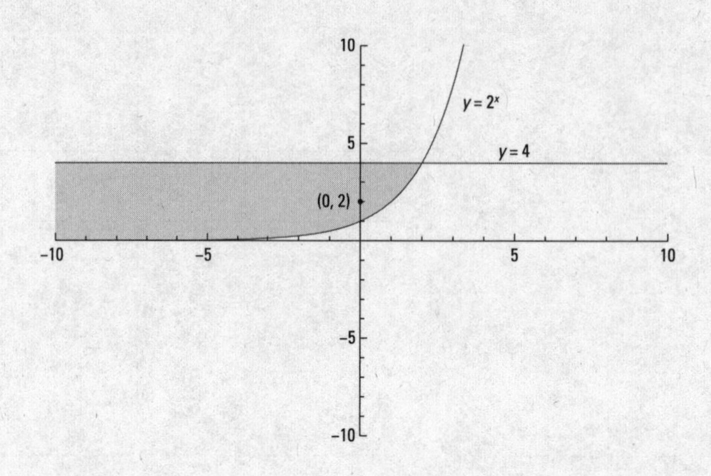

743. Includes (1, 1)

The $x \geq 0$ and $y \geq 0$ indicate that the solution is completely in the first quadrant (including the positive axes and origin).

The graph of $x + y \leq 6$ contains all the points on the corresponding line and everything below and to the left of it.

The graph of $y \leq 5$ contains all the points on the corresponding horizontal line and everything below it.

A test point could be (1, 1).

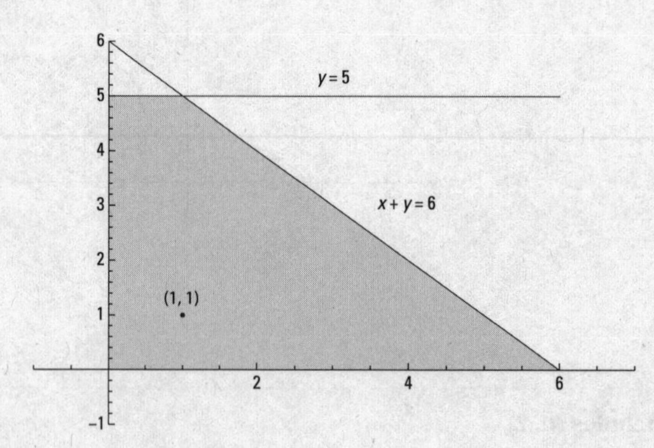

744. Includes (1, 2)

The $x \geq 0$ and $y \geq 0$ indicate that the solution is completely in the first quadrant (including the positive axes and origin).

The graph of $2x + y \leq 12$ contains all the points on the corresponding line and everything below and to the left of it.

The graph of $y \geq x$ contains all the points on the corresponding line and everything above and to the left of it.

A test point could be (1, 2).

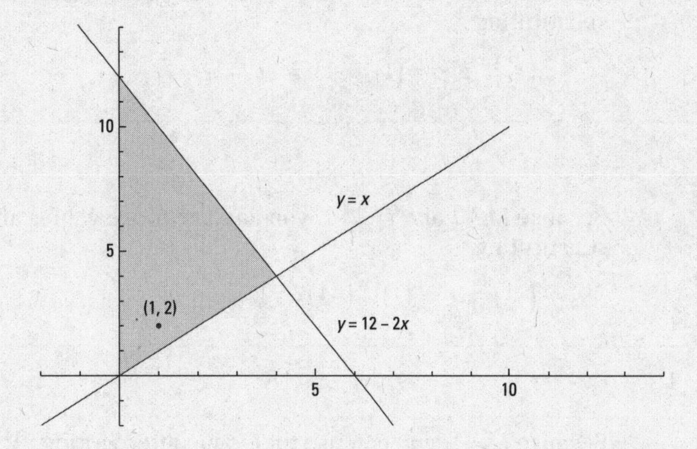

745. Includes (2, 1)

The $x \geq 0$ and $y \geq 0$ indicate that the solution is completely in the first quadrant (including the positive axes and origin).

The graph of $x + y \leq 7$ contains all the points on the corresponding line and everything below and to the left of it.

The graph of $y \leq 2x + 1$ contains all the points on the corresponding line and everything below and to the right of it.

A test point could be (2, 1).

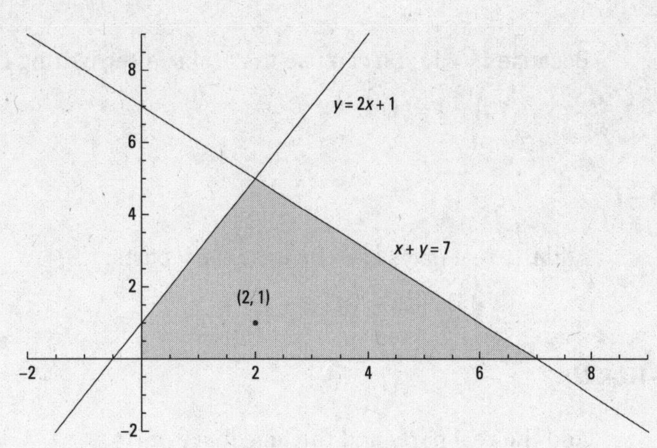

746. $-i$

Because $i^4 = 1$ and $i^2 = -1$, you can use those values after writing i^{15} as $\left(i^4\right)^3 \cdot i^2 \cdot i$ and substituting.

$$\left(i^4\right)^3 \cdot i^2 \cdot i = (1)^3(-1)i = -i$$

747. -1

Because $i^4 = 1$ and $i^2 = -1$, you can use those values after writing i^{402} as $\left(i^4\right)^{100} \cdot i^2$ and substituting.

$$\left(i^4\right)^{100} \cdot i^2 = (1)^{400}(-1) = -1$$

748. 1

Because $i^4 = 1$, you can use that value after writing i^{144} as $\left(i^4\right)^{36}$ and substituting.

$$\left(i^4\right)^{36} = (1)^{36} = 1$$

749. $-i$

Because $i^4 = 1$ and $i^2 = -1$, you can use those values after writing i^{171} as $\left(i^4\right)^{42} \cdot i^2 \cdot i$ and substituting.

$$\left(i^4\right)^{42} \cdot i^2 \cdot i = (1)^{42}(-1) \cdot i = -i$$

750. i

Because $i^4 = 1$, you can use that value after writing i^{57} as $\left(i^4\right)^{14} \cdot i$ and substituting.

$$\left(i^4\right)^{14} \cdot i = (1)^{14} i = i$$

751. $8 - i$

Add the real parts and the imaginary parts.

$$(3 + 5) + (2i - 3i) = 8 - i$$

752. $-10 + 2i$

Add the real parts and the imaginary parts.

$$(-4 - 6) + (-3i + 5i) = -10 + 2i$$

753. **14**

Subtract the real parts and the imaginary parts.

$[9-(-5)]+[-5i-(-5i)]$

$=[9+5]+[-5i+5i]$

$=14$

754. $\mathbf{-4-12i}$

Subtract the real parts and the imaginary parts.

$[4-8]+[-6i-6i]$

$=-4+[-12i]$

$=-4-12i$

755. $\mathbf{10-4i}$

Add and subtract the real parts and the imaginary parts.

$[7-2+5]+[3i-4i+(-3i)]$

$=10+(-4i)$

$=10-4i$

756. **45**

Multiplying the binomials using FOIL, you have $9+18i-18i-36i^2$.

Combine like terms and substitute -1 for i^2.

$=9-36(-1)=9+36=45$

757. **20**

Multiplying the binomials using FOIL, you have $16+8i-8i-4i^2$.

Combine like terms and substitute -1 for i^2.

$=16-4(-1)=16+4=20$

758. $\mathbf{26-2i}$

Multiplying the binomials using FOIL, you have $20+10i-12i-6i^2$.

Combine like terms and substitute -1 for i^2.

$=20-2i-6(-1)=20-2i+6=26-2i$

759. $13 + 40i$

Multiplying the binomials using FOIL, you have $-12 + 30i + 10i - 25i^2$.

Combine like terms and substitute -1 for i^2.

$$= -12 + 40i - 25(-1) = -12 + 40i + 25 = 13 + 40i$$

760. $9 + 18i$

Distribute the factor over the terms in the binomial.

$$18i - 9i^2$$

Substitute -1 for i^2.

$$= 18i - 9(-1) = 18i + 9$$

761. $8 - i$

Multiplying the binomials using FOIL, you have $6 - 4i + 3i - 2i^2$.

Combine like terms and substitute -1 for i^2.

$$= 6 - i - 2(-1) = 6 - i + 2 = 8 - i$$

762. $20 - 90i$

Multiplying the binomials using FOIL, you have $56 - 42i - 48i + 36i^2$.

Combine like terms and substitute -1 for i^2.

$$= 56 - 90i + 36(-1) = 56 - 90i - 36 = 20 - 90i$$

763. $7 - 24i$

The square of the binomial is the same as multiplying the binomial times itself.

$$(4 - 3i)(4 - 3i) = 16 - 12i - 12i + 9i^2$$

Combine like terms and substitute -1 for i^2.

$$= 16 - 24i + 9(-1) = 16 - 24i - 9 = 7 - 24i$$

764. $3 - 4i$

The square of the binomial is the same as multiplying the binomial times itself.

$$(-2 + i)(-2 + i)$$

$$= 4 - 2i - 2i + i^2$$

Combine like terms and substitute -1 for i^2.

$$= 4 - 4i + (-1) = 3 - 4i$$

765. **32*i***

The square of the binomial is the same as multiplying the binomial times itself.

$(-4 - 4i)(-4 - 4i)$

$= 16 + 16i + 16i + 16i^2$

Combine like terms and substitute –1 for i^2.

$= 16 + 32i + 16(-1) = 16 + 32i - 16 = 32i$

766. **8*i***

The square of the binomial is the same as multiplying the binomial times itself.

$(2 + 2i)(2 + 2i)$

$= 4 + 4i + 4i + 4i^2$

Combine like terms and substitute –1 for i^2.

$= 4 + 8i + 4(-1) = 4 + 8i - 4 = 8i$

767. **–5 – *i***

First, find the product of the binomials.

$= i^2(3i + 3 - 2i^2 - 2i)$

$= i^2(i + 3 - 2i^2)$

Now replace i^2 with –1.

$= -1(i + 3 - 2(-1)) = -1(i + 3 + 2) = -1(i + 5)$

$= -i - 5$

768. **–2 – 25*i***

First, find the product of the binomials.

$i^3(24 + 4i - 6i - i^2)$

$= i^3(24 - 2i - i^2)$

Now replace i^3 with $-i$ and i^2 with –1.

$= -i(24 - 2i - (-1))$

$= -i(24 - 2i + 1)$

$= -i(25 - 2i)$

Distribute the $-i$.

$= -25i + 2i^2$

Replace i^2 with –1.

$= -25i + 2(-1) = -25i - 2$

769. **$9 - 19i$**

Find the product of the binomials.

$$15 + 9i^3 - 10i^5 - 6i^8$$

Replace i^3 with $-i$, i^5 with i, and i^8 with 1.

$$= 15 + 9(-i) - 10(i) - 6(1)$$

$$= 15 - 9i - 10i - 6 = 9 - 19i$$

770. **$-4 + 32i$**

First, find the product of the binomials.

$$2i(12 - 6i + 8i - 4i^2)$$

$$= 2i(12 + 2i - 4(-1)) = 2i(12 + 2i + 4)$$

$$= 2i(16 + 2i) = 32i + 4i^2$$

$$= 32i + 4(-1) = 32i - 4$$

771. **$-\frac{10}{17} + \frac{11}{17}i$**

Multiply the numerator and denominator by the conjugate of the denominator.

$$\frac{2+3i}{1-4i} \cdot \frac{1+4i}{1+4i} = \frac{(2+3i)(1+4i)}{(1-4i)(1+4i)}$$

$$= \frac{2+8i+3i+12i^2}{1-16i^2}$$

$$= \frac{2+11i+12(-1)}{1-16(-1)}$$

$$= \frac{2+11i-12}{1+16}$$

$$= \frac{-10+11i}{17}$$

772. **$-\frac{88}{85} - \frac{29}{85}i$**

Multiply the numerator and denominator by the conjugate of the denominator.

$$\frac{1-10i}{2+9i} \cdot \frac{2-9i}{2-9i} = \frac{2-9i-20i+90i^2}{4-81i^2}$$

$$= \frac{2-29i+90(-1)}{4-81(-1)}$$

$$= \frac{2-29i-90}{4+81}$$

$$= \frac{-88-29i}{85}$$

773. $\frac{71}{65}+\frac{17}{65}i$

Multiply the numerator and denominator by the conjugate of the denominator.

$$\begin{aligned}\frac{9+i}{8-i}\cdot\frac{8+i}{8+i}&=\frac{72+9i+8i+i^2}{64-i^2}\\&=\frac{72+17i+(-1)}{64-(-1)}\\&=\frac{72+17i-1}{64+1}\\&=\frac{71+17i}{65}\end{aligned}$$

774. $\frac{2}{13}+\frac{7}{26}i$

Multiply the numerator and denominator by the conjugate of the denominator.

$$\begin{aligned}\frac{2+i}{6-4i}\cdot\frac{6+4i}{6+4i}&=\frac{12+8i+6i+4i^2}{36-16i^2}\\&=\frac{12+14i+4(-1)}{36-16(-1)}\\&=\frac{12+14i-4}{36+16}\\&=\frac{8+14i}{52}=\frac{4+7i}{26}\end{aligned}$$

775. $\frac{13}{10}+\frac{9}{10}i$

Multiply the numerator and denominator by the conjugate of the denominator.

$$\begin{aligned}\frac{7+i}{4-2i}\cdot\frac{4+2i}{4+2i}&=\frac{28+14i+4i+2i^2}{16-4i^2}\\&=\frac{28+18i+2(-1)}{16-4(-1)}\\&=\frac{28+18i-2}{16+4}\\&=\frac{26+18i}{20}=\frac{13+9i}{10}\end{aligned}$$

776. $-\frac{4}{5}-\frac{3}{5}i$

Multiply the numerator and denominator by the conjugate of the denominator.

$$\begin{aligned}\frac{1-3i}{1+3i}\cdot\frac{1-3i}{1-3i}&=\frac{1-6i+9i^2}{1-9i^2}\\&=\frac{1-6i+9(-1)}{1-9(-1)}\\&=\frac{1-6i-9}{1+9}\\&=\frac{-8-6i}{10}=\frac{-4-3i}{5}\end{aligned}$$

777. $\frac{21}{29}+\frac{20}{29}i$

Multiply the numerator and denominator by the conjugate of the denominator.

$$\begin{aligned}\frac{-5-2i}{-5+2i}\cdot\frac{-5-2i}{-5-2i}&=\frac{25+20i+4i^2}{25-4i^2}\\&=\frac{25+20i+4(-1)}{25-4(-1)}\\&=\frac{25+20i-4}{25+4}\\&=\frac{21+20i}{29}\end{aligned}$$

778. $-\frac{24}{25}+\frac{7}{25}i$

Multiply the numerator and denominator by the conjugate of the denominator.

$$\begin{aligned}\frac{1+7i}{1-7i}\cdot\frac{1+7i}{1+7i}&=\frac{1+14i+49i^2}{1-49i^2}\\&=\frac{1+14i+49(-1)}{1-49(-1)}\\&=\frac{1+14i-49}{1+49}\\&=\frac{-48+14i}{50}=\frac{-24+7i}{25}\end{aligned}$$

779. **–1**

Multiply the numerator and denominator by the conjugate of the denominator.

$$\begin{aligned}\frac{6-8i}{-6+8i}\cdot\frac{-6-8i}{-6-8i}&=\frac{-36+64i^2}{36-64i^2}\\&=\frac{-36+64(-1)}{36-64(-1)}\\&=\frac{-36-64}{36+64}\\&=\frac{-100}{100}=-1\end{aligned}$$

This happened because the numerator is –1 times the denominator.

780. **–1**

Multiply the numerator and denominator by the conjugate of the denominator.

$$\begin{aligned}\frac{-3-9i}{3+9i}\cdot\frac{3-9i}{3-9i}&=\frac{-9+81i^2}{9-81i^2}\\&=\frac{-9+81(-1)}{9-81(-1)}\\&=\frac{-9-81}{9+81}\\&=\frac{-90}{90}=-1\end{aligned}$$

This happened because the numerator is –1 times the denominator.

781. $-\frac{8}{29}+\frac{20}{29}i$

Multiply the numerator and denominator by the conjugate of the denominator.

$$\begin{aligned}\frac{4i}{5-2i}\cdot\frac{5+2i}{5+2i}&=\frac{20i+8i^2}{25-4i^2}\\&=\frac{20i+8(-1)}{25-4(-1)}\\&=\frac{20i-8}{25+4}\\&=\frac{20i-8}{29}\end{aligned}$$

782. $\frac{9}{5}+\frac{3}{5}i$

Multiply the numerator and denominator by the conjugate of the denominator.

$$\begin{aligned}\frac{6}{3-i}\cdot\frac{3+i}{3+i}&=\frac{18+6i}{9-i^2}\\&=\frac{18+6i}{9-(-1)}\\&=\frac{18+6i}{10}=\frac{9+3i}{5}\end{aligned}$$

783. **–1 – 3*i***

Multiply the numerator and denominator by the conjugate of the denominator.

$$\begin{aligned}\frac{6-2i}{2i}\cdot\frac{-2i}{-2i}&=\frac{-12i+4i^2}{-4i^2}\\&=\frac{-12i+4(-1)}{-4(-1)}\\&=\frac{-12i-4}{4}\\&=-3i-1\end{aligned}$$

784. $\frac{1}{3}+2i$

Multiply the numerator and denominator by the conjugate of the denominator.

$$\begin{aligned}\frac{-6+i}{3i}\cdot\frac{-3i}{-3i}&=\frac{18i-3i^2}{-9i^2}\\&=\frac{18i-3(-1)}{-9(-1)}\\&=\frac{18i+3}{9}=2i+\frac{1}{3}\end{aligned}$$

785. $\frac{1}{2}+\frac{1}{2}i$

Multiply the numerator and denominator by the conjugate of the denominator.

$$\begin{aligned}\frac{i}{1+i}\cdot\frac{1-i}{1-i}&=\frac{i-i^2}{1-i^2}\\&=\frac{i-(-1)}{1-(-1)}\\&=\frac{i+1}{1+1}=\frac{i+1}{2}\end{aligned}$$

786. $-1\pm 2i$

Using the quadratic formula,

$$\begin{aligned}x&=\frac{-2\pm\sqrt{2^2-4(1)(5)}}{2(1)}\\&=\frac{-2\pm\sqrt{4-20}}{2}=\frac{-2\pm\sqrt{-16}}{2}\\&=\frac{-2\pm\sqrt{16}\sqrt{-1}}{2}=\frac{-2\pm 4i}{2}\\&=-1\pm 2i\end{aligned}$$

787. $2\pm i$

Using the quadratic formula,

$$\begin{aligned}x&=\frac{4\pm\sqrt{4^2-4(1)(5)}}{2(1)}\\&=\frac{4\pm\sqrt{16-20}}{2}=\frac{4\pm\sqrt{-4}}{2}\\&=\frac{4\pm\sqrt{4}\sqrt{-1}}{2}=\frac{4\pm 2i}{2}\\&=2\pm i\end{aligned}$$

788. $\frac{1}{2} \pm \frac{1}{2}i$

Using the quadratic formula,

$$\begin{aligned} x &= \frac{2 \pm \sqrt{2^2 - 4(2)(1)}}{2(2)} \\ &= \frac{2 \pm \sqrt{4-8}}{4} = \frac{2 \pm \sqrt{-4}}{4} \\ &= \frac{2 \pm \sqrt{4}\sqrt{-1}}{4} = \frac{2 \pm 2i}{4} \\ &= \frac{1}{2} \pm \frac{1}{2}i \end{aligned}$$

789. $-3 \pm 2\sqrt{2}i$

Using the quadratic formula,

$$\begin{aligned} x &= \frac{-6 \pm \sqrt{6^2 - 4(1)(17)}}{2(1)} \\ &= \frac{-6 \pm \sqrt{36-68}}{2} = \frac{-6 \pm \sqrt{-32}}{2} \\ &= \frac{-6 \pm \sqrt{16}\sqrt{2}\sqrt{-1}}{2} = \frac{-6 \pm 4\sqrt{2}i}{2} \\ &= -3 \pm 2\sqrt{2}i \end{aligned}$$

790. $\frac{3}{2} \pm \frac{3}{2}i$

Using the quadratic formula,

$$\begin{aligned} x &= \frac{6 \pm \sqrt{(-6)^2 - 4(2)(9)}}{2(2)} \\ &= \frac{6 \pm \sqrt{36-72}}{4} = \frac{6 \pm \sqrt{-36}}{4} \\ &= \frac{6 \pm \sqrt{36}\sqrt{-1}}{4} = \frac{6 \pm 6i}{4} \\ &= \frac{3 \pm 3i}{2} \end{aligned}$$

791. $-4 \pm i$

Using the quadratic formula,

$$\begin{aligned} x &= \frac{-8 \pm \sqrt{8^2 - 4(1)(17)}}{2(1)} \\ &= \frac{-8 \pm \sqrt{64-68}}{2} = \frac{-8 \pm \sqrt{-4}}{2} \\ &= \frac{-8 \pm \sqrt{4}\sqrt{-1}}{2} = \frac{-8 \pm 2i}{2} \\ &= -4 \pm i \end{aligned}$$

792. $-\frac{3}{2} \pm \frac{\sqrt{11}}{2}i$

Using the quadratic formula,

$$x = \frac{-3 \pm \sqrt{3^2 - 4(1)(5)}}{2(1)}$$
$$= \frac{-3 \pm \sqrt{9-20}}{2} = \frac{-3 \pm \sqrt{-11}}{2}$$
$$= \frac{-3 \pm \sqrt{11}\sqrt{-1}}{2} = \frac{-3 \pm \sqrt{11}i}{2}$$

793. $\frac{1}{2} \pm \frac{3\sqrt{3}}{2}i$

Using the quadratic formula,

$$x = \frac{1 \pm \sqrt{(-1)^2 - 4(1)(7)}}{2(1)}$$
$$= \frac{1 \pm \sqrt{1-28}}{2} = \frac{1 \pm \sqrt{-27}}{2}$$
$$= \frac{1 \pm \sqrt{9}\sqrt{3}\sqrt{-1}}{2} = \frac{1 \pm 3\sqrt{3}i}{2}$$

794. $\frac{3}{4} \pm \frac{\sqrt{55}}{4}i$

Using the quadratic formula,

$$x = \frac{3 \pm \sqrt{(-3)^2 - 4(2)(8)}}{2(2)}$$
$$= \frac{3 \pm \sqrt{9-64}}{4} = \frac{3 \pm \sqrt{-55}}{4}$$
$$= \frac{3 \pm \sqrt{55}\sqrt{-1}}{4} = \frac{3 \pm \sqrt{55}i}{4}$$

795. $-\frac{1}{3} \pm \frac{\sqrt{5}}{3}i$

Using the quadratic formula,

$$x = \frac{-2 \pm \sqrt{2^2 - 4(3)(2)}}{2(3)}$$
$$= \frac{-2 \pm \sqrt{4-24}}{6} = \frac{-2 \pm \sqrt{-20}}{6}$$
$$= \frac{-2 \pm \sqrt{4}\sqrt{5}\sqrt{-1}}{6} = \frac{-2 \pm 2\sqrt{5}i}{6}$$
$$= \frac{-1 \pm \sqrt{5}i}{3}$$

796. 2×3

Give the number of rows and then the number of columns.

797. 5×2

Give the number of rows and then the number of columns.

798. 4×1; column matrix

Give the number of rows and then the number of columns. Because the number of columns is 1, it's a column matrix.

799. 3×3; square matrix

Give the number of rows and then the number of columns. Because the number of rows and columns is the same, it's a square matrix.

800. 4×3; zero matrix

Give the number of rows and then the number of columns. Because all the entries are 0, it's a zero matrix.

801. $\begin{bmatrix} -1 & -1 & 13 \\ -1 & 3 & -2 \end{bmatrix}$

When adding matrices, find the sum of the corresponding elements.

$$\begin{bmatrix} 1+(-2) & 0+(-1) & 7+6 \\ -4+3 & 3+0 & 3+(-5) \end{bmatrix} = \begin{bmatrix} -1 & -1 & 13 \\ -1 & 3 & -2 \end{bmatrix}$$

802. $\begin{bmatrix} 5 & 3 \\ -3 & -4 \\ 3 & -8 \end{bmatrix}$

When subtracting matrices, find the difference between the corresponding elements.

$$\begin{bmatrix} 4-(-1) & 3-0 \\ -5-(-2) & 2-6 \\ 0-(-3) & -7-1 \end{bmatrix} = \begin{bmatrix} 5 & 3 \\ -3 & -4 \\ 3 & -8 \end{bmatrix}$$

Another method is to multiply each term in the second matrix by –1 (scalar multiplication) and then add.

$$\begin{bmatrix} 4 & 3 \\ -5 & 2 \\ 0 & -7 \end{bmatrix} - 1\begin{bmatrix} -1 & 0 \\ -2 & 6 \\ -3 & 1 \end{bmatrix}$$

$$= \begin{bmatrix} 4 & 3 \\ -5 & 2 \\ 0 & -7 \end{bmatrix} + \begin{bmatrix} 1 & 0 \\ 2 & -6 \\ 3 & -1 \end{bmatrix} = \begin{bmatrix} 5 & 3 \\ -3 & -4 \\ 3 & -8 \end{bmatrix}$$

803. $\begin{bmatrix} 0 & 0 \\ 0 & 0 \end{bmatrix}$

When adding matrices, find the sum of the corresponding elements.

$$\begin{bmatrix} 6+(-6) & -4+4 \\ -1+1 & 0+0 \end{bmatrix} = \begin{bmatrix} 0 & 0 \\ 0 & 0 \end{bmatrix}$$

All the corresponding elements are opposites (additive inverses), so the sum of the matrices is the zero matrix.

804. $\begin{bmatrix} -6 \\ 0 \\ 5 \\ -2 \end{bmatrix}$

Add the first two matrices.

$$\begin{bmatrix} 1 \\ -1 \\ 7 \\ 0 \end{bmatrix} + \begin{bmatrix} -2 \\ 3 \\ -3 \\ 4 \end{bmatrix} = \begin{bmatrix} 1+(-2) \\ -1+3 \\ 7+(-3) \\ 0+4 \end{bmatrix} = \begin{bmatrix} -1 \\ 2 \\ 4 \\ 4 \end{bmatrix}$$

$$\begin{bmatrix} -1 \\ 2 \\ 4 \\ 4 \end{bmatrix} - \begin{bmatrix} 5 \\ 2 \\ -1 \\ 6 \end{bmatrix} = \begin{bmatrix} -1-5 \\ 2-2 \\ 4-(-1) \\ 4-6 \end{bmatrix} = \begin{bmatrix} -6 \\ 0 \\ 5 \\ -2 \end{bmatrix}$$

805. $\begin{bmatrix} 6 & 3 \end{bmatrix}$

Subtract the second matrix from the first matrix.

$$\begin{bmatrix} 3 & 6 \end{bmatrix} - \begin{bmatrix} -2 & 4 \end{bmatrix} = \begin{bmatrix} 3-(-2) & 6-4 \end{bmatrix} = \begin{bmatrix} 5 & 2 \end{bmatrix}$$

Now subtract the last matrix from the difference.

$$= \begin{bmatrix} 5-(-1) & 2-(-1) \end{bmatrix}$$

$$= \begin{bmatrix} 6 & 3 \end{bmatrix}$$

806. $\begin{bmatrix} -2 & 0 & 8 \\ 4 & -4 & 12 \end{bmatrix}$

Multiply each element in the matrix by 2.

$$\begin{bmatrix} 2(-1) & 2(0) & 2(4) \\ 2(2) & 2(-2) & 2(6) \end{bmatrix}$$

$$= \begin{bmatrix} -2 & 0 & 8 \\ 4 & -4 & 12 \end{bmatrix}$$

807. $\begin{bmatrix} 1 & -4 \\ -5 & 3 \end{bmatrix}$

Multiply each element in the matrix by $-\frac{1}{2}$.

$$\begin{bmatrix} -\frac{1}{2}(-2) & -\frac{1}{2}(8) \\ -\frac{1}{2}(10) & -\frac{1}{2}(-6) \end{bmatrix}$$

$$= \begin{bmatrix} 1 & -4 \\ -5 & 3 \end{bmatrix}$$

808. $\begin{bmatrix} -4 \\ 5 \\ 0 \end{bmatrix}$

Multiply each element in the matrix by –1.

$$\begin{bmatrix} -1(4) \\ -1(-5) \\ -1(0) \end{bmatrix} = \begin{bmatrix} -4 \\ 5 \\ 0 \end{bmatrix}$$

809. $\begin{bmatrix} -0.6 & 0.4 & 0 \\ -0.8 & -0.2 & 0.2 \end{bmatrix}$

Multiply each element in the matrix by –0.2.

$$\begin{bmatrix} -0.2(3) & -0.2(-2) & -0.2(0) \\ -0.2(4) & -0.2(1) & -0.2(-1) \end{bmatrix}$$

$$= \begin{bmatrix} -0.6 & 0.4 & 0 \\ -0.8 & -0.2 & 0.2 \end{bmatrix}$$

810. $\begin{bmatrix} 1 & 0 & 0 \\ 0 & -3 & 0 \\ 0 & 0 & 2 \end{bmatrix}$

Multiply each element in the matrix by $\frac{1}{3}$.

$$\begin{bmatrix} \frac{1}{3}(3) & \frac{1}{3}(0) & \frac{1}{3}(0) \\ \frac{1}{3}(0) & \frac{1}{3}(-9) & \frac{1}{3}(0) \\ \frac{1}{3}(0) & \frac{1}{3}(0) & \frac{1}{3}(6) \end{bmatrix}$$

$$= \begin{bmatrix} 1 & 0 & 0 \\ 0 & -3 & 0 \\ 0 & 0 & 2 \end{bmatrix}$$

811. $\begin{bmatrix} -2 & -4 \\ -17 & 10 \end{bmatrix}$

Each element, a_{ij}, in the product is computed by finding the sum of the products of the elements in the *i*th row times the *j*th column.

$$\begin{bmatrix} 1(-4)+(-2)(-1) & 1(0)+(-2)(2) \\ 3(-4)+5(-1) & 3(0)+5(2) \end{bmatrix}$$
$$= \begin{bmatrix} -2 & -4 \\ -17 & 10 \end{bmatrix}$$

812. $\begin{bmatrix} -6 & 5 \\ 20 & 27 \end{bmatrix}$

Each element, a_{ij}, in the product is computed by finding the sum of the products of the elements in the *i*th row times the *j*th column.

$$\begin{bmatrix} -1(6)+0(-1)+2(0) & -1(3)+0(1)+2(4) \\ 3(6)+(-2)(-1)+5(0) & 3(3)+(-2)(1)+5(4) \end{bmatrix}$$
$$= \begin{bmatrix} -6 & 5 \\ 20 & 27 \end{bmatrix}$$

813. $\begin{bmatrix} 7 & -9 & 1 \\ 6 & -12 & 3 \\ 24 & -28 & 2 \end{bmatrix}$

Each element, a_{ij}, in the product is computed by finding the sum of the products of the elements in the *i*th row times the *j*th column.

$$\begin{bmatrix} 1(5)+(-1)(-2) & 1(-5)+(-1)(4) & 1(0)+(-1)(-1) \\ 0(5)+(-3)(-2) & 0(-5)+(-3)(4) & 0(0)+(-3)(-1) \\ 4(5)+(-2)(-2) & 4(-5)+(-2)(4) & 4(0)+(-2)(-1) \end{bmatrix}$$
$$= \begin{bmatrix} 7 & -9 & 1 \\ 6 & -12 & 3 \\ 24 & -28 & 2 \end{bmatrix}$$

814. $\begin{bmatrix} -1 \\ -21 \\ -17 \\ -23 \end{bmatrix}$

Each element, a_{ij}, in the product is computed by finding the sum of the products of the elements in the ith row times the jth column.

$$\begin{bmatrix} 1(2)+(-1)3+0(-4) \\ 9(2)+(-1)3+9(-4) \\ 2(2)+(1)3+6(-4) \\ 4(2)+(-1)3+7(-4) \end{bmatrix}$$

$$= \begin{bmatrix} -1 \\ -21 \\ -17 \\ -23 \end{bmatrix}$$

815. $\begin{bmatrix} 31 \\ 9 \end{bmatrix}$

Each element, a_{ij}, in the product is computed by finding the sum of the products of the elements in the ith row times the jth column.

$$\begin{bmatrix} -1(1)+4(8) \\ 9(1)+0(8) \end{bmatrix}$$

$$= \begin{bmatrix} 31 \\ 9 \end{bmatrix}$$

816. [–14]

Each element, a_{ij}, in the product is computed by finding the sum of the products of the elements in the ith row times the jth column.

$$\begin{bmatrix} -2(1)+3(-4)+9(0)+0(-3) \end{bmatrix}$$

$$= \begin{bmatrix} -14 \end{bmatrix}$$

When you multiply a 1×4 matrix times a 4×1 matrix, the resulting matrix is 1×1.

817. $\begin{bmatrix} -5 & 4 \\ -4 & 3 \end{bmatrix}$

Each element, a_{ij}, in the product is computed by finding the sum of the products of the elements in the ith row times the jth column.

$$\begin{bmatrix} 2(2)+3(-3) & 2(-1)+3(2) \\ 1(2)+2(-3) & 1(-1)+2(2) \end{bmatrix}$$

$$= \begin{bmatrix} -5 & 4 \\ -4 & 3 \end{bmatrix}$$

818. $\begin{bmatrix} 6 & 0 & 0 \\ 0 & 6 & 0 \\ 0 & 0 & 6 \end{bmatrix}$

Each element, a_{ij}, in the product is computed by finding the sum of the products of the elements in the ith row times the jth column.

$$\begin{bmatrix} 1(2)+1(1)+1(3) & 1(1)+1(0)+1(-1) & 1(-3)+1(1)+1(2) \\ 1(2)+13(1)+(-5)(3) & 1(1)+13(0)+(-5)(-1) & 1(-3)+13(1)+(-5)(2) \\ -1(2)+5(1)+(-1)(3) & -1(1)+(5)(0)+(-1)(-1) & -1(-3)+5(1)+(-1)(2) \end{bmatrix}$$

$$=\begin{bmatrix} 6 & 0 & 0 \\ 0 & 6 & 0 \\ 0 & 0 & 6 \end{bmatrix}$$

819. $\begin{bmatrix} 1 & 0 & 0 \\ 0 & 1 & 0 \\ 0 & 0 & 1 \end{bmatrix}$

Each element, a_{ij}, in the product is computed by finding the sum of the products of the elements in the ith row times the jth column.

$$\begin{bmatrix} \frac{5}{4}(0)+\frac{7}{4}(1)-\frac{1}{4}(3) & -\frac{1}{4}(0)-\frac{3}{4}(1)+\frac{1}{4}(3) & -\frac{7}{4}(0)-\frac{9}{4}(1)+\frac{3}{4}(3) \\ \frac{5}{4}(3)+\frac{7}{4}(-2)-\frac{1}{4}(1) & -\frac{1}{4}(3)-\frac{3}{4}(-2)+\frac{1}{4}(1) & -\frac{7}{4}(3)-\frac{9}{4}(-2)+\frac{3}{4}(1) \\ \frac{5}{4}(-1)+\frac{7}{4}(1)-\frac{1}{4}(2) & -\frac{1}{4}(-1)-\frac{3}{4}(1)+\frac{1}{4}(2) & -\frac{7}{4}(-1)-\frac{9}{4}(1)+\frac{3}{4}(2) \end{bmatrix}$$

$$=\begin{bmatrix} 1 & 0 & 0 \\ 0 & 1 & 0 \\ 0 & 0 & 1 \end{bmatrix}$$

When the product of two matrices results in an identity matrix, this means that the matrices are inverses of one another.

820. $\begin{bmatrix} 6 & 0 & 0 & 0 \\ 0 & 6 & 0 & 0 \\ 0 & 0 & 6 & 0 \\ 0 & 0 & 0 & 6 \end{bmatrix}$

$$\begin{bmatrix} 6 & -12 & 0 & 0 \\ 0 & 18 & 0 & 0 \\ 0 & 0 & 6 & 0 \\ 0 & 12 & 0 & 6 \end{bmatrix}$$

$$\begin{bmatrix} 2(6)+3(4)-1(-6)+4(-6) & 2(12)+3(6)-1(-18)+4(-15) \\ 2(6)-3(4)+0(-6)+0(-6) & 2(12)-3(6)+0(-18)+0(-15) \\ 1(6)+0(4)-1(-6)+2(-6) & 1(12)+0(6)-1(-18)+2(-15) \\ 3(6)-3(4)+1(-6)+0(-6) & 3(12)-3(6)+1(-18)+0(-15) \end{bmatrix}$$

$$\begin{matrix} 2(-12)+3(-8)-1(12)+4(15) & 2(-6)+3(-4)-1(12)+4(9) \\ 2(-12)-3(-8)+0(12)+0(15) & 2(-6)-3(-4)+0(12)+0(9) \\ 1(-12)+0(-8)-1(12)+2(15) & 1(-6)+0(-4)-1(12)+2(9) \\ 3(-12)-3(-8)+1(12)+0(15) & 3(-6)-3(-4)+1(12)+0(9) \end{matrix}\Bigg]$$

$$= \begin{bmatrix} 6 & 0 & 0 & 0 \\ 0 & 6 & 0 & 0 \\ 0 & 0 & 6 & 0 \\ 0 & 0 & 0 & 6 \end{bmatrix}$$

821. $\begin{bmatrix} \mathbf{3} & \mathbf{2} \\ \mathbf{4} & \mathbf{3} \end{bmatrix}$

Use the formula:

$$\begin{bmatrix} a & b \\ c & d \end{bmatrix}^{-1} = \frac{1}{ad-bc}\begin{bmatrix} d & -b \\ -c & a \end{bmatrix}$$

The multiplier is:

$$\frac{1}{3(3)-(-2)(-4)} = \frac{1}{9-8} = 1$$

Switching a and d makes no apparent difference. The values of b and c change from negative to positive.

$$\begin{bmatrix} 3 & -2 \\ -4 & 3 \end{bmatrix}^{-1} = \begin{bmatrix} 3 & 2 \\ 4 & 3 \end{bmatrix}$$

822. $\begin{bmatrix} \mathbf{-5} & \mathbf{7} \\ \mathbf{3} & \mathbf{-4} \end{bmatrix}$

Use the formula:

$$\begin{bmatrix} a & b \\ c & d \end{bmatrix}^{-1} = \frac{1}{ad-bc}\begin{bmatrix} d & -b \\ -c & a \end{bmatrix}$$

The multiplier is:

$$\frac{1}{4(5)-(7)(3)} = \frac{1}{20-21} = -1$$

Switch a and d. The values of b and c change from positive to negative. Multiply the matrix by the scalar –1.

$$\begin{bmatrix} 4 & 7 \\ 3 & 5 \end{bmatrix}^{-1} = -1\begin{bmatrix} 5 & -7 \\ -3 & 4 \end{bmatrix} = \begin{bmatrix} -5 & 7 \\ 3 & -4 \end{bmatrix}$$

823. $\begin{bmatrix} 7 & 5 \\ 3 & 2 \end{bmatrix}$

Use the formula:

$$\begin{bmatrix} a & b \\ c & d \end{bmatrix}^{-1} = \frac{1}{ad-bc}\begin{bmatrix} d & -b \\ -c & a \end{bmatrix}$$

The multiplier is:

$$\frac{1}{-2(-7)-(5)(3)} = \frac{1}{14-15} = -1$$

Switch a and d. The values of b and c change from positive to negative. Multiply the matrix by the scalar –1.

$$\begin{bmatrix} -2 & 5 \\ 3 & -7 \end{bmatrix}^{-1} = -1\begin{bmatrix} -7 & -5 \\ -3 & -2 \end{bmatrix} = \begin{bmatrix} 7 & 5 \\ 3 & 2 \end{bmatrix}$$

824. $\begin{bmatrix} 2 & 5/2 \\ 1 & 3/2 \end{bmatrix}$

Use the formula:

$$\begin{bmatrix} a & b \\ c & d \end{bmatrix}^{-1} = \frac{1}{ad-bc}\begin{bmatrix} d & -b \\ -c & a \end{bmatrix}$$

The multiplier is:

$$\frac{1}{3(4)-(-5)(-2)} = \frac{1}{12-10} = \frac{1}{2}$$

Switch a and d. The values of b and c change from negative to positive. Multiply the matrix by the scalar $\frac{1}{2}$.

$$\begin{bmatrix} 3 & -5 \\ -2 & 4 \end{bmatrix}^{-1} = \frac{1}{2}\begin{bmatrix} 4 & 5 \\ 2 & 3 \end{bmatrix} = \begin{bmatrix} 2 & 5/2 \\ 1 & 3/2 \end{bmatrix}$$

825. $\begin{bmatrix} 0 & 1 & 0 \\ -1 & 2 & 0 \\ 4 & -7 & 1 \end{bmatrix}$

Create a side-by-side matrix with the identity matrix.

$$\left[\begin{array}{ccc|ccc} 2 & -1 & 0 & 1 & 0 & 0 \\ 1 & 0 & 0 & 0 & 1 & 0 \\ -1 & 4 & 1 & 0 & 0 & 1 \end{array}\right]$$

Perform row operations designed to change the left-side matrix into an identity matrix.

$$R_1 \leftrightarrow R_2 \left[\begin{array}{ccc|ccc} 1 & 0 & 0 & 0 & 1 & 0 \\ 2 & -1 & 0 & 1 & 0 & 0 \\ -1 & 4 & 1 & 0 & 0 & 1 \end{array}\right]$$

$$\begin{array}{l} -2R_1 + R_2 \to R_2 \\ R_1 + R_3 \to R_3 \end{array} \left[\begin{array}{ccc|ccc} 1 & 0 & 0 & 0 & 1 & 0 \\ 0 & -1 & 0 & 1 & -2 & 0 \\ 0 & 4 & 1 & 0 & 1 & 1 \end{array}\right]$$

$$-1R_2 \to R_2 \left[\begin{array}{ccc|ccc} 1 & 0 & 0 & 0 & 1 & 0 \\ 0 & 1 & 0 & -1 & 2 & 0 \\ 0 & 4 & 1 & 0 & 1 & 1 \end{array}\right]$$

$$-4R_2 + R_3 \to R_3 \left[\begin{array}{ccc|ccc} 1 & 0 & 0 & 0 & 1 & 0 \\ 0 & 1 & 0 & -1 & 2 & 0 \\ 0 & 0 & 1 & 4 & -7 & 1 \end{array}\right]$$

The inverse is the matrix on the right side.

826. $\begin{bmatrix} \mathbf{23} & \mathbf{-6} & \mathbf{-2} \\ \mathbf{-3} & \mathbf{1} & \mathbf{0} \\ \mathbf{11} & \mathbf{-3} & \mathbf{-1} \end{bmatrix}$

Create a side-by-side matrix with the identity matrix.

$$\left[\begin{array}{ccc|ccc} 1 & 0 & -2 & 1 & 0 & 0 \\ 3 & 1 & -6 & 0 & 1 & 0 \\ 2 & -3 & -5 & 0 & 0 & 1 \end{array}\right]$$

Perform row operations designed to change the left-side matrix into an identity matrix.

$$\begin{array}{l} -3R_1 + R_2 \to R_2 \\ -2R_1 + R_3 \to R_3 \end{array} \left[\begin{array}{ccc|ccc} 1 & 0 & -2 & 1 & 0 & 0 \\ 0 & 1 & 0 & -3 & 1 & 0 \\ 0 & -3 & -1 & -2 & 0 & 1 \end{array}\right]$$

$$3R_2 + R_3 \to R_3 \left[\begin{array}{ccc|ccc} 1 & 0 & -2 & 1 & 0 & 0 \\ 0 & 1 & 0 & -3 & 1 & 0 \\ 0 & 0 & -1 & -11 & 3 & 1 \end{array}\right]$$

$$-1R_3 \to R_3 \left[\begin{array}{ccc|ccc} 1 & 0 & -2 & 1 & 0 & 0 \\ 0 & 1 & 0 & -3 & 1 & 0 \\ 0 & 0 & 1 & 11 & -3 & -1 \end{array}\right]$$

$$2R_3 + R_1 \to R_1 \left[\begin{array}{ccc|ccc} 1 & 0 & 0 & 23 & -6 & -2 \\ 0 & 1 & 0 & -3 & 1 & 0 \\ 0 & 0 & 1 & 11 & -3 & -1 \end{array}\right]$$

The inverse is the matrix on the right side.

827. $\begin{bmatrix} -11 & 28 & -6 \\ 6 & -14 & 3 \\ -2 & 5 & -1 \end{bmatrix}$

Create a side-by-side matrix with the identity matrix.

$$\left[\begin{array}{ccc|ccc} 1 & 2 & 0 & 1 & 0 & 0 \\ 0 & 1 & 3 & 0 & 1 & 0 \\ -2 & 1 & 14 & 0 & 0 & 1 \end{array}\right]$$

Perform row operations designed to change the left-side matrix into an identity matrix.

$$2R_1 + R_3 \to R_3 \left[\begin{array}{ccc|ccc} 1 & 2 & 0 & 1 & 0 & 0 \\ 0 & 1 & 3 & 0 & 1 & 0 \\ 0 & 5 & 14 & 2 & 0 & 1 \end{array}\right]$$

$$\begin{array}{l} -2R_2 + R_1 \to R_1 \\ -5R_2 + R_3 \to R_3 \end{array} \left[\begin{array}{ccc|ccc} 1 & 0 & -6 & 1 & -2 & 0 \\ 0 & 1 & 3 & 0 & 1 & 0 \\ 0 & 0 & -1 & 2 & -5 & 1 \end{array}\right]$$

$$-1R_3 \to R_3 \left[\begin{array}{ccc|ccc} 1 & 0 & -6 & 1 & -2 & 0 \\ 0 & 1 & 3 & 0 & 1 & 0 \\ 0 & 0 & 1 & -2 & 5 & -1 \end{array}\right]$$

$$\begin{array}{l} 6R_3 + R_1 \to R_1 \\ -3R_3 + R_2 \to R_2 \end{array} \left[\begin{array}{ccc|ccc} 1 & 0 & 0 & -11 & 28 & -6 \\ 0 & 1 & 0 & 6 & -14 & 3 \\ 0 & 0 & 1 & -2 & 5 & -1 \end{array}\right]$$

The inverse is the matrix on the right side.

828. $\begin{bmatrix} -1/3 & -1/3 & 1/3 \\ -8/3 & -7/6 & 5/3 \\ -2/3 & -1/6 & 2/3 \end{bmatrix}$

Create a side-by-side matrix with the identity matrix.

$$\left[\begin{array}{ccc|ccc} 3 & -1 & 1 & 1 & 0 & 0 \\ -4 & 0 & 2 & 0 & 1 & 0 \\ 2 & -1 & 3 & 0 & 0 & 1 \end{array}\right]$$

Perform row operations designed to change the left-side matrix into an identity matrix.

$$\frac{1}{3}R_1 \to R_1 \left[\begin{array}{ccc|ccc} 1 & -\frac{1}{3} & \frac{1}{3} & \frac{1}{3} & 0 & 0 \\ -4 & 0 & 2 & 0 & 1 & 0 \\ 2 & -1 & 3 & 0 & 0 & 1 \end{array}\right]$$

$$\begin{array}{l} 4R_1 + R_2 \to R_2 \\ -2R_1 + R_3 \to R_3 \end{array} \left[\begin{array}{ccc|ccc} 1 & -\frac{1}{3} & \frac{1}{3} & \frac{1}{3} & 0 & 0 \\ 0 & -\frac{4}{3} & \frac{10}{3} & \frac{4}{3} & 1 & 0 \\ 0 & -\frac{1}{3} & \frac{7}{3} & -\frac{2}{3} & 0 & 1 \end{array}\right]$$

$$-\frac{3}{4}R_2 \to R_2 \left[\begin{array}{ccc|ccc} 1 & -\frac{1}{3} & \frac{1}{3} & \frac{1}{3} & 0 & 0 \\ 0 & 1 & -\frac{5}{2} & -1 & -\frac{3}{4} & 0 \\ 0 & -\frac{1}{3} & \frac{7}{3} & -\frac{2}{3} & 0 & 1 \end{array}\right]$$

$$\begin{array}{l} \frac{1}{3}R_2 + R_1 \to R_1 \\ \frac{1}{3}R_2 + R_3 \to R_3 \end{array} \left[\begin{array}{ccc|ccc} 1 & 0 & -\frac{1}{2} & 0 & -\frac{1}{4} & 0 \\ 0 & 1 & -\frac{5}{2} & -1 & -\frac{3}{4} & 0 \\ 0 & 0 & \frac{3}{2} & -1 & -\frac{1}{4} & 1 \end{array}\right]$$

$$\frac{2}{3}R_3 \to R_3 \left[\begin{array}{ccc|ccc} 1 & 0 & -\frac{1}{2} & 0 & -\frac{1}{4} & 0 \\ 0 & 1 & -\frac{5}{2} & -1 & -\frac{3}{4} & 0 \\ 0 & 0 & 1 & -\frac{2}{3} & -\frac{1}{6} & \frac{2}{3} \end{array}\right]$$

$$\begin{array}{l} \frac{1}{2}R_3 + R_1 \to R_1 \\ \frac{5}{2}R_3 + R_2 \to R_2 \end{array} \left[\begin{array}{ccc|ccc} 1 & 0 & 0 & -\frac{1}{3} & -\frac{1}{3} & \frac{1}{3} \\ 0 & 1 & 0 & -\frac{8}{3} & -\frac{7}{6} & \frac{5}{3} \\ 0 & 0 & 1 & -\frac{2}{3} & -\frac{1}{6} & \frac{2}{3} \end{array}\right]$$

The inverse is the matrix on the right side.

829.

$$\begin{bmatrix} \frac{3}{4} & -\frac{1}{12} & -\frac{1}{2} \\ -\frac{1}{4} & \frac{1}{4} & \frac{1}{2} \\ -\frac{3}{8} & \frac{5}{24} & \frac{1}{4} \end{bmatrix}$$

Create a side-by-side matrix with the identity matrix.

$$\left[\begin{array}{ccc|ccc} 1 & 2 & -2 & 1 & 0 & 0 \\ 3 & 0 & 6 & 0 & 1 & 0 \\ -1 & 3 & -4 & 0 & 0 & 1 \end{array}\right]$$

Perform row operations designed to change the left-side matrix into an identity matrix.

$$\begin{matrix} -3R_1+R_2 \to R_2 \\ R_1+R_3 \to R_3 \end{matrix} \left[\begin{array}{ccc|ccc} 1 & 2 & -2 & 1 & 0 & 0 \\ 0 & -6 & 12 & -3 & 1 & 0 \\ 0 & 5 & -6 & 1 & 0 & 1 \end{array}\right]$$

$$-\frac{1}{6}R_2 \to R_2 \left[\begin{array}{ccc|ccc} 1 & 2 & -2 & 1 & 0 & 0 \\ 0 & 1 & -2 & \frac{1}{2} & -\frac{1}{6} & 0 \\ 0 & 5 & -6 & 1 & 0 & 1 \end{array}\right]$$

$$\begin{matrix} -2R_2+R_1 \to R_1 \\ -5R_2+R_3 \to R_3 \end{matrix} \left[\begin{array}{ccc|ccc} 1 & 0 & 2 & 0 & \frac{1}{3} & 0 \\ 0 & 1 & -2 & \frac{1}{2} & -\frac{1}{6} & 0 \\ 0 & 0 & 4 & -\frac{3}{2} & \frac{5}{6} & 1 \end{array}\right]$$

$$\frac{1}{4}R_3 \to R_3 \left[\begin{array}{ccc|ccc} 1 & 0 & 2 & 0 & \frac{1}{3} & 0 \\ 0 & 1 & -2 & \frac{1}{2} & -\frac{1}{6} & 0 \\ 0 & 0 & 1 & -\frac{3}{8} & \frac{5}{24} & \frac{1}{4} \end{array}\right]$$

$$\begin{matrix} -2R_3+R_1 \to R_1 \\ 2R_3+R_2 \to R_2 \end{matrix} \left[\begin{array}{ccc|ccc} 1 & 0 & 0 & \frac{3}{4} & -\frac{1}{12} & -\frac{1}{2} \\ 0 & 1 & 0 & -\frac{1}{4} & \frac{1}{4} & \frac{1}{2} \\ 0 & 0 & 1 & -\frac{3}{8} & \frac{5}{24} & \frac{1}{4} \end{array}\right]$$

The inverse is the matrix on the right side.

830.

$$\begin{bmatrix} \mathbf{1} & \mathbf{3} & \mathbf{-1} \\ \mathbf{0} & \mathbf{2} & \mathbf{-2} \\ \mathbf{4} & \mathbf{4} & \mathbf{-10} \end{bmatrix}$$

Create a side-by-side matrix with the identity matrix.

$$\left[\begin{array}{ccc|ccc} \frac{3}{7} & -\frac{13}{14} & \frac{1}{7} & 1 & 0 & 0 \\ \frac{2}{7} & \frac{3}{14} & -\frac{1}{14} & 0 & 1 & 0 \\ \frac{2}{7} & -\frac{2}{7} & -\frac{1}{14} & 0 & 0 & 1 \end{array}\right]$$

The fractions seem a bit daunting. If you multiply by the scalar 14, the values become more manageable.

$$14 * \left[\begin{array}{ccc|ccc} \frac{3}{7} & -\frac{13}{14} & \frac{1}{7} & 1 & 0 & 0 \\ \frac{2}{7} & \frac{3}{14} & -\frac{1}{14} & 0 & 1 & 0 \\ \frac{2}{7} & -\frac{2}{7} & -\frac{1}{14} & 0 & 0 & 1 \end{array}\right]$$

$$= \left[\begin{array}{ccc|ccc} 6 & -13 & 2 & 14 & 0 & 0 \\ 4 & 3 & -1 & 0 & 14 & 0 \\ 4 & -4 & -1 & 0 & 0 & 14 \end{array}\right]$$

Perform row operations designed to change the left-side matrix into an identity matrix.

$$R_1 \leftrightarrow R_3 \left[\begin{array}{ccc|ccc} 4 & -4 & -1 & 0 & 0 & 14 \\ 4 & 3 & -1 & 0 & 14 & 0 \\ 6 & -13 & 2 & 14 & 0 & 0 \end{array}\right]$$

$$\frac{1}{4}R_1 \to R_1 \left[\begin{array}{ccc|ccc} 1 & -1 & -\frac{1}{4} & 0 & 0 & \frac{7}{2} \\ 4 & 3 & -1 & 0 & 14 & 0 \\ 6 & -13 & 2 & 14 & 0 & 0 \end{array}\right]$$

$$\begin{array}{l} -4R_1 + R_2 \to R_2 \\ -6R_1 + R_3 \to R_3 \end{array} \left[\begin{array}{ccc|ccc} 1 & -1 & -\frac{1}{4} & 0 & 0 & \frac{7}{2} \\ 0 & 7 & 0 & 0 & 14 & -14 \\ 0 & -7 & \frac{7}{2} & 14 & 0 & -21 \end{array}\right]$$

$$\frac{1}{7}R_2 \to R_2 \left[\begin{array}{ccc|ccc} 1 & -1 & -\frac{1}{4} & 0 & 0 & \frac{7}{2} \\ 0 & 1 & 0 & 0 & 2 & -2 \\ 0 & -7 & \frac{7}{2} & 14 & 0 & -21 \end{array}\right]$$

$$\begin{array}{l} R_2 + R_1 \to R_1 \\ 7R_2 + R_3 \to R_3 \end{array} \left[\begin{array}{ccc|ccc} 1 & 0 & -\frac{1}{4} & 0 & 2 & \frac{3}{2} \\ 0 & 1 & 0 & 0 & 2 & -2 \\ 0 & 0 & \frac{7}{2} & 14 & 14 & -35 \end{array}\right]$$

$$\frac{2}{7}R_3 \to R_3 \left[\begin{array}{ccc|ccc} 1 & 0 & -\frac{1}{4} & 0 & 2 & \frac{3}{2} \\ 0 & 1 & 0 & 0 & 2 & -2 \\ 0 & 0 & 1 & 4 & 4 & -10 \end{array}\right]$$

$$\frac{1}{4}R_3 + R_1 \to R_1 \left[\begin{array}{ccc|ccc} 1 & 0 & -\frac{1}{4} & 1 & 3 & -1 \\ 0 & 1 & 0 & 0 & 2 & -2 \\ 0 & 0 & 1 & 4 & 4 & -10 \end{array}\right]$$

The inverse is the matrix on the right side.

831. $\begin{bmatrix} \mathbf{1} & \mathbf{1} \\ \mathbf{-1} & \mathbf{-3} \end{bmatrix}$

Matrix "division" is performed by multiplying the numerator (dividend) by the inverse of the denominator (divisor).

First, find the inverse of the divisor.

$$\begin{bmatrix} 3 & 5 \\ -2 & -3 \end{bmatrix}^{-1} = \frac{1}{-9-(-10)}\begin{bmatrix} -3 & -5 \\ 2 & 3 \end{bmatrix} = \begin{bmatrix} -3 & -5 \\ 2 & 3 \end{bmatrix}$$

Now, multiply:

$$\begin{bmatrix} 1 & 2 \\ 3 & 4 \end{bmatrix}\begin{bmatrix} -3 & -5 \\ 2 & 3 \end{bmatrix}$$

$$= \begin{bmatrix} -3+4 & -5+6 \\ -9+8 & -15+12 \end{bmatrix}$$

$$= \begin{bmatrix} 1 & 1 \\ -1 & -3 \end{bmatrix}$$

832. $\begin{bmatrix} 5 & 8 \\ -26 & -42 \end{bmatrix}$

Matrix "division" is performed by multiplying the numerator (dividend) by the inverse of the denominator (divisor).

First, find the inverse of the divisor.

$$\begin{bmatrix} 5 & 8 \\ -3 & -5 \end{bmatrix}^{-1} = \frac{1}{-25-(-24)}\begin{bmatrix} -5 & -8 \\ 3 & 5 \end{bmatrix} = \begin{bmatrix} 5 & 8 \\ -3 & -5 \end{bmatrix}$$

Now, multiply:

$$\begin{bmatrix} 1 & 0 \\ -4 & 2 \end{bmatrix}\begin{bmatrix} 5 & 8 \\ -3 & -5 \end{bmatrix}$$

$$= \begin{bmatrix} 5+0 & 8+0 \\ -20-6 & -32-10 \end{bmatrix}$$

$$= \begin{bmatrix} 5 & 8 \\ -26 & -42 \end{bmatrix}$$

833. $\begin{bmatrix} 2 & 0 \\ 0 & 2 \end{bmatrix}$

Matrix "division" is performed by multiplying the numerator (dividend) by the inverse of the denominator (divisor).

First, find the inverse of the divisor.

$$\begin{bmatrix} 4 & 5 \\ 6 & 8 \end{bmatrix}^{-1} = \frac{1}{32-30}\begin{bmatrix} 8 & -5 \\ -6 & 4 \end{bmatrix} = \begin{bmatrix} 4 & -5/2 \\ -3 & 2 \end{bmatrix}$$

Answers 801–900

Now, multiply:

$$\begin{bmatrix} 8 & 10 \\ 12 & 16 \end{bmatrix}\begin{bmatrix} 4 & -5/2 \\ -3 & 2 \end{bmatrix}$$

$$= \begin{bmatrix} 32-30 & -20+20 \\ 48-(-48) & -30+32 \end{bmatrix}$$

$$= \begin{bmatrix} 2 & 0 \\ 0 & 2 \end{bmatrix}$$

Did you notice that the matrix in the numerator was a scalar multiple of the matrix in the denominator?

834. $\begin{bmatrix} \mathbf{-102} & \mathbf{62} & \mathbf{22} \\ \mathbf{-80} & \mathbf{49} & \mathbf{17} \\ \mathbf{-38} & \mathbf{22} & \mathbf{8} \end{bmatrix}$

Matrix "division" is performed by multiplying the numerator (dividend) by the inverse of the denominator (divisor).

First, find the inverse of the divisor.

$$\begin{bmatrix} 1 & 1 & -1 \\ 1 & 2 & 3 \\ 2 & -1 & -13 \end{bmatrix}^{-1} = \begin{bmatrix} -23 & 14 & 5 \\ 19 & -11 & -4 \\ -5 & 3 & 1 \end{bmatrix}$$

The inverse is obtained through row operations performed on the expanded matrix.

$$\left[\begin{array}{ccc|ccc} 1 & 1 & -1 & 1 & 0 & 0 \\ 1 & 2 & 3 & 0 & 1 & 0 \\ 2 & -1 & -13 & 0 & 0 & 1 \end{array}\right]$$

$$\begin{array}{l} -1R_1 + R_2 \to R_2 \\ -2R_1 + R_3 \to R_3 \end{array} \left[\begin{array}{ccc|ccc} 1 & 1 & -1 & 1 & 0 & 0 \\ 0 & 1 & 4 & -1 & 1 & 0 \\ 0 & -3 & -11 & -2 & 0 & 1 \end{array}\right]$$

$$\begin{array}{l} -1R_2 + R_1 \to R_1 \\ 3R_2 + R_3 \to R_3 \end{array} \left[\begin{array}{ccc|ccc} 1 & 0 & -5 & 2 & -1 & 0 \\ 0 & 1 & 4 & -1 & 1 & 0 \\ 0 & 0 & 1 & -5 & 3 & 1 \end{array}\right]$$

$$\begin{array}{l} 5R_3 + R_1 \to R_1 \\ -4R_3 + R_2 \to R_2 \end{array} \left[\begin{array}{ccc|ccc} 1 & 0 & 0 & -23 & 14 & 5 \\ 0 & 1 & 0 & 19 & -11 & -4 \\ 0 & 0 & 1 & -5 & 3 & 1 \end{array}\right]$$

Now, multiply:

$$\begin{bmatrix} 4 & 0 & 2 \\ 3 & 1 & 6 \\ 0 & -2 & 0 \end{bmatrix}\begin{bmatrix} -23 & 14 & 5 \\ 19 & -11 & -4 \\ -5 & 3 & 1 \end{bmatrix}$$

$$= \begin{bmatrix} -92+0-10 & 56+0+6 & 20+0+2 \\ -69+19-30 & 42-11+18 & 15-4+6 \\ 0-38+0 & 0+22+0 & 0+8+0 \end{bmatrix}$$

$$= \begin{bmatrix} -102 & 62 & 22 \\ -80 & 49 & 17 \\ -38 & 22 & 8 \end{bmatrix}$$

835. $\begin{bmatrix} \mathbf{14} & -\mathbf{31/2} & -\mathbf{11/2} \\ \mathbf{16} & \mathbf{-20} & \mathbf{-8} \\ \mathbf{-5} & \mathbf{7} & \mathbf{3} \end{bmatrix}$

Matrix "division" is performed by multiplying the numerator (dividend) by the inverse of the denominator (divisor).

First, find the inverse of the divisor.

$$\begin{bmatrix} 1 & 0 & 3 \\ 0 & 1 & 4 \\ 2 & -3 & -4 \end{bmatrix}^{-1} = \begin{bmatrix} 4 & -9/2 & -3/2 \\ 4 & -5 & -2 \\ -1 & 3/2 & 1/2 \end{bmatrix}$$

The inverse is obtained through row operations performed on the expanded matrix.

$$\left[\begin{array}{ccc|ccc} 1 & 0 & 3 & 1 & 0 & 0 \\ 0 & 1 & 4 & 0 & 1 & 0 \\ 2 & -3 & -4 & 0 & 0 & 1 \end{array}\right]$$

$$-2R_1+R_3 \to R_3 \left[\begin{array}{ccc|ccc} 1 & 0 & 3 & 1 & 0 & 0 \\ 0 & 1 & 4 & 0 & 1 & 0 \\ 0 & -3 & -10 & -2 & 0 & 1 \end{array}\right]$$

$$3R_2+R_3 \to R_3 \left[\begin{array}{ccc|ccc} 1 & 0 & 3 & 1 & 0 & 0 \\ 0 & 1 & 4 & 0 & 1 & 0 \\ 0 & 0 & 2 & -2 & 3 & 1 \end{array}\right]$$

$$\frac{1}{2}R_3 \to R_3 \left[\begin{array}{ccc|ccc} 1 & 0 & 3 & 1 & 0 & 0 \\ 0 & 1 & 4 & 0 & 1 & 0 \\ 0 & 0 & 1 & -1 & 3/2 & 1/2 \end{array}\right]$$

$$\begin{array}{l} -3R_3+R_1 \to R_1 \\ -4R_3+R_2 \to R_2 \end{array} \left[\begin{array}{ccc|ccc} 1 & 0 & 0 & 4 & -9/2 & -3/2 \\ 0 & 1 & 0 & 4 & -5 & -2 \\ 0 & 0 & 1 & -1 & 3/2 & 1/2 \end{array}\right]$$

Now, multiply:

$$\begin{bmatrix} 3 & 1 & 2 \\ 0 & 4 & 0 \\ 1 & -2 & 1 \end{bmatrix} \begin{bmatrix} 4 & -9/2 & -3/2 \\ 4 & -5 & -2 \\ -1 & 3/2 & 1/2 \end{bmatrix}$$

$$= \begin{bmatrix} 12+4-2 & -27/2-5+3 & -9/2-2+1 \\ 0+16+0 & 0-20+0 & 0-8+0 \\ 4-8-1 & -9/2+10+3/2 & -3/2+4+1/2 \end{bmatrix}$$

$$= \begin{bmatrix} 14 & -31/2 & -11/2 \\ 16 & -20 & -8 \\ -5 & 7 & 3 \end{bmatrix}$$

836. $x = 1, y = -2$

Write the coefficient matrix and the constant matrix.

$$A = \begin{bmatrix} 1 & -2 \\ 3 & -5 \end{bmatrix}, C = \begin{bmatrix} 5 \\ 13 \end{bmatrix}$$

Multiply the inverse of the coefficient matrix times the constant matrix.

$$A^{-1} = \begin{bmatrix} 1 & -2 \\ 3 & -5 \end{bmatrix}^{-1} = \begin{bmatrix} -5 & 2 \\ -3 & 1 \end{bmatrix}$$

$$A^{-1} * C = \begin{bmatrix} -5 & 2 \\ -3 & 1 \end{bmatrix} \begin{bmatrix} 5 \\ 13 \end{bmatrix}$$

$$= \begin{bmatrix} -25+26 \\ -15+13 \end{bmatrix} = \begin{bmatrix} 1 \\ -2 \end{bmatrix}$$

The solution is $x = 1, y = -2$.

837. $x = -2, y = 3$

Write the coefficient matrix and the constant matrix.

$$A = \begin{bmatrix} 2 & 3 \\ 5 & 7 \end{bmatrix}, C = \begin{bmatrix} 5 \\ 11 \end{bmatrix}$$

Multiply the inverse of the coefficient matrix times the constant matrix.

$$A^{-1} = \begin{bmatrix} 2 & 3 \\ 5 & 7 \end{bmatrix}^{-1} = \begin{bmatrix} -7 & 3 \\ 5 & -2 \end{bmatrix}$$

$$A^{-1} * C = \begin{bmatrix} -7 & 3 \\ 5 & -2 \end{bmatrix} \begin{bmatrix} 5 \\ 11 \end{bmatrix}$$

$$= \begin{bmatrix} -35+33 \\ 25-22 \end{bmatrix} = \begin{bmatrix} -2 \\ 3 \end{bmatrix}$$

The solution is $x = -2, y = 3$.

838. $x = 3, y = -1$

Write the coefficient matrix and the constant matrix.

$$A = \begin{bmatrix} 4 & 7 \\ 3 & 5 \end{bmatrix}, C = \begin{bmatrix} 5 \\ 4 \end{bmatrix}$$

Multiply the inverse of the coefficient matrix times the constant matrix.

$$A^{-1} = \begin{bmatrix} 4 & 7 \\ 3 & 5 \end{bmatrix}^{-1} = \begin{bmatrix} -5 & 7 \\ 3 & -4 \end{bmatrix}$$

$$A^{-1} * C = \begin{bmatrix} -5 & 7 \\ 3 & -4 \end{bmatrix} \begin{bmatrix} 5 \\ 4 \end{bmatrix}$$

$$= \begin{bmatrix} -25+28 \\ 15-16 \end{bmatrix} = \begin{bmatrix} 3 \\ -1 \end{bmatrix}$$

The solution is $x = 3$, $y = -1$.

839. $x = 4, y = 2, z = -1$

Write the coefficient matrix and the constant matrix.

$$A = \begin{bmatrix} 1 & 1 & 3 \\ 1 & 2 & 0 \\ 2 & 3 & 2 \end{bmatrix}, C = \begin{bmatrix} 3 \\ 8 \\ 12 \end{bmatrix}$$

Multiply the inverse of the coefficient matrix times the constant matrix.

$$A^{-1} = \begin{bmatrix} 1 & 1 & 3 \\ 1 & 2 & 0 \\ 2 & 3 & 2 \end{bmatrix}^{-1} = \begin{bmatrix} -4 & -7 & 6 \\ 2 & 4 & -3 \\ 1 & 1 & -1 \end{bmatrix}$$

$$A^{-1} * C = \begin{bmatrix} -4 & -7 & 6 \\ 2 & 4 & -3 \\ 1 & 1 & -1 \end{bmatrix} \begin{bmatrix} 3 \\ 8 \\ 12 \end{bmatrix}$$

$$= \begin{bmatrix} -12-56+72 \\ 6+32-36 \\ 3+8-12 \end{bmatrix} = \begin{bmatrix} 4 \\ 2 \\ -1 \end{bmatrix}$$

The solution is $x = 4$, $y = 2$, $z = -1$.

840. $x = 3, y = -2, z = 4$

Write the coefficient matrix and the constant matrix.

$$A = \begin{bmatrix} 1 & 2 & 1 \\ 1 & -3 & 2 \\ 1 & 1 & 2 \end{bmatrix}, C = \begin{bmatrix} 3 \\ 17 \\ 9 \end{bmatrix}$$

Multiply the inverse of the coefficient matrix times the constant matrix.

$$A^{-1} = \begin{bmatrix} 1 & 2 & 1 \\ 1 & -3 & 2 \\ 1 & 1 & 2 \end{bmatrix}^{-1} = \begin{bmatrix} 2 & 3/4 & -7/4 \\ 0 & -1/4 & 1/4 \\ -1 & -1/4 & 5/4 \end{bmatrix}$$

$$A^{-1} * C = \begin{bmatrix} 2 & 3/4 & -7/4 \\ 0 & -1/4 & 1/4 \\ -1 & -1/4 & 5/4 \end{bmatrix} \begin{bmatrix} 3 \\ 17 \\ 9 \end{bmatrix}$$

$$= \begin{bmatrix} 6 + 51/4 - 63/4 \\ 0 - 17/4 + 9/4 \\ -3 - 17/4 + 45/4 \end{bmatrix} = \begin{bmatrix} 3 \\ -2 \\ 4 \end{bmatrix}$$

The solution is $x = 3$, $y = -2$, $z = 4$.

841. **3, 5, 7, 9, 11**

$$a_1 = 2(1) + 1 = 2 + 1 = 3$$
$$a_2 = 2(2) + 1 = 4 + 1 = 5$$
$$a_3 = 2(3) + 1 = 6 + 1 = 7$$
$$a_4 = 2(4) + 1 = 8 + 1 = 9$$
$$a_5 = 2(5) + 1 = 10 + 1 = 11$$

This is one way to describe the odd integers.

842. **1, 3, 7, 13, 21**

$$b_1 = 1^2 - 1 + 1 = 1$$
$$b_2 = 2^2 - 2 + 1 = 3$$
$$b_3 = 3^2 - 3 + 1 = 7$$
$$b_4 = 4^2 - 4 + 1 = 13$$
$$b_5 = 5^2 - 5 + 1 = 21$$

Notice that the difference between the terms is increasingly larger even numbers.

843. $\frac{1}{2}, 1, \frac{5}{4}, \frac{7}{5}, \frac{3}{2}$

$$c_1 = \frac{2(1)-1}{1+1} = \frac{2-1}{2} = \frac{1}{2}$$
$$c_2 = \frac{2(2)-1}{2+1} = \frac{4-1}{3} = \frac{3}{3} = 1$$
$$c_3 = \frac{2(3)-1}{3+1} = \frac{6-1}{4} = \frac{5}{4}$$
$$c_4 = \frac{2(4)-1}{4+1} = \frac{8-1}{5} = \frac{7}{5}$$
$$c_5 = \frac{2(5)-1}{5+1} = \frac{10-1}{6} = \frac{9}{6} = \frac{3}{2}$$

The terms will keep increasing in size — eventually coming very close to 2.

844. $\frac{1}{2}, \frac{1}{2}, \frac{3}{4}, \frac{3}{2}, \frac{15}{4}$

$$d_1 = \frac{1!}{2^1} = \frac{1}{2} = \frac{1}{2}$$
$$d_2 = \frac{2!}{2^2} = \frac{2}{4} = \frac{1}{2}$$
$$d_3 = \frac{3!}{2^3} = \frac{6}{8} = \frac{3}{4}$$
$$d_4 = \frac{4!}{2^4} = \frac{24}{16} = \frac{3}{2}$$
$$d_5 = \frac{5!}{2^5} = \frac{120}{32} = \frac{15}{4}$$

Notice that the numerator is growing faster than the denominator.

845. 3, 6, 12, 24, 48

$$f_1 = 3\left(2^{1-1}\right) = 3\left(2^0\right) = 3(1) = 3$$
$$f_2 = 3\left(2^{2-1}\right) = 3\left(2^1\right) = 3(2) = 6$$
$$f_3 = 3\left(2^{3-1}\right) = 3\left(2^2\right) = 3(4) = 12$$
$$f_4 = 3\left(2^{4-1}\right) = 3\left(2^3\right) = 3(8) = 24$$
$$f_5 = 3\left(2^{5-1}\right) = 3\left(2^4\right) = 3(16) = 48$$

This is a geometric series with each term 2 times the one before it.

846. $a_n = 2n$

The difference between the terms is 2. The first term is 2.

Using $a_n = a_1 + (n-1)d$, replace d with 2 and a_1 with 2.

$$a_n = 2 + (n-1)2 = 2 + 2n - 2 = 2n$$

847. $a_n = 5n - 4$

The difference between the terms is 5. The first term is 1.

Using $a_n = a_1 + (n-1)d$, replace d with 5 and a_1 with 1.

$$a_n = 1 + (n-1)5 = 1 + 5n - 5 = 5n - 4$$

848. $a_n = 3n + 2$

The difference between the terms is 3. The first term is 5.

Using $a_n = a_1 + (n-1)d$, replace d with 3 and a_1 with 5.

$$a_n = 5 + (n-1)3 = 5 + 3n - 3 = 3n + 2$$

849. $a_n = 6n - 12$

The difference between the terms is 6. The first term is –6.

Using $a_n = a_1 + (n-1)d$, replace d with 6 and a_1 with –6.

$$a_n = -6 + (n-1)6 = -6 + 6n - 6 = 6n - 12$$

850. $a_n = \frac{1}{8}n + \frac{1}{8}$

The difference between the terms is $\frac{1}{8}$. The first term is $\frac{1}{4}$.

Using $a_n = a_1 + (n-1)d$, replace d with $\frac{1}{8}$ and a_1 with $\frac{1}{4}$.

$$a_n = \frac{1}{4} + (n-1)\left(\frac{1}{8}\right) = \frac{1}{4} + \frac{1}{8}n - \frac{1}{8} = \frac{1}{8}n + \frac{1}{8}$$

851. $g_n = 2^{n-1}$

The ratio between the terms is 2, and the first term is 1.

Using $g_n = g_1\left(r^{n-1}\right)$, replace r with 2 and g_1 with 1.

$$g_n = 1\left(2^{n-1}\right) = 2^{n-1}$$

852. $g_n = 2\left(3^{n-1}\right)$

The ratio between the terms is 3, and the first term is 2.

Using $g_n = g_1\left(r^{n-1}\right)$, replace r with 3 and g_1 with 2.

$$g_n = 2\left(3^{n-1}\right)$$

853. $g_n = 2^{7-n}$

The ratio between the terms is $\frac{1}{2}$, and the first term is 64.

Using $g_n = g_1(r^{n-1})$, replace r with $\frac{1}{2}$ and g_1 with 64.

$$g_n = 64\left(\left(\frac{1}{2}\right)^{n-1}\right) = 2^6\left(\left(2^{-1}\right)^{n-1}\right)$$
$$= 2^6\left(2^{-1(n-1)}\right) = 2^6\left(2^{-n+1}\right) = 2^{6+(-n)+1}$$
$$= 2^{7-n}$$

854. $g_n = 3\left((-2)^{n-1}\right)$

The ratio between the terms is –2, and the first term is 3.

Using $g_n = g_1(r^{n-1})$, replace r with –2 and g_1 with 3.

$$g_n = 3\left((-2)^{n-1}\right)$$

855. $g_n = \left(\frac{3}{2}\right)^n$

The ratio between the terms is $\frac{3}{2}$, and the first term is $\frac{3}{2}$.

Using $g_n = g_1(r^{n-1})$, replace r with $\frac{3}{2}$ and g_1 with $\frac{3}{2}$.

$$g_n = \frac{3}{2}\left(\left(\frac{3}{2}\right)^{n-1}\right) = \left(\frac{3}{2}\right)^1\left(\frac{3}{2}\right)^{n-1}$$
$$= \left(\frac{3}{2}\right)^{1+n-1} = \left(\frac{3}{2}\right)^n$$

856. $n^2 - 1$; 35, 48, 63, 80

The difference between the terms is:

$$\begin{array}{ccccccccc} 0 & & 3 & & 8 & & 15 & & 24 \\ & \vee & & \vee & & \vee & & \vee & \\ & 3 & & 5 & & 7 & & 9 & \end{array}$$

The differences aren't the same, so this isn't an arithmetic sequence.

Because the "second differences" are the same, in this case 2, this is a sequence involving a squared term.

Using $n^2 - 1$ gives you:

$$n = 1,\ 1^2 - 1 = 0$$
$$n = 2,\ 2^2 - 1 = 3$$
$$n = 3,\ 3^2 - 1 = 8$$
$$n = 4,\ 4^2 - 1 = 15$$
$$n = 5,\ 5^2 - 1 = 24$$
$$n = 6,\ 6^2 - 1 = 35$$
$$n = 7,\ 7^2 - 1 = 48$$
$$n = 8,\ 8^2 - 1 = 63$$
$$n = 9,\ 9^2 - 1 = 80$$

So the next four terms are 35, 48, 63, 80.

857. **$2^n - 1$; 63, 127, 255, 511**

There is no common difference or ratio, so the sequence isn't arithmetic or geometric. Look at the difference between the terms:

1		3		7		15		31
	V		V		V		V	
	2		4		8		16	

You find the differences to be consecutive powers of 2. Using a formula involving powers of 2, $2^n - 1$, you have:

$n = 1, 2^1 - 1 = 0$

$n = 2, 2^2 - 1 = 3$

$n = 3, 2^3 - 1 = 7$

$n = 4, 2^4 - 1 = 15$

$n = 5, 2^5 - 1 = 31$

$n = 6, 2^6 - 1 = 63$

$n = 7, 2^7 - 1 = 127$

$n = 8, 2^8 - 1 = 255$

$n = 9, 2^9 - 1 = 511$

So the next four terms are 63, 127, 255, 511.

858. **$n!$; 720, 5,040, 40,320, 362,880**

There is no common difference or ratio, so the sequence isn't arithmetic or geometric.

1		2		6		24		120
	V		V		V		V	
	1		4		18		96	

That pattern isn't particularly helpful, either. Instead, look at the factorizations of the terms in the sequence.

$n = 1, 1 = 1$

$n = 2, 2 = 2 \cdot 1$

$n = 3, 6 = 3 \cdot 2 \cdot 1$

$n = 4, 24 = 4 \cdot 3 \cdot 2 \cdot 1$

$n = 5, 120 = 5 \cdot 4 \cdot 3 \cdot 2 \cdot 1$

The numbers are factorials. So the next four terms are 6!, 7!, 8!, 9! = 720, 5,040, 40,320, 362,880.

859. **1, 243, 1, 729**

This sequence has two patterns in one. The even-numbered terms are all "1," and the odd-numbered terms are powers of 3.

$a_{2n-1} = 3^{(n+1)/2}$

$a_{even} = 1$

So the next four terms are:

$n = 8, a_8 = a_{4(2)} = 1$

$n = 9, a_9 = 3^{(9+1)/2} = 3^5 = 243$

$n = 10, a_{10} = a_{5(2)} = 1$

$n = 11, a_{11} = 3^{(11+1)/2} = 3^6 = 729$

860. $n^3 + 2$; **218, 345, 514, 731**

There is no common difference or ratio, so the sequence isn't arithmetic or geometric. Look at the difference between the terms:

```
3       10      29      66      127
    V       V       V       V
    7       19      37      61
        V       V       V
        12      18      24
```

Looking at the second differences, you see a common difference between them. This suggests a pattern with a perfect cube. Using $n^3 + 2$,

$n = 1, 1^3 + 2 = 3$

$n = 2, 2^3 + 2 = 10$

$n = 3, 3^3 + 2 = 29$

$n = 4, 4^3 + 2 = 66$

$n = 5, 5^3 + 2 = 127$

$n = 6, 6^3 + 2 = 218$

$n = 7, 7^3 + 2 = 345$

$n = 8, 8^3 + 2 = 514$

$n = 9, 9^3 + 2 = 731$

So the next four terms are 218, 345, 514, 731.

861. **5, –1, 4, 3, 7**

The nth term is the sum of the two previous terms.

$a_1 = 5$

$a_2 = -1$

$a_3 = 5 + (-1) = 4$

$a_4 = -1 + 4 = 3$

$a_5 = 4 + 3 = 7$

862. **1, 1, 3, 5, 11**

The *n*th term is the sum of the previous term plus twice the one before that.

$$a_1 = 1$$
$$a_2 = 1$$
$$a_3 = 1 + 2(1) = 3$$
$$a_4 = 3 + 2(1) = 5$$
$$a_5 = 5 + 2(3) = 11$$

863. **2, 3, –1, 4, –5**

The *n*th term is the difference between the two previous terms.

$$a_1 = 2$$
$$a_2 = 3$$
$$a_3 = 2 - 3 = -1$$
$$a_4 = 3 - (-1) = 4$$
$$a_5 = -1 - 4 = -5$$

864. **0, 2, –4, 16, –48**

The *n*th term is 4 times the term two previous minus twice the term directly before.

$$a_1 = 0$$
$$a_2 = 2$$
$$a_3 = 4(0) - 2(2) = -4$$
$$a_4 = 4(2) - 2(-4) = 16$$
$$a_5 = 4(-4) - 2(16) = -48$$

865. **1, 4, 3, 8, 15**

The *n*th term is the sum of the three previous terms.

$$a_1 = 1$$
$$a_2 = 4$$
$$a_3 = 3$$
$$a_4 = 1 + 4 + 3 = 8$$
$$a_5 = 4 + 3 + 8 = 15$$

866. **21, 34, 55, 89**

The *n*th term is the sum of the two previous terms. This is the famous Fibonacci sequence.

$$a_8 = 8 + 13 = 21$$
$$a_9 = 13 + 21 = 34$$
$$a_{10} = 21 + 34 = 55$$
$$a_{11} = 34 + 55 = 89$$

867. $\frac{1}{120}, \frac{1}{720}, \frac{1}{5040}, \frac{1}{40,320}$

$$a_6 = \frac{1}{(6-1)!} = \frac{1}{5!} = \frac{1}{120}$$

$$a_7 = \frac{1}{(7-1)!} = \frac{1}{6!} = \frac{1}{720}$$

$$a_8 = \frac{1}{(8-1)!} = \frac{1}{7!} = \frac{1}{5040}$$

$$a_9 = \frac{1}{(9-1)!} = \frac{1}{8!} = \frac{1}{40,320}$$

868. $\frac{1}{4}(x-1)^4, \frac{1}{5}(x-1)^5, \frac{1}{6}(x-1)^6, \frac{1}{7}(x-1)^7$

$$a_4 = \frac{1}{4}(x-1)^4$$

$$a_5 = \frac{1}{5}(x-1)^5$$

$$a_6 = \frac{1}{6}(x-1)^6$$

$$a_7 = \frac{1}{7}(x-1)^7$$

869. **28, 36, 45, 55**

$$a_7 = 21+7 = 28$$

$$a_8 = 28+8 = 36$$

$$a_9 = 36+9 = 45$$

$$a_{10} = 45+10 = 55$$

These are the Triangular numbers.

870. **312211, 13112221, 1113213211, 31131211131221**

This is the “look and say” sequence. You create the next number by reading off the previous number.

1 is read: “one 1”

11 is read: “two 1s”

21 is read: “one 2 then one 1”

1211 is read: “one 1 then one 2 then two 1s”

111221 is read: “three 1s then two 2s then one 1”

312211 is read: “one 3 then one 1 then two 2s then two 1s”

13112221 is read: “one 1 then one 3 then two 1s then three 2s then one 1”

1113213211 is read: “three 1s then one 3 then one 2 then one 1 then one 3 then one 2 then two 1s”

31131211131221 is read: “one 3 then two 1s then one 3 then one 1 then one 2 then three 1s then one 3 then one 1 then two 2s then one 1”

871. **14**

The summation asks for:

$$1^2 + 2^2 + 3^2 = 1 + 4 + 9 = 14$$

872. **24**

The summation asks for:

$$(2 \cdot 1 + 1) + (2 \cdot 2 + 1) + (2 \cdot 3 + 1) + (2 \cdot 4 + 1)$$
$$= (2 + 1) + (4 + 1) + (6 + 1) + (8 + 1)$$
$$= 3 + 5 + 7 + 9 = 24$$

873. **45**

The summation asks for:

$$3(2^0)+3(2^1)+3(2^2)+3(2^3)$$
$$=3(1)+3(2)+3(4)+3(8)$$
$$=3+6+12+24=45$$

874. $\mathbf{\frac{31}{2}}$

The summation asks for:

$$8\left(\frac{1}{2}\right)^0+8\left(\frac{1}{2}\right)^1+8\left(\frac{1}{2}\right)^2+8\left(\frac{1}{2}\right)^3+8\left(\frac{1}{2}\right)^4$$
$$=8(1)+8\left(\frac{1}{2}\right)+8\left(\frac{1}{4}\right)+8\left(\frac{1}{8}\right)+8\left(\frac{1}{16}\right)$$
$$=8+4+2+1+\frac{1}{2}=15\frac{1}{2}=\frac{31}{2}$$

875. **63**

The summation asks for:

$$2^{1-1}+2^{2-1}+2^{3-1}+2^{4-1}+2^{5-1}+2^{6-1}$$
$$=2^0+2^1+2^2+2^3+2^4+2^5$$
$$=1+2+4+8+16+32=63$$

876. **49**

The first term is 1, and the *n*th term is the seventh term, which is 13.

$$a_1=1$$
$$a_7=13$$
$$n=7$$
$$S_7=\frac{7(1+13)}{2}=\frac{7(14)}{2}=7\cdot 7=49$$

877. **36**

This is an arithmetic sequence beginning with 2(1) – 1 = 1 and then increasing by 2 for each subsequent term. The *n*th term is the sixth term, which is 2(6) – 1 = 11.

$$a_1 = 1$$

$$a_6 = 11$$

$$n = 6$$

$$S_6 = \frac{6(1+11)}{2} = \frac{6(12)}{2} = 6 \cdot 6 = 36$$

878. **5,050**

This is an arithmetic sequence of the first 100 natural numbers. The first term is 1, and the *n*th term is the hundredth term, which is 100.

$$a_1 = 1$$

$$a_{100} = 100$$

$$n = 100$$

$$S_{100} = \frac{100(1+100)}{2}$$

$$= \frac{100(101)}{2} = 50(101) = 5{,}050$$

879. **392**

This is an arithmetic sequence beginning with 3(0) + 2 = 2 and then increasing by 3 for each subsequent term. The *n*th term is the 16th term, which is 3(15) + 2 = 47. Note that the terms start with the "0" term, which is why there are 16 terms.

$$a_1 = 2$$

$$a_{16} = 47$$

$$n = 16$$

$$S_{16} = \frac{16(2+47)}{2} = \frac{16(49)}{2} = 8 \cdot 49 = 392$$

880. **400**

The first odd integer is 1, the second is 3, and the general term is $a_n = 2n - 1$. The 20th term is then $a_{20} = 2(20) - 1 = 39$. So the sum is found with:

$$a_1 = 1$$

$$a_{20} = 39$$

$$n = 20$$

$$S_{20} = \frac{20(1+39)}{2} = \frac{20(40)}{2} = 10(40) = 400$$

An alternative formula for the sum of the first *n* odd integers is $\sum_{i=1}^{n} 2i - 1 = n^2$.

881. **10,100**

The first even integer is 2, the second is 4, and the general term is $a_n = 2n$. The 100th term is then $a_{100} = 2(100) = 200$. So the sum is found with:

$$a_1 = 2$$
$$a_{100} = 200$$
$$n = 100$$
$$S_{100} = \frac{100(2+200)}{2}$$
$$= \frac{100(202)}{2} = 100(101) = 10{,}100$$

882. **136**

You need the sum of the integers from 1 through 16.

$$a_1 = 1$$
$$a_{16} = 16$$
$$n = 16$$
$$S_{16} = \frac{16(1+16)}{2} = \frac{16(17)}{2} = 8(17) = 136$$

883. **75**

This is the sum of 19, 17, 15, 13, and 11.

You can just add them up, but to create a more general solution (in case the next time you see this type of problem there are hundreds of rows), consider this to be a sequence of five numbers, starting with 19, and decreasing by 2 for each subsequent term. The general term is $a_n = 21 - 2n$.

For this particular problem,

$$a_1 = 19$$
$$a_5 = 11$$
$$n = 5$$
$$S_5 = \frac{5(19+11)}{2} = \frac{5(30)}{2} = 5(15) = 75$$

884. **2,460**

The number of trees per row is 3, 6, 9, 12, and so on for a total of 40 rows. The general term for the sequence is $a_n = 3n$. So,

$$a_1 = 3$$
$$a_{40} = 3(40) = 120$$
$$n = 40$$
$$S_{40} = \frac{40(3+120)}{2} = \frac{40(123)}{2}$$
$$= 20(123) = 2{,}460$$

885. **15,229**

The difference between the terms is 3, and the first term is 13, so the general term is:

$$a_n = a_1 + (n-1)d = 13 + (n-1)3$$
$$= 13 + 3n - 3 = 10 + 3n$$

Use the general term to determine which term the 301 is in the sequence.

$301 = 10 + 3n$

$291 = 3n, n = 97$

So the sum of the terms is found by:

$a_1 = 13$

$a_{97} = 301$

$n = 97$

$$S_{97} = \frac{97(13+301)}{2} = \frac{97(314)}{2}$$
$$= 97(157) = 15,229$$

886. **2,046**

The summation asks for:

$2^1 + 2^2 + 2^3 + \cdots + 2^{10}$

The ratio, r, is 2.

Using the formula,

$g_1 = 2^1 = 2$

$r = 2$

$n = 10$

$$S_{10} = \frac{2\left(1-2^{10}\right)}{1-2}$$
$$= \frac{2(1-1,024)}{-1} = -2(-1,023) = 2,046$$

887. **3,906**

The summation asks for:

$5^0 + 5^1 + 5^2 \cdots + 5^5$

The ratio, r, is 5.

Using the formula,

$$g_1 = 5^0 = 1$$

$$r = 5$$

$$n = 6$$

$$S_6 = \frac{1(1-5^6)}{1-5}$$

$$= \frac{1(1-15{,}625)}{-4} = \frac{-15{,}624}{-4}$$

$$= 3{,}906$$

888. **3,279**

The summation asks for:

$$3^{0+1} + 3^{1+1} + 3^{2+1} \cdots + 3^{6+1}$$

$$= 3^1 + 3^2 + 3^3 \cdots + 3^7$$

The ratio, r, is 3.

Using the formula,

$$g_1 = 3^1 = 3$$

$$r = 3$$

$$n = 7$$

$$S_7 = \frac{3(1-3^7)}{1-3}$$

$$= \frac{3(1-2{,}187)}{-2} = \frac{3(-2{,}186)}{-2}$$

$$= 3(1{,}093) = 3{,}279$$

889. **630**

The summation asks for:

$$5(2^{0+1}) + 5(2^{1+1}) + 5(2^{2+1}) \cdots + 5(2^{5+1})$$

$$= 5(2) + 5(4) + 5(8) \cdots + 5(64)$$

$$= 10 + 20 + 40 + \cdots + 320$$

The ratio, r, is 2.

Using the formula,

$$g_1 = 10$$

$$r = 2$$

$$n = 6$$

$$S_6 = \frac{10(1-2^6)}{1-2}$$

$$= \frac{10(1-64)}{-1} = -10(-63) = 630$$

890. **11,111.1**

The terms are powers of 10 with the general term:

$$g_n = 10^{n-2}$$

The ratio is 10.

Using the formula,

$$g_1 = \frac{1}{10}$$

$$r = 10$$

$$n = 6$$

$$S_6 = \frac{\frac{1}{10}\left(1-10^6\right)}{1-10} = \frac{\frac{1}{10}(1-1{,}000{,}000)}{-9}$$

$$= \frac{\frac{1}{10}(-999{,}999)}{-9} = \frac{1}{10}(111{,}111) = 11{,}111.1$$

Of course, it would have been simpler to just add the numbers from the original problem, but this shows the technique needed if there were more terms.

891. **170**

The terms are alternating powers of 2 with $g_n = (-2)^n$.

If you couldn't count the terms easily,, you could solve for the number of terms using:

$$256 = (-2)^n$$

$$2^8 = (-1)^n\left(2^n\right)$$

$$n = 8$$

There are 8 terms.

The ratio is –2.

Using the formula,

$$g_1 = -2$$

$$r = -2$$

$$n = 8$$

$$S_8 = \frac{-2\left(1-(-2)^8\right)}{1-(-2)} = \frac{-2(1-256)}{3}$$

$$= \frac{-2(-255)}{3} = -2(-85) = 170$$

892. $40\frac{40}{81}$

The terms are powers of 3 with a ratio of $\frac{1}{3}$. The general term is $g_n = 3^{4-n}$. To find the number of terms, solve:

$$\frac{1}{81} = 3^{4-n}$$
$$\frac{1}{3^4} = 3^{4-n}$$
$$3^{-4} = 3^{4-n}$$

Setting the exponents equal to one another,

$$-4 = 4 - n$$
$$-8 = -n,\ n = 8$$

Using the sum formula,

$$S_8 = \frac{27\left(1-\left(\frac{1}{3}\right)^8\right)}{1-\left(\frac{1}{3}\right)} = \frac{3^3\left(1-\frac{1}{3^8}\right)}{2\left(\frac{1}{3}\right)}$$
$$= \frac{3^3 - \frac{3^3}{3^8}}{2\left(3^{-1}\right)} = \frac{3^3 - 3^{-5}}{2\left(3^{-1}\right)}$$
$$= \frac{3^3 - 3^{-5}}{2\left(3^{-1}\right)} \cdot \frac{3^1}{3^1} = \frac{3^4 - 3^{-4}}{2}$$
$$= \frac{81 - \frac{1}{81}}{2} = \frac{80\frac{80}{81}}{2} = 40\frac{40}{81}$$

893. $47\frac{5}{8}$

The terms are powers of 2 multiplied by 3 with a ratio of 2. The general term is $g_n = 3\left(2^{n-4}\right)$. To find the number of terms, solve:

$$24 = 3\left(2^{n-4}\right)$$
$$3(8) = 3\left(2^{n-4}\right)$$
$$8 = 2^{n-4}$$
$$2^3 = 2^{n-4}$$

Setting the exponents equal to one another,

$$3 = n - 4$$
$$7 = n$$

Using the sum formula,

$$S_7 = \frac{\frac{3}{8}\left(1-2^7\right)}{1-2} = \frac{\frac{3}{8}(1-128)}{-1}$$
$$= \frac{\frac{3}{8}(-127)}{-1} = \frac{3}{8}(127) = \frac{381}{8} = 47\frac{5}{8}$$

894. **2,147,483,647**

$1 + 2 + 4 + 8 + \ldots$ is the sum of powers of 2 from 2^0 to 2^{30}.

Using the sum formula and $g_1 = 1$, $r = 2$, and $n = 31$,

$$S_{31} = \frac{1(1-2^{31})}{1-2} = \frac{1-2{,}147{,}483{,}648}{-1}$$
$$= \frac{-2{,}147{,}483{,}647}{-1} = 2{,}147{,}483{,}647$$

895. **6 hours**

You want to determine the number of hours, which translates into the number of terms needed to add up to 364.

The pieces of candy, $1 + 3 + 9 + 27 + \ldots$ are the sum of powers of 3 beginning with 3^0. The ratio is 3.

Using the sum formula,

$$364 = \frac{1(1-3^n)}{1-3} = \frac{1-3^n}{-2} = \frac{3^n-1}{2}$$

Multiply each side of the equation by 2.

$$2 \cdot 364 = 3^n - 1$$
$$728 = 3^n - 1$$
$$729 = 3^n$$

The number 729 is the 6th power of 3, so $n = 6$. It took 6 hours to eat all the candy, with the last helping being 3^5 or 243 pieces.

896. $2\frac{43}{60}$

The first six terms are:

$$a_1 = \frac{1}{(1-1)!} = \frac{1}{0!} = \frac{1}{1} = 1$$
$$a_2 = \frac{1}{(2-1)!} = \frac{1}{1!} = \frac{1}{1} = 1$$
$$a_3 = \frac{1}{(3-1)!} = \frac{1}{2!} = \frac{1}{2}$$
$$a_4 = \frac{1}{(4-1)!} = \frac{1}{3!} = \frac{1}{6}$$
$$a_5 = \frac{1}{(5-1)!} = \frac{1}{4!} = \frac{1}{24}$$
$$a_6 = \frac{1}{(6-1)!} = \frac{1}{5!} = \frac{1}{120}$$

The sum of the terms is:

$$1+1+\frac{1}{2}+\frac{1}{6}+\frac{1}{24}+\frac{1}{120}$$
$$=\frac{163}{60}=2\frac{43}{60}$$

This sum is approximately 2.7167. Keep adding more terms, and you get closer and closer to the value of *e*.

897. $7\frac{4}{15}$

$$a_1=\frac{2^{1-1}}{(1-1)!}=\frac{2^0}{0!}=\frac{1}{1}=1$$
$$a_2=\frac{2^{2-1}}{(2-1)!}=\frac{2^1}{1!}=\frac{2}{1}=2$$
$$a_3=\frac{2^{3-1}}{(3-1)!}=\frac{2^2}{2!}=\frac{4}{2}=2$$
$$a_4=\frac{2^{4-1}}{(4-1)!}=\frac{2^3}{3!}=\frac{8}{6}=\frac{4}{3}$$
$$a_5=\frac{2^{5-1}}{(5-1)!}=\frac{2^4}{4!}=\frac{16}{24}=\frac{2}{3}$$
$$a_6=\frac{2^{6-1}}{(6-1)!}=\frac{2^5}{5!}=\frac{32}{120}=\frac{4}{15}$$

The sum of the terms is:

$$1+2+2+\frac{4}{3}+\frac{2}{3}+\frac{4}{15}$$
$$=\frac{109}{15}=7\frac{4}{15}$$

The sum is approximately 7.2667. Keep adding more terms, and you get closer to the value of e^2.

898. About 2.9760

$$a_1=\frac{(-1)^{1+1}}{2(1)-1}=\frac{1}{2-1}=\frac{1}{1}=1$$
$$a_2=\frac{(-1)^{2+1}}{2(2)-1}=\frac{-1}{4-1}=-\frac{1}{3}$$
$$a_3=\frac{(-1)^{3+1}}{2(3)-1}=\frac{1}{6-1}=\frac{1}{5}$$
$$a_4=\frac{(-1)^{4+1}}{2(4)-1}=\frac{-1}{8-1}=-\frac{1}{7}$$
$$a_5=\frac{(-1)^{5+1}}{2(5)-1}=\frac{1}{10-1}=\frac{1}{9}$$
$$a_6=\frac{(-1)^{6+1}}{2(6)-1}=\frac{-1}{12-1}=-\frac{1}{11}$$

The sum of the terms is:

$$1-\frac{1}{3}+\frac{1}{5}-\frac{1}{7}+\frac{1}{9}-\frac{1}{11}$$
$$=\frac{2,578}{3,465}$$

Multiply this by 4 to get $\frac{10,312}{3,465}$.

The decimal approximation of this fraction is 2.9760. It'll take a few more terms to get close to the expected 3.14.

899. **128**

This is a geometric series with the first term of 64 and a ratio of $\frac{1}{2}$. Infinite geometric series with a ratio whose absolute value is less than 1 can be summed using the formula $S_n=\frac{a_1}{1-r}$. In this case,

$$S_n=\frac{64}{1-\frac{1}{2}}=\frac{64}{\frac{1}{2}}=128$$

900. **1,792 feet**

Write some of the terms of this series.

Down	Up
256	$\frac{3}{4}(256)=192$
192	$\frac{3}{4}(192)=144$
144	$\frac{3}{4}(144)=108$
108	$\frac{3}{4}(108)=81$
81	$\frac{3}{4}(81)=60.75$

Except for the initial drop of 256 feet, each of the other heights occur twice (up, then down). The *doubled* distances are a geometric series with an initial value of 192 and a ratio of $\frac{3}{4}$. Infinite geometric series with a ratio whose absolute value is less than 1 can be summed using the formula $S_n=\frac{a_1}{1-r}$.

This problem is solved with:

$$256+2\left(\frac{192}{1-\frac{3}{4}}\right)=256+2\left(\frac{192}{\frac{1}{4}}\right)$$
$$=256+2(768)=256+1,536=1,792$$

901. $A = \{2, 4, 6, 8, 10\}$

The name of the set is A, and the roster (list) of the elements can be written in any order between the braces.

902. $2 \in B$

The element 2 could be the only element in B, or it could be one of many elements of the set B.

903. $C = \{7, 14, 21, 28, \ldots\}$

The positive multiples of seven are all the numbers that can be written as the product of a natural number times seven.

904. $D = \{\ \}$ or $D = \varnothing$

The empty set is written with either empty braces or the symbol ∅. Do not put the symbol inside the braces — that would be a set that has something in it: the empty set!

905. $E = F$

When two sets contain exactly the same elements, they're equal. The elements can be listed in a different order in a roster; that doesn't change the fact that the sets are equal.

906. $3 \notin G$

The number three is not an element of the set G; it is not contained in that set.

907. $H \subset J$

When a subset has fewer elements than the *superset,* the two are not equal. When a subset has fewer elements than the *superset,* it's called a *proper* subset.

908. K'

This is also sometimes written $\overline{K}$. It means that no elements in set K are to be considered.

909. $M = P \cap Q$

The elements in the *intersection* of sets P and Q are all the elements shared by the two sets.

910. $M = P \cup Q$

The elements in the *union* of sets P and Q are all the elements in P plus all the elements in Q.

911. **{0, 2, 3, 4, 6, 8, 9}**

The *union* of sets B and C contains all the elements in B plus all the elements in C. You don't list a common element more than once.

912. **{3, 9}**

The *intersection* of sets B and E consists of only the elements that they both contain.

913. **{2, 3, 5, 7} = D**

Set D is a subset of A, so the intersection of the two sets is D.

914. ∅

The intersection has no elements — the two sets have nothing in common.

915. **{0, 1, 2, 3, 5, 6, 7, 9}**

The union of the three sets contains all the elements in the three sets.

916. **{3, 6}**

First find the union of sets C and D.

$$C \cup D = \{2, 3, 4, 5, 6, 7, 8\}$$

Now find the intersection of B and this union.

$$B \cap (C \cup D) = \{3, 6\}$$

917. **{2, 3, 5, 6, 7}**

First find the intersection of sets B and C.

$$(B \cap C) = \{6\}$$

Now find the union of this set and the set D.

$$(B \cap C) \cup D = \{2, 3, 5, 6, 7\}$$

918. **4**

The number of elements in set B is 4.

919. **7**

First find the union of sets B and C.

$$B \cup C = \{0, 2, 3, 4, 6, 8, 9\}$$

The number of elements in this set is 7.

920. **7**

Find the number of elements in the intersection of B and C.

$(B \cap C) = \{6\}$, so the number of elements is 1.

The number of elements in set B is 4, and the number of elements in set C is 4.

$$n(B) + n(C) - n(B \cap C) = 4 + 4 - 1 = 7$$

921. **U**

The union of all even integers and odd integers is all the integers.

922. **∅**

The intersection of even and odd integers is the empty set; they have nothing in common.

923. **C**

The multiples of 6 are a subset of the even integers.

924. **A**

Set C is a subset of A, so everything in C is also in A. Their union is set A.

925. **{multiples of 24}**

The intersection of the multiples of 6 and the multiples of 8 form a new set containing the multiples of 24. The prime factorization of 24 is $2^3 \cdot 3$. This factorization is the smallest number containing all the factors of both 6 and 8, so its multiples will also be multiples of both 6 and 8.

926. **A**

Integers that are not odd must be even.

927. **B**

Integers that are not even must be odd.

928. **∅**

All the multiples of 6 are even numbers, so no elements are shared with the set of odd numbers.

929. A

If you combine all the elements in A with the set that has no elements, you have set A.

930. ∅

Because the empty set has no elements, there are none to share with those in set A.

931. {7, 9, 11, 13}

The numbers between 1 and 14, inclusive, are 1, 2, 3, 4, 5, 6, 7, 8, 9, 10, 11, 12, 13, 14.

The factors of 30 between 1 and 14 are 1, 2, 3, 5, 6, 10.

The elements not in B are 4, 7, 8, 9, 11, 12, 13, 14.

And the elements not in B that are also in A are 7, 9, 11, 13.

The solution is the shaded area in the figure.

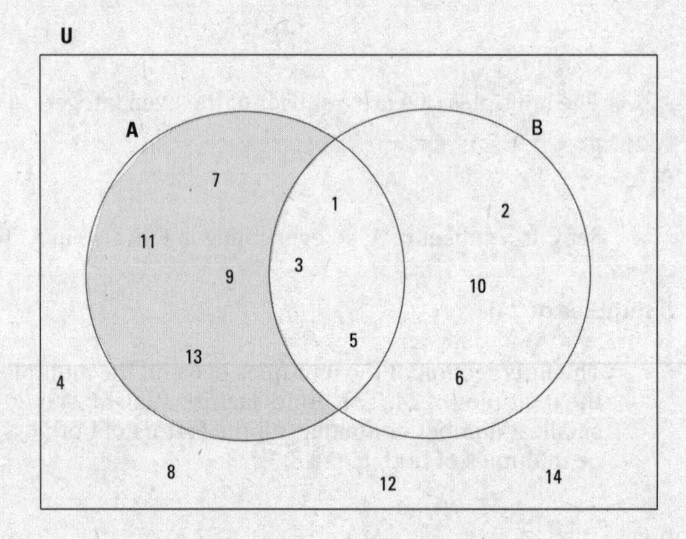

932. {c, e, f, i, o, s, t, u}

The only element shared by both A and B is the letter *a*, so the solution is everything else in U.

The solution is everything *except* the intersection, which contains the letter *a*.

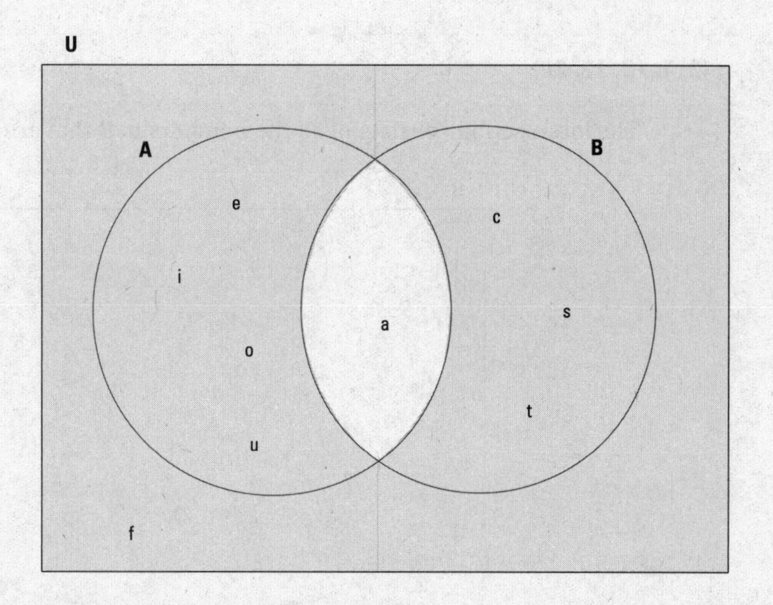

933. {1, 3, 7, 9, 11, 13, 17, 19}

The union of A and B contains all the even numbers between 1 and 19 plus the numbers 5 and 15.

So the elements *not* in the union are the odd numbers that are not multiples of 5.

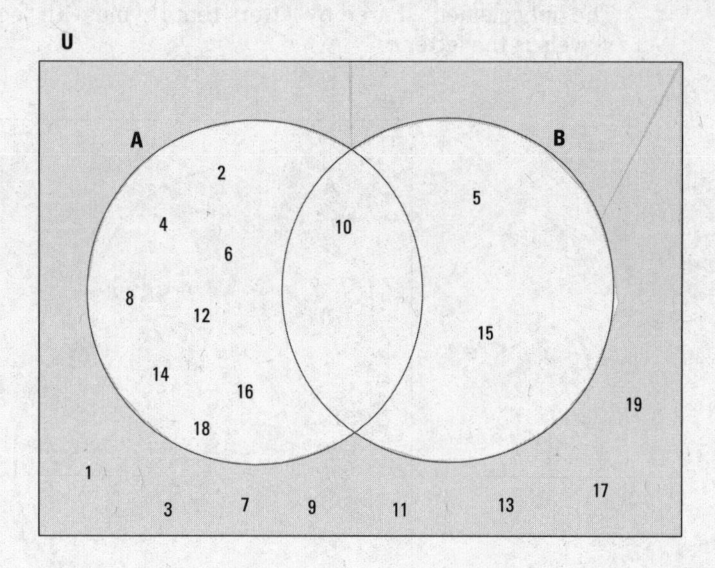

934. {2, 5, 13, 17, 29}

The intersection consists of all the numbers in B that are not in A.

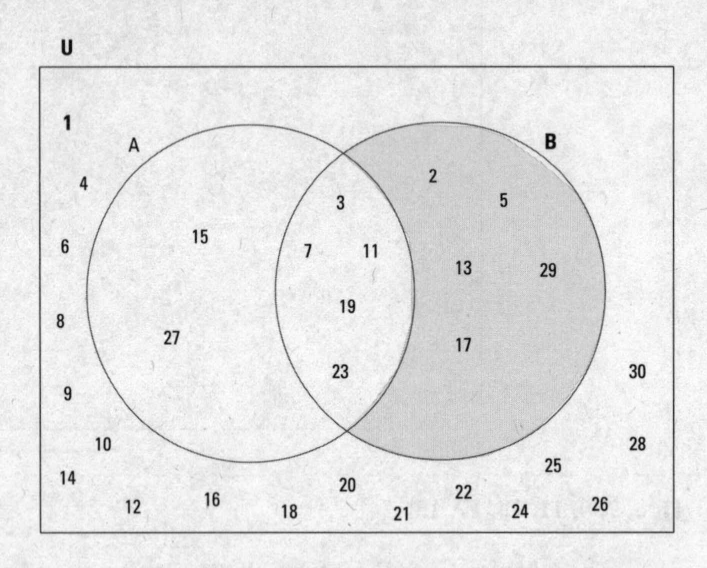

935. {e}

The only element shared by letters that rhyme with "eee", the letters in Egypt, and the vowels is the letter *e*.

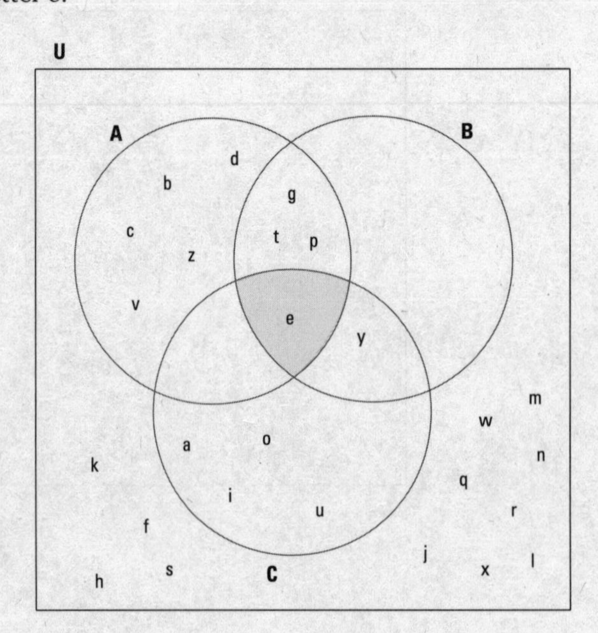

936. **{g, o, n, r}**

First, find the union of A and B. Then find what that union shares with C.

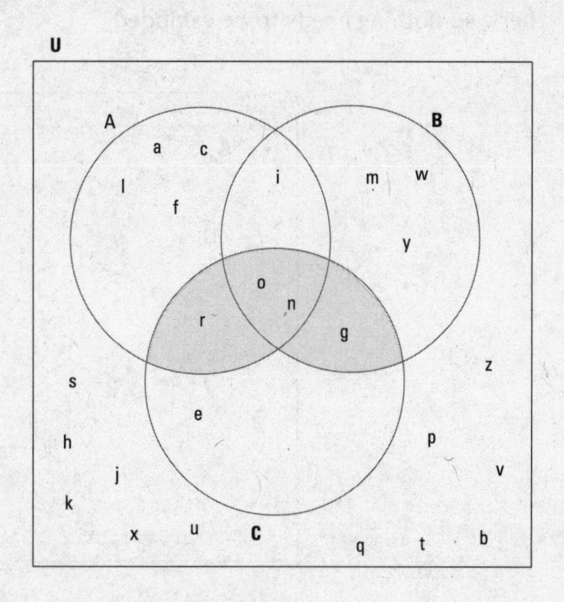

937. **{p, l, a, t, i, n, u, m, e, r}**

The intersection of A and B is {e, r, i}. Combine those with the elements in C, and the two additions are *e* and *r*.

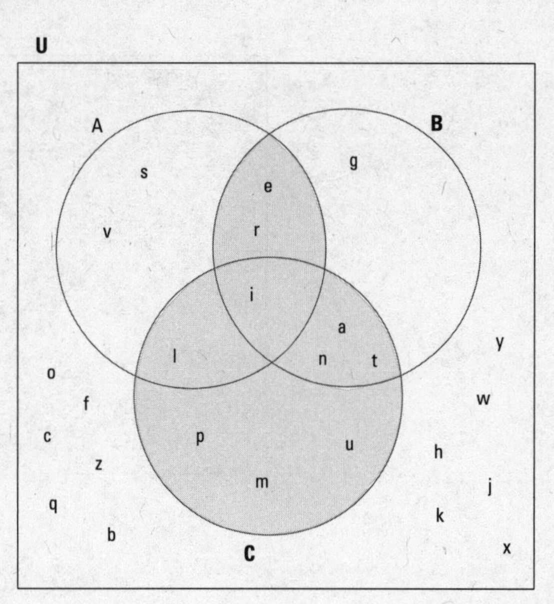

938. **{20}**

The intersection of A and B contains only the number 20. Set C only contains odd numbers, so nothing needs to be excluded.

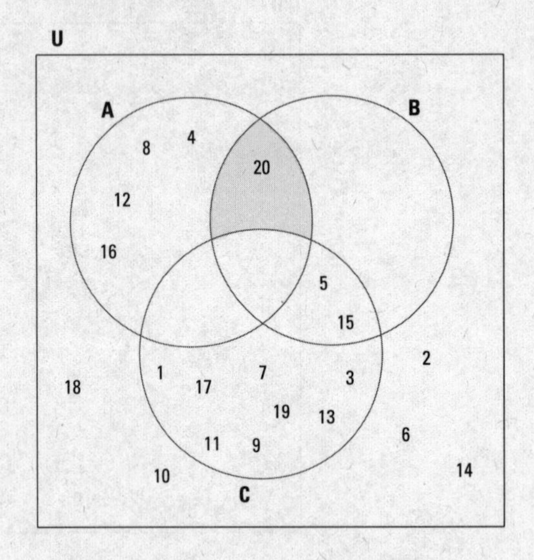

939. **{b, f, g}**

The union of the three sets is everything in all three sets. The only letters not in A, B, or C are *b*, *f* and *g*.

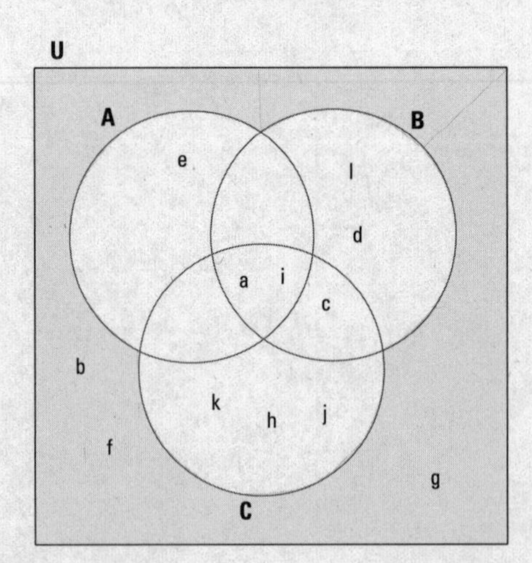

940. {21}

The union of sets A and B contains 12, 15, 24, 30, and 36. The elements *not* in that intersection are all the other elements in U. Because the only element in C is 21, its intersection with what's not in A or B is just the element 21 — or the set C.

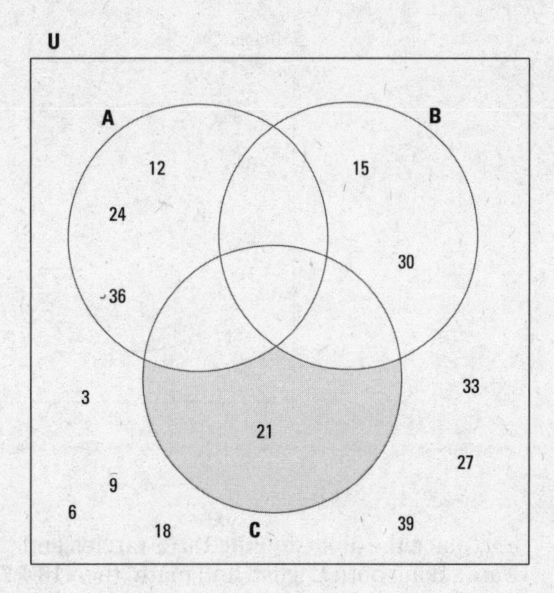

941. 15

The intersection of wood and brick contains 10 homes; that leaves 35 with only wood and 40 with only brick. The sum of the numbers in the circles is 85, leaving 15 homes with neither.

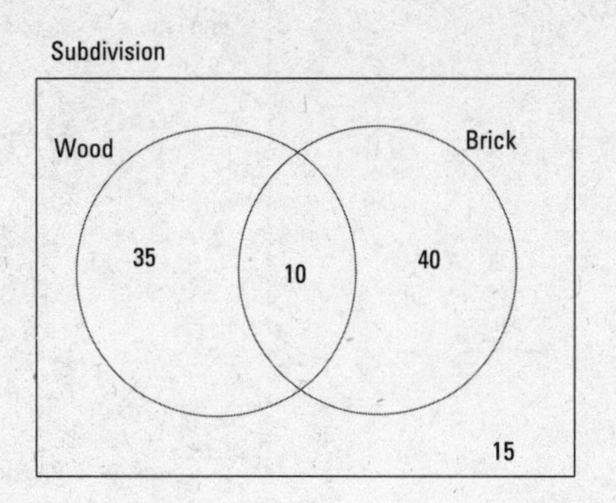

942. 63

If you add 52 + 84 + 15, the sum is 151, which is 21 more than the total number of orders. Put 21 in the intersection of the pie and ice cream circles. That leaves 52 – 21 = 31 eating pie only and 84 – 21 = 63 eating ice cream only.

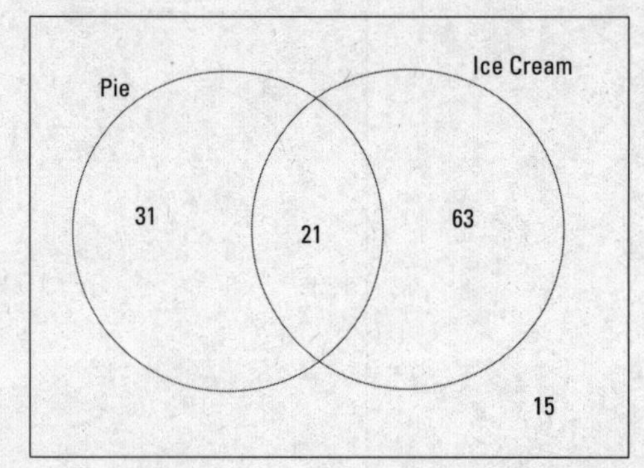

943. 4

First, place the 5 outside the three circles and the 2 in the intersection of all 3. Because 10 are taking both English and math, then 10 – 2 = 8 for the rest of the intersection. Because 7 are taking both math and physics, then 7 – 2 = 5 for the rest of the intersection. Because 3 are taking both English and physics, then 3 – 2 = 1 for the rest of the intersection. So far there are 8 + 2 + 5 = 15 in the math circle, which leaves 19 – 15 = 4 who are taking math but not English or physics.

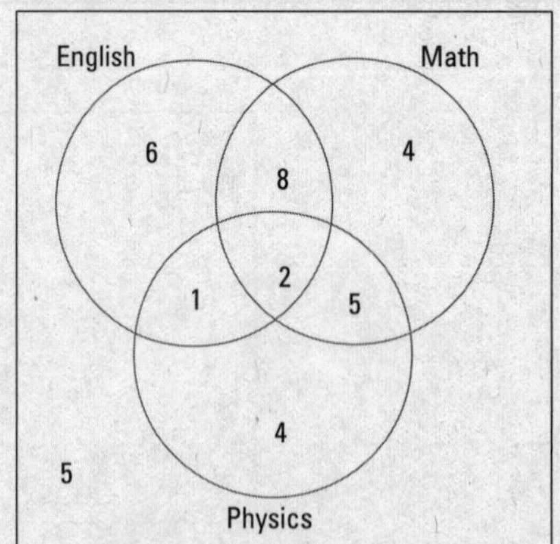

944. 6

Put 2 in the intersection of all three toppings. Then 3 – 2 = 1 had sausage and anchovies but no pepperoni, 2 – 2 = 0 had pepperoni and anchovies but no sausage, and 10 – 2 = 8 had sausage and pepperoni but no anchovies. A total of 31 had sausage, so put 20 for those with just sausage; a total of 20 had pepperoni, so put 10 for those with just pepperoni; and 6 had anchovies, so put 3 for those with just anchovies. The total in the circles is now 44, so 50 – 44 = 6 who had none of the toppings.

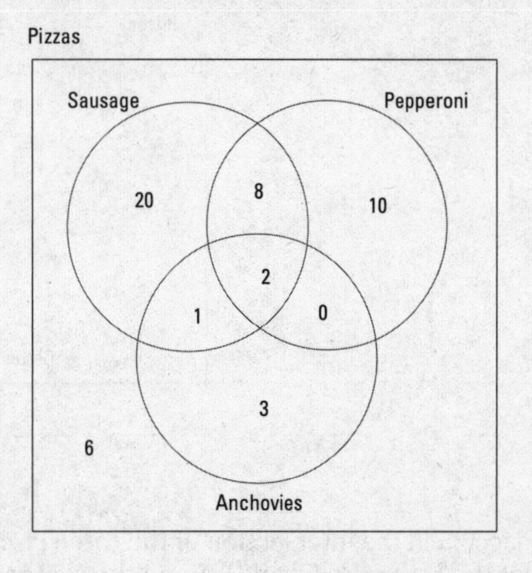

945. 20

Put 2 in the intersection of all three media. Then 12 – 2 = 10 who got it on television and the Internet but not the radio, 5 – 2 = 3 who got it on the radio and the Internet but not television, and 3 – 2 = 1 who got it on television and the radio but not the Internet. If 33 got the weather on television, then 33 – 13 = 20 who got it on television only. If 9 got the weather on the radio, then 9 – 6 = 3 who got it on the radio only. And if 56 got the weather on the Internet, then 56 – 15 = 41 who got it on the Internet only. The sum of all the numbers in the circle is 80, so that leaves 100 – 80 = 20 who found out what the weather was with none of the media given.

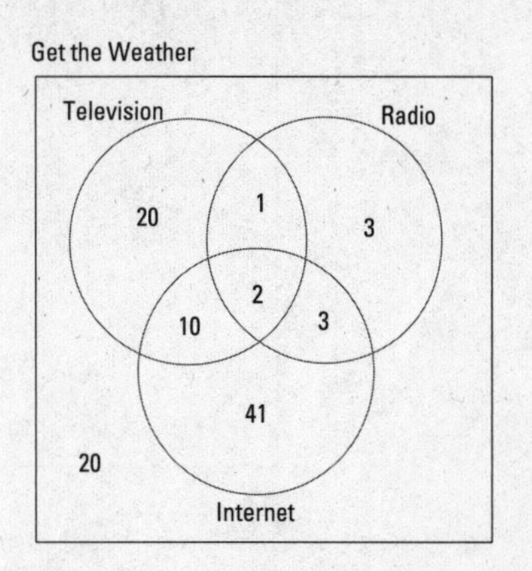

946. 5

Subtract 61 from 71 roses and mums to get 10, the number of roses and mums but not carnations. Subtract 61 from 67 roses and carnations to get roses and carnations but no mums. And subtract 61 from 66 mums and carnations to get mums and carnations but no roses. So 66 – 61 = 5.

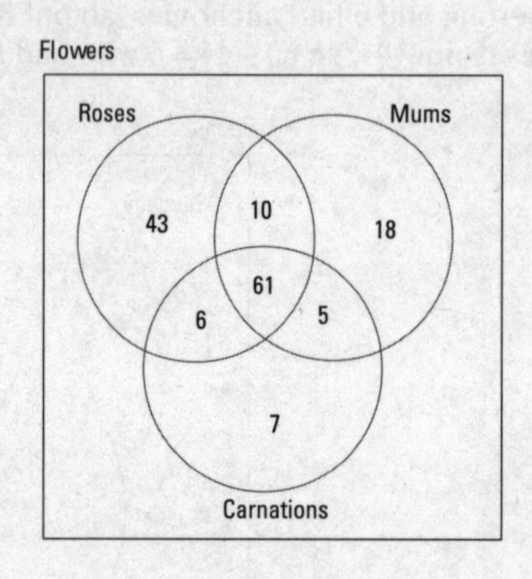

947. 25

Place 25 in the intersection of the three circles. Then 5 had GPS and 20-inch wheels but not the roof rack, 5 had GPS and the roof rack but not 20-inch wheels, and 20 had the roof rack and 20-inch wheels but not GPS. Twenty had only GPS, 10 had only the roof rack, and 10 had only the 20-inch wheels. Adding up the numbers in the circles, you get 95 cars. That means 5 cars had none of the features. So 20 + 5 = 25 had either GPS alone or none of the features.

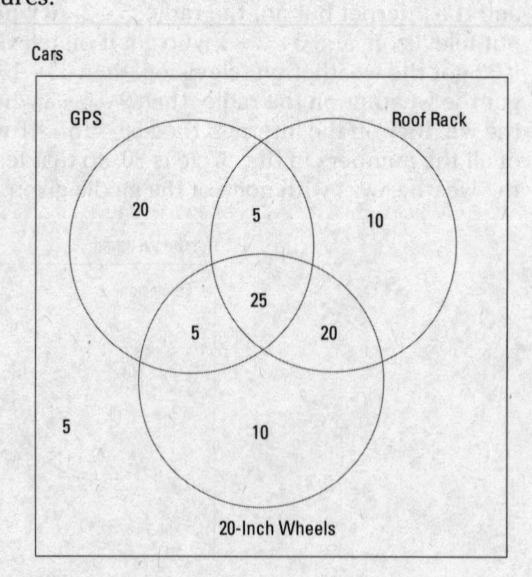

948. 47

Place 2 in the intersection of the three circles. Then 1 is on the top floor with an ocean view, 20 are a block from the beach with an ocean view, and 10 are on the top floor and a block from the beach. Forty have only an ocean view, 20 are only on the top floor, and 50 are a block from the beach but don't have the other two features. Add up all the numbers, and you have 143 units, leaving 7 to have none of these features. So 7 + 40 = 47 with none of the features or just the ocean view.

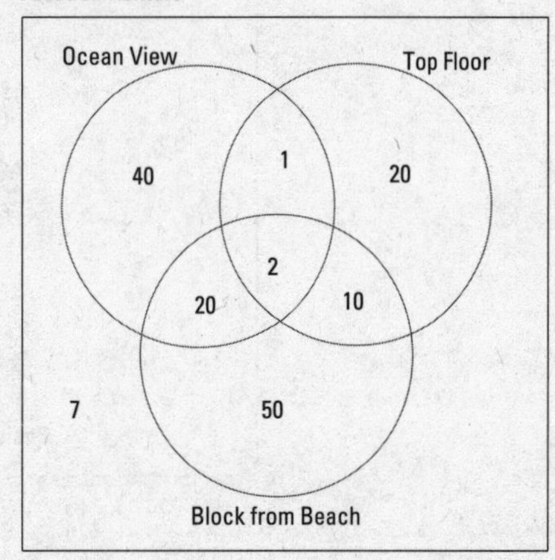

949. AB+

Blood contains either the A antigen or the B antigen (or neither) and is either positive or negative. Label the three circles as A, B, and Positive. All blood types outside the positive circle are negative. And, because O negative blood contains neither A nor B nor the positive, it is outside all the circles.

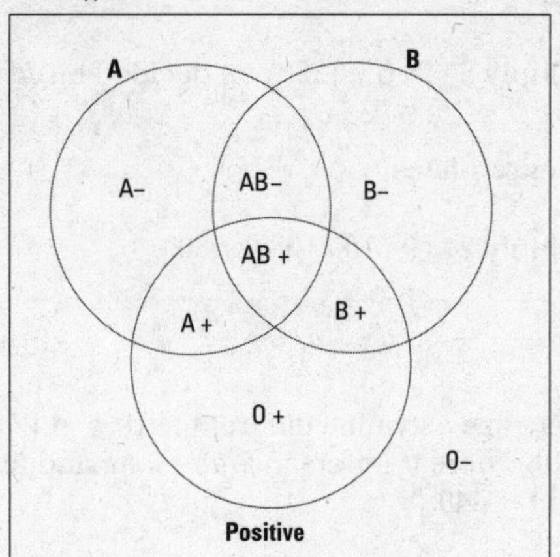

950. 4

Start with 6 in the AB+ section, and then add 1 to the AB– section to make 7 total AB type. Put 20 in the A+ section and 8 in the B+ section. Put 32 – 27 = 5 in the A– section, and put 17 – 15 = 2 in the B– section. Put 58 – 34 = 24 in the O+ section. The sum of the numbers in the circles is 66, so that leaves 70 – 66 = 4 who had O– blood.

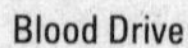

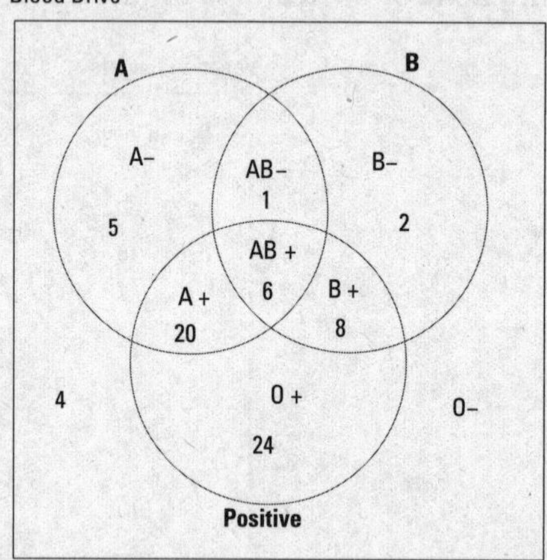

951. 1,680 outfits

Multiply 10 · 7· 6 · 4 = 1,680.

952. 60 desserts

Multiply 2 · 6 · 5 = 60.

953. 125 names

Multiply 5 · 5 · 5 = 125. You decided on *Jaz*.

954. 21,600 license plates

Multiply 24 · 9 · 10 · 10 = 21,600.

955. 1,440 codes

There are only nine different letters in *Washington*, four different letters in *Alabama*, eight different letters in *Minnesota*, and five different letters in *Illinois*. Multiplying 9 · 4 · 8 · 5 = 1,440.

956. 24

This is a permutation of four things taken four at a time:

$$P(4,4)=\frac{4!}{(4-4)!}=\frac{4!}{0!}=\frac{24}{1}=24$$

957. 5,040

This is a permutation of seven things taken seven at a time:

$$P(7,7)=\frac{7!}{(7-7)!}=\frac{7!}{0!}=\frac{5{,}040}{1}=5{,}040$$

958. 120

Since Doc has to be first and Sleepy last, that leaves just five of the Dwarfs to be arranged in the line. That's five Dwarfs taken all five at a time.

$$\frac{n!}{(n-r)!}=\frac{5!}{(5-5)!}=\frac{5!}{0!}=\frac{120}{1}=120$$

959. 24,360

There are thirty people, and three will be chosen.

$$\frac{n!}{(n-r)!}=\frac{30!}{(30-3)!}=\frac{30!}{27!}$$
$$=\frac{30\cdot 29\cdot 28\cdot 27!}{27!}=\frac{30\cdot 29\cdot 28\cdot \cancel{27!}}{\cancel{27!}}$$
$$=30\cdot 29\cdot 28=24{,}360$$

960. 3,024

With nine digits available, he will need four of them.

$$\frac{n!}{(n-r)!}=\frac{9!}{(9-4)!}=\frac{9!}{5!}$$
$$=\frac{9\cdot 8\cdot 7\cdot 6\cdot 5!}{5!}=\frac{9\cdot 8\cdot 7\cdot 6\cdot \cancel{5!}}{\cancel{5!}}$$
$$=9\cdot 8\cdot 7\cdot 6=3{,}024$$

961. 2,520

With seven tiles to choose from, he'll need five of them.

$$\frac{n!}{(n-r)!}=\frac{7!}{(7-5)!}=\frac{7!}{2!}$$
$$=\frac{7\cdot 6\cdot 5\cdot 4\cdot 3\cdot 2!}{2!}=\frac{7\cdot 6\cdot 5\cdot 4\cdot 3\cdot \cancel{2!}}{\cancel{2!}}$$
$$=7\cdot 6\cdot 5\cdot 4\cdot 3=2{,}520$$

962. **19,958,400**

With twelve products to choose from, she'll put up eight of them.

$$\frac{n!}{(n-r)!}=\frac{12!}{(12-8)!}=\frac{12!}{4!}$$

$$=\frac{12\cdot 11\cdot 10\cdot 9\cdot 8\cdot 7\cdot 6\cdot 5\cdot 4!}{4!}=\frac{12\cdot 11\cdot 10\cdot 9\cdot 8\cdot 7\cdot 6\cdot 5\cdot \cancel{4!}}{\cancel{4!}}$$

$$=12\cdot 11\cdot 10\cdot 9\cdot 8\cdot 7\cdot 6\cdot 5=19{,}958{,}400$$

963. **3,276,000**

This requires both permutations and the multiplication property.

First, find the number of ways to choose the two letters. Twenty-six things taken two at a time is as follows:

$$\frac{n!}{(n-r)!}=\frac{26!}{(26-2)!}=\frac{26!}{24!}$$

$$=\frac{26\cdot 25\cdot 24!}{24!}=\frac{26\cdot 25\cdot \cancel{24!}}{\cancel{24!}}$$

$$=26\cdot 25=650$$

Now find the number of ways of choosing four different digits from a possible ten.

$$\frac{n!}{(n-r)!}=\frac{10!}{(10-4)!}=\frac{10!}{6!}$$

$$=\frac{10\cdot 9\cdot 8\cdot 7\cdot 6!}{6!}=\frac{10\cdot 9\cdot 8\cdot 7\cdot \cancel{6!}}{\cancel{6!}}$$

$$=10\cdot 9\cdot 8\cdot 7=5{,}040$$

Finally, multiply the number of ways of choosing the two letters times the number of ways of choosing the four digits.

$$650\cdot 5{,}040=3{,}276{,}000$$

964. **2,903,040**

This requires the multiplication property and a permutation.

First, find the number of ways to choose the two letters. There are 24 to choose from if you eliminate the *O* and the *I*. Because the letter can be repeated, there are 24 choices for the first letter and 24 choices for the second letter. Using the multiplication property, the product is $24\cdot 24=576$.

Now find the number of ways of choosing four different digits from a possible ten.

$$\frac{n!}{(n-r)!}=\frac{10!}{(10-4)!}=\frac{10!}{6!}$$

$$=\frac{10\cdot 9\cdot 8\cdot 7\cdot 6!}{6!}=\frac{10\cdot 9\cdot 8\cdot 7\cdot \cancel{6!}}{\cancel{6!}}$$

$$=10\cdot 9\cdot 8\cdot 7=5{,}040$$

Finally, multiply the number of ways of choosing the two letters times the number of ways of choosing the four digits.

$$576\cdot 5{,}040=2{,}903{,}040$$

965. **116,280**

The number of ways of choosing four different countries from a possible twenty is:

$$\frac{n!}{(n-r)!}=\frac{20!}{(20-4)!}=\frac{20!}{16!}$$
$$=\frac{20\cdot 19\cdot 18\cdot 17\cdot 16!}{16!}=\frac{20\cdot 19\cdot 18\cdot 17\cdot \cancel{16!}}{\cancel{16!}}$$
$$=20\cdot 19\cdot 18\cdot 17=116{,}280$$

966. **210**

The number of ways of choosing six people from a possible ten is:

$$\frac{n!}{(n-r)!r!}=\frac{10!}{(10-6)!6!}=\frac{10!}{4!6!}$$
$$=\frac{10\cdot 9\cdot 8\cdot 7\cdot 6!}{4!6!}=\frac{10\cdot 9\cdot 8\cdot 7\cdot \cancel{6!}}{4!\cancel{6!}}$$
$$=\frac{10\cdot 9\cdot 8\cdot 7}{4\cdot 3\cdot 2\cdot 1}=\frac{10\cdot \cancel{9}^{3}\cdot \cancel{8}\cdot 7}{\cancel{4}\cdot \cancel{3}\cdot \cancel{2}\cdot 1}=10\cdot 3\cdot 7=210$$

967. **1,140**

The number of ways of choosing three scouts from twenty is:

$$\frac{n!}{(n-r)!r!}=\frac{20!}{(20-3)!3!}=\frac{20!}{17!3!}$$
$$=\frac{20\cdot 19\cdot 18\cdot 17!}{17!3!}=\frac{20\cdot 19\cdot 18\cdot \cancel{17!}}{\cancel{17!}3!}$$
$$=\frac{20\cdot 19\cdot 18}{3\cdot 2\cdot 1}=\frac{20\cdot 19\cdot \cancel{18}^{3}}{\cancel{3}\cdot \cancel{2}\cdot 1}=20\cdot 19\cdot 3=1{,}140$$

968. **658,008**

The number of ways of choosing five numbers from forty is:

$$\frac{n!}{(n-r)!r!}=\frac{40!}{(40-5)!5!}=\frac{40!}{35!5!}$$
$$=\frac{40\cdot 39\cdot 38\cdot 37\cdot 36\cdot 35!}{35!5!}$$
$$=\frac{40\cdot 39\cdot 38\cdot 37\cdot 36\cdot \cancel{35!}}{\cancel{35!}5!}$$
$$=\frac{40\cdot 39\cdot 38\cdot 37\cdot 36}{5\cdot 4\cdot 3\cdot 2\cdot 1}$$
$$=\frac{\cancel{40}\cdot \cancel{39}^{13}\cdot 38\cdot 37\cdot 36}{\cancel{5}\cdot \cancel{4}\cdot \cancel{3}\cdot \cancel{2}\cdot 1}$$
$$=13\cdot 38\cdot 37\cdot 36=658{,}008$$

If you could buy one ticket in a second, it would take you over seven and a half days to buy that many tickets.

969. 2,703,448,440

This requires finding a combination and then using the multiplication property.

The number of ways of choosing six numbers from sixty is:

$$\frac{n!}{(n-r)!r!}=\frac{60!}{(60-6)!6!}=\frac{60!}{54!6!}$$

$$=\frac{60\cdot 59\cdot 58\cdot 57\cdot 56\cdot 55\cdot 54!}{54!6!}$$

$$=\frac{60\cdot 59\cdot 58\cdot 57\cdot 56\cdot 55\cdot \cancel{54!}}{\cancel{54!}6!}$$

$$=\frac{60\cdot 59\cdot 58\cdot 57\cdot 56\cdot 55}{6\cdot 5\cdot 4\cdot 3\cdot 2\cdot 1}$$

$$=\frac{\cancel{60}\cdot 59\cdot 58\cdot \cancel{57}^{19}\cdot \cancel{56}^{14}\cdot 55}{\cancel{6}\cdot \cancel{5}\cdot \cancel{4}\cdot \cancel{3}\cdot \cancel{2}\cdot 1}$$

$$=59\cdot 58\cdot 19\cdot 14\cdot 55=50{,}063{,}860$$

Now there are 54 numbers remaining from which to choose the "bonus" number. Multiply the number of combinations times 54.

$$50{,}063{,}860\cdot 54=2{,}703{,}448{,}440$$

That's almost 3 billion!

970. 3,150

This requires two different combinations and the multiplication property.

First, choose the "fling in air" rides:

$$\frac{n!}{(n-r)!r!}=\frac{10!}{(10-4)!4!}=\frac{10!}{6!4!}$$

$$=\frac{10\cdot 9\cdot 8\cdot 7\cdot 6!}{6!4!}=\frac{10\cdot 9\cdot 8\cdot 7\cdot \cancel{6!}}{\cancel{6!}4!}$$

$$=\frac{10\cdot 9\cdot 8\cdot 7}{4\cdot 3\cdot 2\cdot 1}=\frac{10\cdot \cancel{9}^{3}\cdot \cancel{8}\cdot 7}{\cancel{4}\cdot \cancel{3}\cdot \cancel{2}\cdot 1}=10\cdot 3\cdot 7=210$$

Next, choose the "water" rides:

$$\frac{n!}{(n-r)!r!}=\frac{6!}{(6-4)!4!}=\frac{6!}{2!4!}$$

$$=\frac{6\cdot 5\cdot 4!}{2!4!}=\frac{6\cdot 5\cdot \cancel{4!}}{2!\cancel{4!}}$$

$$=\frac{6\cdot 5}{2\cdot 1}=\frac{30}{2}=15$$

Multiply the number of "air" rides times the number of "water" rides: $210\cdot 15=3{,}150$

971. **630**

This requires two different combinations and the multiplication property.

First, choose the six books from the ten in the first group:

$$\frac{n!}{(n-r)!r!}=\frac{10!}{(10-6)!6!}=\frac{10!}{4!6!}$$

$$=\frac{10\cdot 9\cdot 8\cdot 7\cdot 6!}{4!6!}=\frac{10\cdot 9\cdot 8\cdot 7\cdot \not{6!}}{4!\not{6!}}$$

$$=\frac{10\cdot 9\cdot 8\cdot 7}{4\cdot 3\cdot 2\cdot 1}=\frac{10\cdot \not{9}^{3}\cdot \not{8}\cdot 7}{\not{4}\cdot \not{3}\cdot \not{2}\cdot 1}=10\cdot 3\cdot 7=210$$

Next, choose the last two books:

$$\frac{n!}{(n-r)!r!}=\frac{3!}{(3-2)!2!}=\frac{3!}{1!2!}$$

$$=\frac{6}{2}=3$$

Multiply the number of books from the first group times the number from the second group: $210\cdot 3=630$

972. **5,760**

This requires two different combinations and the multiplication property.

First, choosing the vase, there are eight different choices, and she'll pick one: 8

Next, choosing the flowers, there are ten different choices, and she'll pick three:

$$\frac{n!}{(n-r)!r!}=\frac{10!}{(10-3)!3!}=\frac{10!}{7!3!}$$

$$=\frac{10\cdot 9\cdot 8\cdot 7!}{7!3!}=\frac{10\cdot 9\cdot 8\cdot \not{7!}}{\not{7!}3!}$$

$$=\frac{10\cdot 9\cdot 8}{3\cdot 2\cdot 1}=\frac{\not{10}^{4}\cdot \not{9}^{3}\cdot 8}{\not{3}\cdot \not{2}\cdot 1}=4\cdot 3\cdot 8=96$$

Now, to choose the greenery, she picks two out of the four:

$$\frac{n!}{(n-r)!r!}=\frac{4!}{(4-2)!2!}=\frac{4!}{2!2!}$$

$$=\frac{\not{4}\cdot 3\cdot 2\cdot 1}{\not{2}\cdot 1\cdot \not{2}\cdot 1}=6$$

Multiply the number of vase choices times the number of flower choices times the number of greenery choices:

$8\cdot 120\cdot 6=5,760$

973. 1,800

This requires three different combinations and the multiplication property.

Meat selections are two out of four choices:

$$\frac{n!}{(n-r)!r!}=\frac{4!}{(4-2)!2!}=\frac{4!}{2!2!}$$

$$=\frac{\cancel{4}\cdot 3\cdot 2\cdot 1}{\cancel{2}\cdot 1\cdot \cancel{2}\cdot 1}=6$$

Veggie/non-meat selections are two out of five choices:

$$\frac{n!}{(n-r)!r!}=\frac{5!}{(5-2)!2!}=\frac{5!}{3!2!}$$

$$=\frac{5\cdot \cancel{4}\cdot \cancel{3}\cdot 2\cdot 1}{\cancel{3}\cdot \cancel{2}\cdot 1\cdot \cancel{2}\cdot 1}=10$$

Cheese selections are two out of five choices:

$$\frac{n!}{(n-r)!r!}=\frac{5!}{(5-2)!2!}=\frac{5!}{3!2!}$$

$$=\frac{5\cdot \cancel{4}\cdot \cancel{3}\cdot 2\cdot 1}{\cancel{3}\cdot \cancel{2}\cdot 1\cdot \cancel{2}\cdot 1}=10$$

Multiply the meat times veggie times cheese times three (crust choices), and you get $6 \cdot 10 \cdot 10 \cdot 3 = 1{,}800$ different pizzas.

974. 635,013,559,600

Being dealt 13 cards from 52 choices looks like this:

$$\frac{n!}{(n-r)!r!}=\frac{52!}{(52-13)!13!}=\frac{52!}{39!13!}$$

$$=\frac{52\cdot 51\cdot 50\cdot 49\cdot 48\cdot 47\cdot 46\cdot 45\cdot 44\cdot 43\cdot 42\cdot 41\cdot 40\cdot \cancel{39!}}{\cancel{39!}13!}$$

$$=\frac{52\cdot 51\cdot 50\cdot 49\cdot 48\cdot 47\cdot 46\cdot 45\cdot 44\cdot 43\cdot 42\cdot 41\cdot 40}{13\cdot 12\cdot 11\cdot 10\cdot 9\cdot 8\cdot 7\cdot 6\cdot 5\cdot 4\cdot 3\cdot 2\cdot 1}$$

$$=\frac{\cancel{52}\cdot \cancel{51}^{17}\cdot 50\cdot 49\cdot \cancel{48}\cdot 47\cdot 46\cdot \cancel{45}\cdot \cancel{44}^{4}\cdot 43\cdot \cancel{42}\cdot 41\cdot \cancel{40}}{\cancel{13}\cdot \cancel{12}\cdot \cancel{11}\cdot \cancel{10}\cdot \cancel{9}\cdot \cancel{8}\cdot \cancel{7}\cdot \cancel{6}\cdot \cancel{5}\cdot \cancel{4}\cdot \cancel{3}\cdot \cancel{2}\cdot 1}$$

$$=17\cdot 50\cdot 49\cdot 47\cdot 46\cdot 4\cdot 43\cdot 41=635{,}013{,}559{,}600$$

There are many ways to reduce the fraction; what's shown is just one way. Shall we count the number of ways? No, thank you.

975. 4,368

In a standard deck, there are four suits. In each suit, there are three face cards (king, queen, jack) and one ace. So, of the 52 cards, you have 16 cards that are either a face card or an ace.

The number of 5-card hands that can be created from the 16 cards are:

$$\frac{n!}{(n-r)!r!}=\frac{16!}{(16-5)!5!}=\frac{16!}{11!5!}$$
$$=\frac{16\cdot 15\cdot 14\cdot 13\cdot 12\cdot 11!}{11!5!}=\frac{16\cdot 15\cdot 14\cdot 13\cdot 12\cdot \cancel{11!}}{\cancel{11!}5!}$$
$$=\frac{16\cdot 15\cdot 14\cdot 13\cdot 12}{5\cdot 4\cdot 3\cdot 2\cdot 1}=\frac{\cancel{16}^{2}\cdot \cancel{15}\cdot 14\cdot 13\cdot 12}{\cancel{5}\cdot \cancel{4}\cdot \cancel{3}\cdot \cancel{2}\cdot 1}$$
$$=2\cdot 14\cdot 13\cdot 12=4{,}368$$

976. $x^3+3x^2y+3xy^2+y^3$

The coefficients of the terms are:

$$\binom{3}{0}\quad\binom{3}{1}\quad\binom{3}{2}\quad\binom{3}{3}$$
$$1\qquad 3\qquad 3\qquad 1$$

Using the expansion formula, first insert the coefficients.

$$1\quad 3\quad 3\quad 1$$

Next insert the decreasing powers of x.

$$1x^3\quad 3x^2\quad 3x^1\quad 1x^0$$

Now insert the increasing powers of y.

$$1x^3y^0\quad 3x^2y^1\quad 3x^1y^2\quad 1x^0y^3$$

Put in the + signs and simplify.

977. $x^4+4x^3y+6x^2y^2+4xy^3+y^4$

The coefficients of the terms are:

$$\binom{4}{0}\quad\binom{4}{1}\quad\binom{4}{2}\quad\binom{4}{3}\quad\binom{4}{4}$$
$$1\qquad 4\qquad 6\qquad 4\qquad 1$$

Using the expansion formula, first insert the coefficients.

$$1\quad 4\quad 6\quad 4\quad 1$$

Next insert the decreasing powers of x.

$$1x^4\quad 4x^3\quad 6x^2\quad 4x^1\quad 1x^0$$

Now insert the increasing powers of y.

$$1x^4y^0\quad 4x^3y^1\quad 6x^2y^2\quad 4x^1y^3\quad 1x^0y^4$$

Put in the + signs and simplify.

978. $x^5 + 5x^4y + 10x^3y^2 + 10x^2y^3 + 5xy^4 + y^5$

The coefficients of the terms are:

$$\binom{5}{0} \binom{5}{1} \binom{5}{2} \binom{5}{3} \binom{5}{4} \binom{5}{5}$$

1 5 10 10 5 1

Using the expansion formula, first insert the coefficients.

1 5 10 10 5 1

Next insert the decreasing powers of x.

$1x^5 \quad 5x^4 \quad 10x^3 \quad 10x^2 \quad 5x^1 \quad 1x^0$

Now insert the increasing powers of y.

$1x^5y^0 \quad 5x^4y^1 \quad 10x^3y^2 \quad 10x^2y^3 \quad 5x^1y^4 \quad 1x^0y^5$

Put in the + signs and simplify.

979. $x^6 + 6x^5y + 15x^4y^2 + 20x^3y^3 + 15x^2y^4 + 6xy^5 + y^6$

The coefficients of the terms are:

$$\binom{6}{0} \binom{6}{1} \binom{6}{2} \binom{6}{3} \binom{6}{4} \binom{6}{5} \binom{6}{6}$$

1 6 15 20 15 6 1

Using the expansion formula, first insert the coefficients.

1 6 15 20 15 6 1

Next insert the decreasing powers of x.

$1x^6 \quad 6x^5 \quad 15x^4 \quad 20x^3 \quad 15x^2 \quad 6x^1 \quad 1x^0$

Now insert the increasing powers of y.

$1x^6y^0 \quad 6x^5y^1 \quad 15x^4y^2 \quad 20x^3y^3 \quad 15x^2y^4 \quad 6x^1y^5 \quad 1x^0y^6$

Put in the + signs and simplify.

980. $x^7 + 7x^6y + 21x^5y^2 + 35x^4y^3 + 35x^3y^4 + 21x^2y^5 + 7xy^6 + y^7$

The coefficients of the terms are:

$$\binom{7}{0} \binom{7}{1} \binom{7}{2} \binom{7}{3} \binom{7}{4} \binom{7}{5} \binom{7}{6} \binom{7}{7}$$

Using the expansion formula, first insert the coefficients.

1 7 21 35 35 21 7 1

Next insert the decreasing powers of x.

$1x^7 \quad 7x^6 \quad 21x^5 \quad 35x^4 \quad 35x^3 \quad 21x^2 \quad 7x^1 \quad 1x^0$

Now insert the increasing powers of y.

$$1x^7y^0 \quad 7x^6y^1 \quad 21x^5y^2 \quad 35x^4y^3 \quad 35x^3y^4 \quad 21x^2y^5 \quad 7x^1y^6 \quad 1x^0y^7$$

Put in the + signs and simplify.

981. $\frac{5}{26}$

There are 26 letters in the alphabet. Divide 5 by 26.

982. $\frac{2}{5}$

The prime numbers between 1 and 20 are: 2, 3, 5, 7, 11, 13, 17, 19. There are 8 prime numbers, so divide 8 by 20.

$$\frac{8}{20} = \frac{2}{5}$$

983. $\frac{1}{5}$

There are ten multiples of 5: 5, 10, 15, 20, 25, 30, 35, 40, 45, 50. Divide 10 by 50.

$$\frac{10}{50} = \frac{1}{5}$$

984. $\frac{13}{40}$

There are ten multiples of 4: 4, 8, 12, 16, 20, 24, 28, 32, 36, 40 and six multiples of 6: 6, 12, 18, 24, 30, 36. But the numbers 12, 24, and 36 are shared by the two numbers, so the total number of multiples is 10 + 6 – 3 = 13.

Divide 13 by 40.

985. $\frac{1}{25}$

There are two states beginning with the letter *T:* Tennessee and Texas.

Divide 2 by 50.

$$\frac{2}{50} = \frac{1}{25}$$

986. **0**

There are no states beginning with the letter *B.* Divide 0 by 50, and you get 0.

987. $\frac{1}{24}$

Use a permutation to count the number of ways that you can arrange the four letters.

$${}_4P_4 = \frac{4!}{(4-4)!} = \frac{4!}{0!} = 24$$

Because there's only one way to draw them in the order *MATH,* you divide 1 by 24.

988. $\frac{1}{8}$

Use a permutation to count the number of ways that you can arrange the three letters after drawing from the four available.

$$_4P_3 = \frac{4!}{(4-3)!} = \frac{4!}{1!} = 24$$

Divide the three choices by 24.

$$\frac{3}{24} = \frac{1}{8}$$

989. $\frac{1}{870}$

First, use a permutation to determine how many different ways the offices of president and vice president can be filled; use thirty members, choosing two of them.

$$_{30}P_2 = \frac{30!}{(30-2)!} = \frac{30!}{28!} = 870$$

There's only one way to choose Adam and Betty, in that order, so you divide 1 by 870.

990. $\frac{1}{27{,}405}$

Because the order doesn't matter, use a combination to count all the different ways that four people can be chosen out of thirty.

$$_{30}C_4 = \frac{30!}{(30-4)!4!} = \frac{30!}{26!4!} = 27{,}405$$

Just one of these arrangements will contain all four people mentioned.

991. $\frac{1}{4}$

Make a list of the different possibilities for the two tosses: {HH, HT, TH, TT}. Out of the four possibilities, only one has two heads.

992. $\frac{3}{8}$

Make a list of the different possibilities for the three tosses: {HHH, HHT, HTH, HTT, THH, THT, TTH, TTT}. Out of the eight possibilities, three of the results have two heads: HHT, HTH, and THH.

Divide 3 by 8.

993. $\frac{7}{8}$

Make a list of the different possibilities for three children in a family: {BBB, BBG, BGB, BGG, GBB, GBG, GGB, GGG}. Of the eight arrangements, seven of them have at least one girl. So divide 7 by 8.

994. $\frac{5}{16}$

Make a list of the different possibilities for four children in a family: {BBBB, BBBG, BBGB, BBGG, BGBB, BGBG, BGGB, BGGG, GBBB, GBBG, GBGB, GBGG, GGBB, GGBG, GGGB, GGGG}. Of the 16 possibilities, the ones with more boys than girls are BBBB, BBBG, BBGB, BGBB, and GBBB. So divide 5 by 16.

(It's often helpful to make this list either with a chart or a tree — to be sure you get all the different arrangements.)

995. $\frac{2}{15}$

The first drawing has an effect on the second drawing.

The probability of drawing the first red marble is $\frac{4}{10}$, because four of the ten marbles are red.

If a red marble is drawn, then there are only three red marbles remaining. The probability of drawing a red marble now is $\frac{3}{9}$.

Multiply: $\frac{4}{10} \cdot \frac{3}{9} = \frac{2}{5} \cdot \frac{1}{3} = \frac{2}{15}$

996. $\frac{4}{25}$

The two drawings are completely independent of one another.

The probability of drawing the first red marble is $\frac{4}{10}$, because four of the ten marbles are red. The same is true of the second drawing.

So, multiplying, $\frac{4}{10} \cdot \frac{4}{10} = \frac{2}{5} \cdot \frac{2}{5} = \frac{4}{25}$.

997. $\frac{1}{26!}$

Because the order of the tiles matters, you need a permutation of 26 tiles — drawing all 26.

$$_{26}P_{26} = \frac{26!}{(26-26)!} = \frac{26!}{0!} = 26!$$

The chance of drawing the tiles in alphabetical order is 1 out of 26! Not much of a chance (about 1 in 400 septillion).

998. $\frac{1}{49!}$

Count the total number of ways of visiting the first 49 states — using a permutation.

$$_{49}P_{49} = \frac{49!}{(49-49)!} = \frac{49!}{0!} = 49!$$

The probability of visiting Wyoming last will be 1 divided by all the different ways you can visit the other 49 states.

999. $\frac{1,287}{65,780}$

If the cards are all red, they're either all hearts, all diamonds, or a mixture of hearts and diamonds. To determine the probability that they're all hearts, count the number of ways that you can have 5 out of the 13 hearts. Then count the number of ways you can have 5 red cards out of the 26 red cards. Divide the first number by the second. Use combinations, because the order doesn't matter.

Number of ways to get 5 hearts out of 13:

$$\frac{n!}{(n-r)!r!}=\frac{13!}{(13-5)!5!}=\frac{13!}{8!5!}=1,287$$

Number of ways to get 5 red cards out of 26:

$$\frac{n!}{(n-r)!r!}=\frac{26!}{(26-5)!5!}=\frac{26!}{21!5!}=65,780$$

Divide:

$$\frac{1,287}{65,780}\approx 2\%$$

1000. $\frac{48}{2,598,960}$

Count the number of ways you can be dealt four aces. Then divide that by the total number of ways you can be dealt 5 cards out of 52.

The number of ways to get four aces is 48.

If you have four aces, then you need just one more card. There are 48 left to choose from in the deck.

The number of ways to be dealt 5 cards from a deck of 52 is

$$\frac{n!}{(n-r)!r!}=\frac{52!}{(52-5)!5!}=\frac{52!}{47!5!}=2,598,960$$

1001. $\frac{3}{8}$

The prime factors of 1001 are 7, 11, and 13. Other factors of 1001 are products of these prime numbers: 77, 91, 143, and 1,001. And then there's the number 1. So the total number of divisors of 1001 is 8. Three of them are prime. Divide 3 by 8.

Index

• A •

• B •

• C •

• D •

• E •

• F •

• G •

• H •

• I •

• L •

• T •

• U •

• V •

• W •

Workspace

Workspace

Workspace

Workspace

Workspace

Workspace